Eagle Over the Ice

EAGLE OVER THE ICE

The U.S. in the Antarctic

Christopher C. Joyner and Ethel R. Theis

University Press of New England / Hanover and London

University Press of New England, Hanover, NH 03755
© 1997 by University Press of New England
All rights reserved
Printed in the United States of America   5   4   3   2   1
CIP data appear at the end of the book

For
Nancy, Kristin, Clayton
and
Walter

Contents

Preface		ix
List of Acronyms		xiii
Chapter 1	*Introduction*	1
Chapter 2	*The Antarctic Setting*	10
Chapter 3	*The Antarctic Treaty*	20
Chapter 4	*Making U.S. Antarctic Policy*	45
Chapter 5	*Freedom of Scientific Research*	78
Chapter 6	*Environmental Interests*	100
Chapter 7	*Geostrategic Interests*	133
Chapter 8	*Ideological Interests and Ecopolitics*	161
Conclusion		188
Appendixes		195
A.	The Antarctic Treaty	197
B.	Interview Questions	203
C.	Government and Nongovernment Officials Interviewed for this Study	204
D.	Antarctic Treaty Consultative Meetings	205
E.	Antarctic Treaty Consultative Parties and Acceding States	206
F.	List of Selected Congressional Hearings/Reports on Antarctic Issues 1939–1994	208
G.	U.S. Antarctic Program Funding	210
Notes		212
Selected Bibliography		274
Index		297

Preface

ANTARCTICA IS the frozen continent. It is the only continent without native human inhabitants or states, and the only continent whose mineral and hydrocarbon resources remain essentially unexplored and untapped. The superabundant living marine resources in the circumpolar Antarctic seas are practically untouched. Today, on the eve of the twenty-first century, in an era of wondrous planetary probes, subatomic investigations and achievements in telecommunications, Antarctica remains a vast pristine wilderness that is used almost exclusively for scientific research.

Though Antarctica is the coldest, driest, highest, most desolate, and least hospitable of the continents, human activities have become increasingly evident there during the last half-century. Since 1970, major changes have occurred in how Antarctica is perceived, used, and governed by the international community. The future may witness even more changes, particularly in terms of environmental conservation and resource management. Antarctica thus has become the focus of growing international attention, not only from those states with the economic wherewithal to conduct scientific research there, but also from developing countries, international organizations, and environmental interest groups. While these players have done much to influence the course of recent Antarctic events, the most prominent actor on the Antarctic scene throughout this century has remained the same; namely, the United States.

The purpose of this book is to investigate and assess the political, strategic, and legal aspects of the United States' relationship to the Antarctic. While this study began as a critical examination of U.S. national interests in the Antarctic, it gradually evolved into a broader history, intended to lend depth and perspective to American foreign policy in the region. As a history of U.S. involvement in the polar south, this study traverses several fields, including international and domestic politics, international law and diplomacy, science and national security, and, occasionally, developmental economics and ideological rivalries. In the final analysis, however, it is a study of the means and ends that have determined U.S. policy attitudes toward the Antarctic.

This volume is arranged to be of use to the general reader as well as the scholarly investigator. Chapter 1 sets the stage by examining the role of the national interest as a guide to U.S. Antarctic policy and, to this end, asks a number of penetrating research questions that this study seeks to answer. Chapter 2 turns to an examination of the geophysical characteristics of the Antarctic, while chapter 3 evaluates the Antarctic Treaty as the legal mechanism governing the polar south. Next, chapter 4 examines the ways and means that national policy is made by the U.S. government for Antarctica. Chapters 5 through 8 critically assess preeminent national interests held by the United States in formulating and carrying out policies for Antarctica. Chapter 5 explores the U.S. insistence on free scientific research as a cardinal principle governing activities in the Antarctic. Issues that have historically determined American policy considerations toward resource conservation and environmental protection in the Antarctic are treated in chapter 6. Geostrategic interests, especially matters affecting U.S. perceptions of national security and economic priorities in the polar south, are assessed in chapter 7. During the 1980s, certain international ideologies arose that were perceived as being contrary to United States' national interests and policy ambitions for the Antarctic; these are analyzed in chapter 8. Finally, the conclusion provides an assessment of how national interests have historically guided U.S. policy toward the Antarctic, and how the situation of Antarctica has more recently affected U.S. foreign policy interests.

Many people contributed to this work in many ways. Several U.S. officials and policymakers were interviewed in the course of this study, and we are indebted to their patience and generosity in providing information, as well as their valuable and candid insights into and explanation of the rationale behind U.S. policy. In this respect, a fundamental debt must be acknowledged to R. Tucker Scully, Raymond Arnaudo, Thomas Laughlin, Lawrence Rudolph, and Anton Inderbitzen. The U.S. environmental community was also extremely helpful in furnishing documents and reports elaborating the conservationists' position on various Antarctic issues. Particularly helpful were Beth Marks with the Antarctica Project, Jim Barnes of the Antarctic and Southern Ocean Coalition, and Sue Sabella of Greenpeace. Lee Kimball provided much-appreciated perspective on the nuances earmarking U.S. foreign policy on various Antarctic issues. As to errors of fact and interpretation in this study, the standard disclaimer applies to these officials and to other colleagues, who are hereby absolved of all responsibility.

Personal thanks also go to the National Science Foundation for use of their library facilities over the years. We are greatly indebted to the staff in the Office of Polar Programs, and particularly Guy Guthridge and Polly Penhale, for their considerable assistance and courtesy. We also thank David Friscic for his tireless energy and patience in responding to our requests for materials.

The authors are no less grateful for comments and suggestions made by persons who read earlier drafts of this manuscript. In this regard, special appreciation is owed to Benjamin Nimer and Burton Sapin. Similarly, the authors are immensely grateful to David J. Bederman and the anonymous reviewers whose suggestions enhanced the scholarly quality of the finished product.

We are also deeply grateful to our editor, Paul Schnee at the University Press of New England, who facilitated and expedited the publication of this study, and to Mary Crittendon, who managed its way through the publication process in a professional and most competent manner.

Finally, we wish to express deep appreciation to our families for their love, patience, and understanding over the past four years. Nancy, Kristin, Clayton, and especially Walter provided constant support and devotion that facilitated completion of this volume. As a measure of our love and appreciation, we dedicate this volume to them.

C.C.J.
E.R.T.

Acronyms

AJUS	Antarctic Journal of the United States
AMLRCA	Antarctic Marine Living Resources Convention Act
APG	Antarctic Policy Group
ASPAs	Antarctic Specially Protected Areas
ASMAs	Antarctic Specially Managed Areas
ASOC	Antarctic and Southern Ocean Coalition
ATCM	Antarctic Treaty Consultative Meeting
ATCPs	Antarctic Treaty Consultative Parties
ATS	Antarctic Treaty System
BIOMASS	Biological Investigations of Marine Antarctic Systems and Stocks
CCAMLR	Convention for the Conservation of Antarctic Marine Living Resources
CCAS	Convention for the Conservation of Antarctic Seals
CEQ	Council on Environmental Quality
C.F.R.	Code of Federal Regulations
CHM	Common heritage of mankind
CIA	Central Intelligence Agency
COMNAP	Council of Managers of National Antarctic Programs
CRAMRA	Convention on the Regulation of Antarctic Mineral Resource Activities
DPP	Division of Polar Programs
DVDP	Dry Valley Drilling Project
E.O.	Executive Order
EAM	Environmental Action Memorandum
EDF	Environmental Defense Fund
EIA	Environmental Impact Assessment
EISs	Environmental Impact Statements
EPA	Environmental Protection Agency
FIBEX	First International Biological Experiment
FRUS	Foreign Relations of the United States
FSO	Foreign Service Officer
ICSU	International Council of Scientific Unions
IGY	International Geophysical Year
IO	Bureau of International Organization Affairs
MARPOL 73/78	International Convention for the Prevention of Pollution from Ships
MMC	Marine Mammal Commission
NASA	National Air and Space Administration
NEPA	National Environmental Policy Act
NESDIS	National Environmental Satellite Data and Information Service

NIEO	New International Economic Order
NMFS	National Marine Fisheries Service
NOAA	National Oceanic and Atmospheric Administration
NSC	National Security Council
NSDM	National Security Decision Memorandum
NSF	National Science Foundation
NSFA	Naval Support Force, Antarctica
OCB	Operations Coordinating Board
OES	Bureau for Oceans and International Environment and Scientific Affairs
OGC	Office of General Counsel
OMB	Office of Management and Budget
OPP	Office of Polar Programs (former DPP)
OTA	Office of Technology Assessment
PCC	Policy Coordinating Council
PEIS	Programmatic Environmental Impact Statement
POLEX	Polar Experiment
SCALOP	Standing Committee on Antarctic Logistics and Operations
SCAR	Scientific Committee on Antarctic Research
SEHI	Safety, Environment and Health Initiative
SEIS	Supplemental Environmental Impact Statement
SIBEX	Second International BIOMASS Experiment
SPAs	Specially Protected Areas
SSSIs	Sites of Special Scientific Interest
TIAS	Treaties and Other International Acts Series
UN	United Nations
USAP	United States Antarctic Program
USARP	United States Antarctic Research Program
USAS	United States Antarctic Service
USGS	United States Geological Survey
UST	United States Treaties and Other International Agreements

Eagle Over the Ice

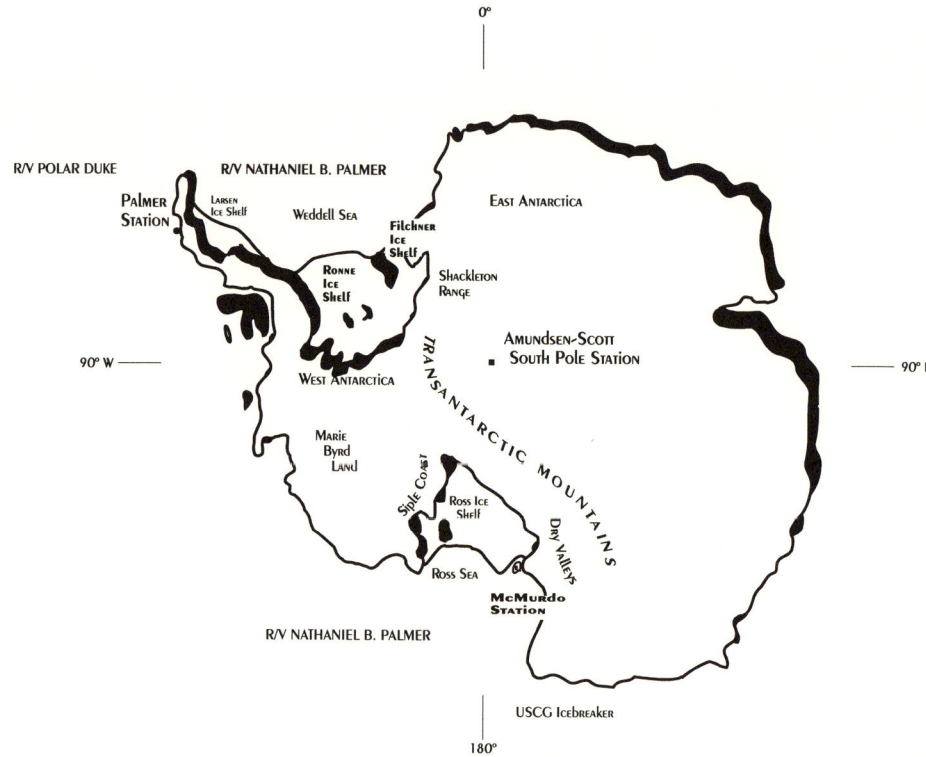

U.S. Antarctic Program locations of major activities.
Courtesy National Science Foundation, Office of Polar Programs.

1 / *Introduction*

THE UNITED States is the chief architect of law and policy for the Antarctic. For over three decades the United States has exerted the political and diplomatic clout necessary to enhance and expand the legal regime governing the Antarctic continent and Southern Ocean. In fact, the legal basis for U.S. involvement in Antarctica stems from the 1959 Antarctic Treaty, an agreement originally called for and substantially forged by the United States.

United States interests in Antarctica, which are inextricably tied to the preservation of the Antarctic Treaty and its auxiliary instruments, are long-standing. The Antarctic Treaty itself evolved from an American initiative. At the conclusion of the 1957/1958 International Geophysical Year, the United States proposed to those states that had participated in Antarctic research that they should all enter into "a treaty designed to preserve the continent as an international laboratory for scientific research and insure that it be used only for peaceful purposes."[1] On December 1, 1959, the United States and eleven other states signed the Antarctic Treaty. Since that instrument entered into force in June 1961, the United States has invested more financial and physical resources than any other government in Antarctic logistics, scientific activities, and environmental programs. Similarly, the United States has been among the most active governments in fostering international cooperation in scientific research about and protection of the fragile Antarctic environment. United States foreign policymakers have consistently viewed the Antarctic Treaty System as the political/legal framework most conducive for promoting U.S. goals and objectives in the region. Recent U.S. policy positions suggest a sustained commitment to the purposes and functions of this family of international instruments.

This study examines U.S. foreign policy in the Antarctic. The focus falls on interests or goals of the United States, especially as they set objectives for policy in Antarctica:[2] What those goals are; whether they can be explained by reference to the national interest; how they are manifested in actual policy decisions; and what factors or influences lie behind them. A number of central questions are consequently posed in the study: What interests or objectives guide U.S. Antarctic policy? How is the U.S. government organized to carry them out? To what extent have U.S. officials been able to realize these objectives through international diplomacy? What constraints inhibit realization of these goals, and what factors facilitate their accomplishment? And, is the pursuit of U.S. national interests in Antarctica compatible or incompatible with global interests, and in what ways? In examining U.S. Antarctic policy, this study challenges a key traditional assumption concerning the national interest and its consideration in policy formulation, namely that pursuit of a state's national interest is often irreconcilable with the pursuit of global interests.[3] The overriding theme here runs counter to that proposition: While U.S. Antarctic policy explicitly serves national interests, it simultaneously promotes active global interests. Scientific research by the United States in the region, for example, has increased knowledge and understanding of basic geological, geophysical, biological, and oceanographic processes, thus helping to address or overcome problems of planetary importance, for example, ozone depletion and global climate change.

Improved understanding of U.S. foreign policy initiatives in the Antarctic involves recognition of the interlocking relationship not only between political and security interests but also among legal, scientific, environmental, and economic interests. United States security interests in the Antarctic have embraced military and strategic concerns; but the Antarctic experience reveals that legal and economic dimensions can overshadow military and strategic components of U.S. security interests in a region. In this regard, international law and diplomacy have functioned as policy instruments for structuring order and building confidence in the governing arrangement. Stability and order in the polar south have benefitted the parties to the Antarctic Treaty and general global interests as well. A core purpose of this study is to examine how these forces have affected and been affected by U.S. policy concerns.

The Notion of National Interest

States chart their courses in international affairs in accordance with their national priorities, which in turn reflect their main stakes in international politics—national security, self-preservation (inclusive of political independence and territorial integrity), economic sufficiency, national prestige, ideological ambitions, and power considerations. The "national interest" of the United States comprises those priorities that guide U.S. actions in international politics. The U.S. national interest, then, encompasses the collective goals sought by the government. Pursuit of the U.S. national interest depends upon the resources available to the government to induce or persuade other states to cooperate in areas where interests coincide; it also depends upon the state's capabilities. The United States has a national interest that benefits the polity as a whole, and statesmen tend to think and act in terms of that interest.[4]

Critical for explaining the foreign policy behavior of the United States in terms of the "national interest" is a clear definition of what the concept means. National interest refers to the collective long-term goals and objectives specified by the government of a state—that is, foreign policymakers and institutions—that are directed toward the external environment, are pursued consistently, and seek to benefit the whole polity of a state rather than just some part of it. This definition emphasizes several important conceptual distinctions. For one, the national interest functions as the source of a government's foreign policy when pursued consistently over time. For another, the national interest seeks to benefit the whole of society rather than merely individual or group interests. In addition, the definition distinguishes between "national interest" and "public interest." Whereas "public interest" refers to the domestic realm—the internal conditions of a polity—"national interest" concerns the international domain, that is, the impact that external relations have upon a state.

Important in the calculus of U.S. national interest are various criteria that can be applied to define the concept. Economic criteria are often held in this regard. Policies that enhance the economic position of the United States are viewed to be in the national interest, for example, when they improve the balance of trade, strengthen the industrial base, or guarantee access to energy or nonfuel mineral resources. Ideological considerations have also been salient for determining U.S. national interest. Ideologies may be often used to legitimize U.S. foreign policies, a strategy often seen during the Cold War era in the tensions between the Western liberal democracies and Marxist-Leninist states. Factors enhancing America's national power are also frequently perceived to be in the U.S. national interest. Chief among these are policies designed to improve U.S. military capabilities, strategic advantage, or economic might.

In recent decades the foreign affairs agenda of the United States has expanded and diversified such that economic, legal, ecological, and other issues have become more prominent and, at times, tend to eclipse military and strategic security issues. This phenomenon has intensified since 1989 with the collapse of communism. The stunning disintegration of the Soviet Union as a global rival presented the Western democracies with a less threatening international environment. For the United States, national survival will nonetheless remain a vital, constant, but less consuming national interest.[5] This trend rings plainly true in the Antarctic context. To some extent, the Antarctic experience shows how a variety of non-vital interests, such as scientific, environmental, and law and order concerns, have displaced those on the more conventional foreign affairs agenda to the benefit of the United States.

Examining U.S. foreign policy through the prism of national interests requires consideration of other factors. An obvious realization is that the U.S. government manages not just one but actually many foreign policies. It is proper to speak of a single, overriding national interest when the physical, cultural, and political survival of the Unites States is threatened, as during the Japanese attack at Pearl Harbor on December 7, 1941; but far more numerous situations arise where U.S. foreign policy for a given geographic region or a specific issue does not involve interests directly related to the survival of the state. Moreover, not only does the United States have multiple foreign policies but each foreign policy issue may have a cluster of objectives or interests attached to it. Similarly, not all interests concerned with a specific geographical region or issue hold equal import for American national leaders, nor do those leaders pursue all interests with the same vigor. National interests relating to a given issue (e.g., resource exploitation or conservation) or region (e.g., the ocean floor or Antarctica) can thus be seen in terms of relative degrees of primacy and permanence.

For the Antarctic situation, this distinction is especially useful. United States interests in Antarctica generally take on non-vital importance; that is, the United States does not have core interests at risk in the Antarctica. The country's stakes in the Antarctic are not likely to endanger the survival of the United States as a corporate polity. It is worth mentioning, however, that this view is not shared by all other Antarctic Treaty parties. Certain claimant states, Argentina and Chile, for example, have often viewed threats to their Antarctic claims as endangering their territorial integrity. Those governments therefore elevate their perceptions of foreign policy in the Antarctic to the level of vital national interests.

Conceptual Strengths and Limitations

Analysts have had difficulty grappling with the concept of national interest, and for good reasons. Doubts have been expressed about its analytic utility and the difficulty of operationalizing the concept for assessing the behavior of foreign policymakers.[6] Perhaps the most serious criticism of the national interest concept is its ambiguity. The term is often used without sound definition or clear, precise criteria for its application.

A related obstacle in conceptually clarifying the term is that the national interest has often been depicted in terms of power, primarily conceived as military capabilities. This visualization creates the problem of discerning what "power" means, what it consists of, and how it can be measured. Such a task in the study of international relations eludes quick or easy solution. This early association of national interest with national security has impeded conceptual development of the term by relegating its use to conflict situations involving threats to a state's territorial or political integrity.

A second limitation pertains to use of the national interest both as a manipulative political symbol and as an intellectual construct.[7] Too often the national interest has been applied loosely and generally, especially to justify foreign policy actions already taken. Still a third limitation mirrors the debate over whether the national interest should be viewed as objective or subjective. With reference to the United States, the question becomes whether the national interest can be ascertained objectively and systematically, or whether it should be regarded as the subjective perceptions and preferences of individuals and groups. Because systematic inquiry does not provide an answer, the national interest of the United States is best viewed as a set of subjective preferences that change with the aspirations of American policymakers.

Despite these conceptual limitations, the notion of "national interest" retains enduring appeal for U.S. government officials, which can be explained by the uses it serves. For foreign policymakers, the "national interest" represents a direction for Antarctic policymaking—an intellectual compass, so to speak, that helps restrain and structure the choice of goals and objectives.[8] By employing the national interest as a prism for policy decisions affecting the Antarctic, policymakers are encouraged to scrutinize their intentions should they be required to justify policies formulated or actions taken. This recourse to the U.S. national interest involves distinguishing between foreign policies that seek to benefit the polity as a whole and those intended to benefit particular or special interests. Rather than being based on the notion of the national interest, foreign policies responsive to special interests may reflect preferences of a dominant group or class in society or result from the competitive struggle among affected groups for influence over certain policy decisions.

The enduring currency of the notion of national interest among American foreign policymakers is of particular note. Leaders in the United States have long felt the need to articulate concepts that substantiate their foreign policies. Though imprecise, the concept of a national interest is deeply rooted in the American political experience.[9] For American policymakers, "national interest" is far more than a slogan; it remains an expression of commitment reflecting both capabilities and intentions. Policymakers have often perceived national interest as a guide to foreign policy decisions and as a limit on the range of policy choices available.[10]

The resiliency of this concept is reflected in the roles it has played in the conduct of American foreign policy in the Antarctic.[11] The notion of national interest has been pivotal in shaping U.S. foreign policy in the polar south. It has also reduced uncertainty about motivations for state behavior in the region and provided constraints on actions taken there by national leaders. All these are significant considerations for assessing the ways, means, and rationales for U.S. foreign policy in the Antarctic.

Actors in U.S. Antarctic Policymaking

The United States remains a key player in Antarctic politics. As a state, the United States seeks certain objectives or goals, decides among them, and finds ways to realize them.[12] In this context, the "United States" actually refers to the high-level foreign policy officials and central decision-making institutions in the government, in other words, the executive branch.[13] International conduct thus refers mainly to actions decided upon by the individuals occupying foreign policymaking positions in government. The goals that the U.S. government formulates and pursues in the Antarctic do not, however, simply reflect demands from or interests of individuals or organized groups. If they did, these preferences could not be construed as "national" interests. We thus assume that foreign policymakers' preferences are distinct from the narrow concerns of domestic groups and are based on the former's perceptions of what is "best" for the nation, that is, what is "in the national interest." These broad policy preferences, often associated with political objectives or ideological goals, are properly labeled "national interests" to indicate their benefit to the national community as a whole rather than mere satisfaction of certain individual or particular group ambitions.

Focus on the United States in the Antarctic suggests that the state remains the preeminent actor in world politics, which indeed is the case. This unique position endows a government's officials with a special legitimacy in the formulation and implementation of foreign policy. Inherent in this analytical approach is the presumption that U.S. foreign policy behavior can be assessed as though states were rational actors.[14] In other words, the model asserts that in reaching decisions, decision makers (1) select the objectives they wish to achieve and the values they expect to maximize; (2) consider the alternative means of achieving these ends; (3) calculate the likely consequences of each of the courses of action they may adopt; and, (4) select the method that has the best prospect of attaining the desired aims.

Foreign policy decision making, however, may not actually correspond to a simple model of rational behavior. Given the complex character of U.S. foreign policy, it is unreasonable to expect that officials make policy and carry out choices that always conform to a rational model. Limited information, possible misperceptions, and inadequate resources for considering possible alternatives circumscribe the ability of policymakers to make decisions in a fully "rational" manner. As a result, they choose from a universe of possible responses to reach a "satisfying" rather than an "optimizing" solution.

This situation is no less true for U.S. policy in the Antarctic than for U.S. policy in general. That U.S. government policymakers cannot make decisions for the Antarctic that coincide with specifications suggested by the rational model does not negate the rationality model's utility as an analytical tool, however.[15] Assumptions of rationality are conceptually useful simplifications about individual behavior, even though they can only approximate reality in Antarctic policy formulation.[16] Consequently, the main utility of a "rational actor" approach for appraising U.S. policy rests in its depiction of international behavior as purposeful acts of a government.

In gauging the role of the United States in Antarctic affairs, specific attention falls on goal-oriented behavior of government officials and politicians responsible for the conduct of American foreign policy. This study thus treats the "state" as consisting of high-level executive branch officials and members of congressional committees with close involvement in Antarctic affairs. The chief individual actors include the president, secretary of state, director of the National Science Foundation, senior National Security Council officials responsible for Antarctic affairs, the Antarctic Policy Group, and other senior executive branch officials intimately involved in Antarctic policymaking. Most important among governmental institutions are the White House, the National Security Council, the Department of State, and the National Science Foundation. Other executive branch agencies that play a less central role in Antarctic policymaking include the Department of the Navy, the National Oceanic and Atmospheric Administration, the U.S. Geological Survey, the Environmental Protection Agency, and the Department of the Interior.

Members of congressional committees charged with oversight for Antarctic affairs also constitute important actors since, on occasion, they have exerted notable influence in Antarctic affairs. Nevertheless, in the Antarctic context, the principal arena of policymaking remains the executive branch. Although congressional input at times has been influential, the insulated position that foreign policymakers enjoy has permitted them to establish and guide Antarctic policy in tandem with a fairly constant set of goals.

Environmental groups have also been active and varied participants in Antarctic policymaking, particularly during the debate over the development of an Antarctic minerals regime. Environmentalists have used the legislature and, to a lesser extent, the courts to attain their objective. At the same time, the executive branch has often sought to exercise a restraining influence on environmentalists' efforts to shape U.S. Antarctic policy. Although environmentalists have not been fully successful in carrying out their pronounced conservationist agenda, their persistence has contributed to a fundamental reworking of the relationship between government and environmentalists in the context of Antarctic policy.

Conclusion

The concept of the national interest remains central to American Antarctic policy. Similarly, the need exists to reexamine national interest considerations in the south polar region in light of changing world conditions. Notwithstanding limitations as a theoretical construct, the "national interest" retains critical significance for American decision makers. It helps guide executive branch policymakers in their choice of future action and serves as a yardstick for assessing those policies followed in the past. Statesmen, in sum, define foreign policy goals and objectives in terms of the interests of their government. This is as true for U.S. policy in the Antarctic as it is for any issue in any other geographical region.

Use of national interest as a prism through which to assess U.S. Antarctic policy sustains the conclusion that the precept is pervasive in international relations. Clearly, the concept has not lost its saliency as the basis for measuring policy results among U.S. decision makers. Resort to "national interests" as a gauge for and justification of foreign policy decisions is ubiquitous to governments throughout the international community. This enduring prominence of the concept highlights the need to sustain its utility in analyzing foreign policy behavior.

In forming an ordered view of American national interests in Antarctica, the importance of consistency and continuity in U.S. policymaking must be underscored. The Antarctic experience, however, also demonstrates the importance of exercising flexibility in domestic decision making such that policy priorities can be shifted in response to changing international circumstances. This flexibility can be seen in the ascendancy of environmental interests over economic interests in United States Antarctic policy during the late 1980s. Maintaining continuity in U.S. Antarctic policy, coupled with the ability to adapt to perceived threats against U.S. national interests in the region, should permit American foreign policymakers to meet the new challenges the Antarctic may present in the future. It is this fundamental theme that the present study explores.

2 / The Antarctic Setting

ANTARCTICA's uniqueness is highlighted by the world's most frigid climate, an extraordinarily fragile ecosystem, and extreme remoteness.[1] A former director of the National Science Foundation put it well when he observed that "Antarctica is a natural laboratory of extreme conditions not found anywhere else on the surface of the earth."[2] The hostile environment severely limits human activity and influences a state's definition and pragmatic pursuit of its national interests on the frozen continent. Paradoxically, the forbidding geophysical characteristics of the Antarctic and its pristine environment offer unparalleled opportunities for scientific research. This aspect of Antarctica has been critical to the U.S. decision to embrace freedom of scientific research as a national interest. Moreover, the fragility of the Antarctic environment, along with uncertainty over the potential damage from mineral extraction activities to the region's ecosystem, led during the late 1980s to a shift in the priority of U.S. national interest in the polar south. A disciplined analysis of U.S. interests in the Antarctic therefore requires examination of the geophysical characteristics of the Antarctic.

The Mainland

Only superlatives can convey a vivid picture of the harsh Antarctic environment and depict the opportunities and limitations for man's activities in the region. Antarctica is the remotest, coldest, highest, driest, windiest, least inhabited, and most barren of all continents. The South Pole marks the geographical center of this high, perennially ice-covered continent surrounded and isolated from the world's six other continents by broad, frigid expanses of the Atlantic, Indian, and Pacific Oceans.[3] Antarctica is larger than the United States and Mexico combined.[4] It covers about 5.4 million square miles and is geographically and geologically divided into East Antarctica and West Antarctica, separated by the Transantarctic Mountains.[5] Whereas the east side consists largely of a high, ice-covered plateau, West Antarctica consists of an archipelago of mountainous islands bonded together by ice.[6] Except for exposed peaks of mountain chains and some patches of bare ground near the coast, about 3,000 square miles in all, the land is solidly encased in permanent ice. The nearest continental mainland, the southern tip of South America, lies about 600 miles distant, separated from Antarctica by the Drake Passage, one of the world's stormiest stretches of ocean.[7] Antarctica thus remains the most isolated of all the continents. South Africa lies almost 2,500 miles away, and Australia, over 1,500 miles.[8]

Ice dominates all other features of Antarctica. About 98 percent of the continent is covered by a mantle of ice averaging 6,600 feet in thickness.[9] The Antarctic ice sheet contains 90 percent of the world's freshwater reserves, and at some depths the ice attains a thickness of 15,580 feet.[10] Over one-third of Antarctica lies below sea level, and radar soundings have revealed that the tremendous weight of the ice has depressed other areas of the continent's landmass below sea level as well.[11] The ice extends offshore in the form of huge ice shelves up to 1,000 feet thick. These ice shelves are floating slabs of land ice ranging in thickness from 600 to 3,900 feet. The Ross Ice Shelf, the largest, is roughly twice the size of New Zealand. Antarctic ice shelves continually discharge the vast tubular icebergs characteristic of the region, which eventually drift with the currents and take years to melt. Annually, Antarctica produces some five thousand icebergs, about 6.5 times the production of the Arctic.[12]

The surface area of the continent varies with the seasonal expansion and retreat of the pack ice formed by freezing sea water. During the austral winter the pack ice may extend from a few hundred to 1,500 miles out from the coastline, with the average being 700 to 800 miles.[13] Antarctic ice thus occurs in three forms: pack ice, ice shelves, and ice sheets. Pack ice is generally ice formed by the freezing of sea water. Shelf ice is very much like pack ice but is formed by the ice sheet pushing past the land edge, over the surface of the sea, normally in bays or other sheltered areas. Ice sheets are generally considered land ice and are formed by the freezing of fresh water or the compacting of snow.[14] In all, approximately 13 million square miles, including the continent, are covered by ice during the winter months, effectively isolating Antarctica from access by ship from mid-March to November.[15] In contrast, ice area in the summer is approximately 6 million square miles less than in winter, and its outer limits average 200 miles seaward from the coast.[16]

The climate of Antarctica is the most extreme on earth. Survival in Antarctica means battling sub-zero temperatures and fierce winds. Even in the summer the temperature rarely climbs above freezing.[17] Although the sun never sets during the summer, only the areas along the coast experience a warming effect. The rest of the continent remains locked in glacial ice. Richard S. Lewis remarked that, although meteorologists were aware in 1912 that the Antarctic was a colder region than the Arctic,

they had no means of determining the actual difference. IGY measurements showed the extent of the contrast. At the south pole monthly average temperatures range from -13 to -80°F, while at the north pole the range is 32 to -31°F. The Arctic grounded ice, including the Greenland ice cap, is barely one-seventh of the Antarctic ice sheet. Hence, it reflects less solar radiation back into space and permits more to heat the ground.[18]

Inevitably, the world's lowest temperatures have been recorded in Antarctica. On June 1982, wintering personnel at the U.S. South Pole station recorded a low temperature of -117.4°F. This low broke the station's previous record of -113.3°F set in July 1956.[19] Vostok, the Soviet base, registered a record low of -126.9°F on August 24, 1960,[20] but surpassed it in July 1983 when the station recorded a low of -128.6°F.[21] At such low temperatures fuel oil turns to jelly and boiling water thrown into the air will burst into ice crystals before reaching the ground.[22]

Antarctica is the windiest place on earth. It is by no means unusual for winds of fifty miles per hour with gusts in excess of two hundred miles per hour to persist for days, or even weeks.[23] Winds can be so strong and persistent that the Australian explorer Sir Douglas Mawson observed at Camp Denison that the weather "proved to be little more than one continuous blizzard the year round; a hurricane of wind roaring for weeks together, pausing for breath only at odd hours."[24] J. R. Dudeney notes that, with a mean annual wind speed of fifty miles per hour being recorded during 1912 and gusts in excess of two hundred miles per hour, Mawson can hardly be accused of exaggeration.[25] Experience with the forceful, steady Antarctic winds has been described by others. Ian Cameron has written:

> When we wintered in Adelie Land the wind on 5th July blew non-stop for eight hours at an average speed of 107 miles per hour. In these conditions it was possible to stand for no more than a few seconds, and then only by leaning forward at an angle of 45 degrees.[26]

Considering Antarctica's sub-zero temperatures, the short, cold summers, the dark, long winters, the high winds, and the lack of water, it is not surprising that the number of land-based flora and fauna is quite limited.[27] The extreme conditions are antagonistic to most organisms, and there is very little life of any kind on the continent. Inland, "except for birds occasionally sighted by field parties in the summer months, no life meets the eye, no sound is heard."[28] Antarctica does not have any land vertebrates. The same impoverishment characterizes continental plant life. No trees or plants such as shrubs, grasses, or herbs grow there.[29] The only plants able to withstand the extreme conditions of the Antarctic mainland, apart from microscopic soil fungi and algae, are lichens and mosses.[30] Mites, springtails, lice, and midges are the main Antarctic land fauna, and they generally live under rocks. Only two flowering plants, "both dwarfed and primitive and rare have been detected in Antarctica, and then only in the less rigorous climate of the sub-Antarctic islands."[31] Penguins, sea birds, and seals that come ashore on Antarctica are all regarded ocean creatures because the sea is the source of their food. It is no surprise, then, that Antarctica's interior essentially consists of a frozen plateau.

Light precipitation and sub-freezing temperatures severely depress the availability of moisture and explain why Antarctica is typically classified as a polar desert. Precipitation falls almost wholly in the form of snow.[32] Because occasional melting occurs only in coastal areas, most of the snowfall on Antarctica accumulates from year to year to nourish the ice sheet. The annual average precipitation for the entire continent, however, is hardly more than four inches.[33] Blizzards do not bring much fresh precipitation but rather result from the fierce Antarctic winds blowing around old snow.[34] As Paul A. Siple aptly put it during his sojourn to the South Pole, "Truly, we were on one of the great deserts of the world as far as the amount of precipitation was concerned. One snowfall in New York City produced far more snow than would be produced in several years at the South Pole."[35]

Antarctic Waters

Despite the distinctive environment of the Antarctic, no universal definition is used to delineate its boundaries. To some observers the totality of the Antarctic is the continent of Antarctica, while others consider the Antarctic Treaty perimeter (60° south latitude) to be a satisfactory boundary.[36] The scientific community, however, generally supports the Antarctic Convergence as a meandering but definite boundary. The Convergence is a belt of water some twenty to thirty miles wide. It forms a biological line of demarcation separating organisms adapted to the warmer and saltier surface from those adapted to the colder and fresher waters.[37] At the Convergence, the cold, low saline, dense Antarctic waters meet the warmer, saltier waters of the Atlantic, Pacific, and Indian Oceans.[38] This clash of different water temperatures and salinities forms a barrier to numerous living species.[39] Although there are fewer species in the waters south of the Antarctic Convergence, those species that do exist are present in enormous quantities.[40] The 1980 Conservation of Antarctic Marine Living Resources Convention (CCAMLR) defines the Antarctic Convergence as the limit of the Antarctic marine ecosystem.[41]

Hardly detectable from the ice-locked land is the Southern Ocean, which contracts and expands with the seasons. The Antarctic Treaty recognized this expandable ecosystem by extending its purview up to 60° south latitude.[42] In contrast to the barren continent, the Southern Ocean supports a rich marine life. It is from this ocean that animals readily associated with Antarctica—especially seals, whales, and penguins—take their food. Paul A. Siple put it eloquently when he observed that the cold seas surrounding Antarctica "offer a strong contrast to the nearly sterile snow-covered continent. The seas virtually teem with life of all sorts."[43] The basis for this life lies in the fundamental process of photosynthesis. Cold water can hold larger concentrations of carbon dioxide, and the long hours of summer sunshine provide the energy required for microscopic plants to thrive.[44] This food source initiates a chain that extends to the largest animals in the world—the blue whales, which attain lengths of over one hundred feet and weights in excess of one hundred tons.[45]

Indeed, the most striking feature of the Southern Ocean is its rich fertility. According to Sayed El-Sayed, the productivity of the Antarctic coastal waters is 400 percent greater than the average productivity of the world's oceans.[46] The Southern Ocean produces phytoplankton at rates as high as nine times the average for the other oceans.[47] Phytoplankton supports rapidly increasing numbers of krill, the key ingredient in the circumpolar marine cycle.[48] The incessant, turbulent mixing of warm waters from more northerly latitudes and icy currents from the pole thrust plankton, algae, and microorganisms up toward the surface; and pack ice extending hundreds of miles beyond the shore suspends algae plants throughout the winter, then releases them in astonishing numbers upon melting.[49] Together, these rich sources of nutrients feed billions of Antarctic krill, which, in turn, nourish perhaps 65 million penguins and 35 million seals, as well as 300 species of sponges, 126 kinds of fishes, and whales of various description.[50]

This one organism, krill, is critical to the food chain in that it supports nearly all larger components of the ecosystem. Although the Southern Ocean ecosystem is bountiful, its relatively few species and short food chain make it very fragile, not yet well understood, and deserving of protection.[51] The dependence of so many predators on one principal prey, krill, is extremely unusual. The central position that krill holds in the Antarctic food chain highlights the fragility of the Southern Ocean's ecosystem, since threats to this single species could produce negative repercussions throughout the entire region.[52]

The Environment and Human Activities

The environment dominates Antarctic operations and constrains human activities in the area.[53] Fierce winds, prolonged periods of darkness, temperatures that rarely rise above freezing, and a hostile frozen terrain present formidable hazards, impose physical hardships, and produce profound psychological stresses on individuals. Only by great effort and expense can humans survive in Antarctica.[54] The extreme low temperatures require special care for equipment. At very low temperatures most metals become brittle. Bulldozers, graders, and similar equipment normally do not function well below -30°F because exposed parts, such as bulldozer blades, are subjected to severe strain and crack easily. Trucks and heavy equipment cannot be shut down for any length of time unless special heaters are available to restart them. Wind chill compounds the effect of cold on humans by disrupting the layer of warmth trapped by skin pores and hair.[55]

Prolonged winter confinement in small campsites contributes to a stressful situation. Ship transport with the aid of icebreakers is restricted to summer months, and air transport is costly. The fierce environment and long periods of winter darkness not only hamper operations of aircraft, ships, and land vehicles, but they also impose psychological stress on their operators. According to the great Antarctic explorer, Admiral Richard E. Byrd, the weather is the greatest hazard and obstacle to flying in the polar regions.[56] Flying can be particularly difficult during a polar "whiteout," an atmospheric condition peculiar to the polar regions when the snow surface and sky appear to blend together, leaving no horizon.[57] Snowdrifts are also extremely hazardous.[58] Blizzards frequently fill the air with powdery snow that buries equipment and structures and reduces visibility to zero.[59] At the now-closed U.S. Byrd Station, for example, visibility was restricted for nearly half the days of the year, and in August, the stormiest month, for two days out of three.[60] Snow accumulates around buildings, which can be buried in a relatively short time. The South Pole station occupied in 1956 has long since been buried and sealed off as hazardous; the station was rebuilt in 1975.[61] Byrd Station was completely covered by snow within five years of construction in 1957,[62] and the British station at Halley Bay was buried in thirty feet of snow in less than a decade.[63]

Visitors to Antarctica will agree with D. H. Walton that "in few spheres of human endeavor are the demands upon a supply organization as onerous as in Antarctica. To ensure both survival and scientific activities supply procedures need to be as near foolproof as possible."[64] The scientific stations in Antarctica must painstakingly import everything essential for human survival from great distances.[65] Inland stations are supported almost entirely by air. As soon as an expedition arrives in Antarctica, its members are made acutely aware of the problems posed by topography and snow surfaces over which they have to travel. These vary as one moves from the coast to the interior, ranging from sea ice thickness, cracks in the ice, snow density, and temperature problems to sastrugi fields, crevasses, and precipitous areas of rocky mountains.

Sastrugi—sharp ridges cut by the wind in the snow surface—are packed by the winds until the snow is as hard as marble. Sastrugi range from a few inches to five feet in height. The smaller types cause little trouble, but those twelve inches or more create problems for all forms of transport.[66] Crevasses—fissures in thick ice or hardened snow—can pose considerable danger.[67] Some crevasses are only a few inches wide at the top, yet hundreds of feet wide below. Some are wide enough to swallow huge pieces of equipment. Blowing snow can form bridges over crevasses, concealing them from view. In the white landscape it is difficult to identify the bridges, which can break under the weight of a sled.[68] During his Antarctic expeditions, Admiral Byrd and his party found crevasses of gigantic proportions. In describing one of these Admiral Byrd observed that "a New York office building could have been quite easily dumped into one of these chasms with enough room for a good-sized oceanliner."[69]

Ice presents, of course, a major navigational obstacle to ships, which must be specifically strengthened to work in ice. Offshore pack ice affects the relative accessibility of shipping to the coastline. Pack ice is never still, for it expands and contracts according to temperatures and season. Even in the austral summer, icebreakers may need to aid other vessels in the area.[70] Icebergs dominate the seascape in the Southern Ocean.[71] Antarctic pack ice is mixed with icebergs, and the differential movement of these two formations can create unpredictable disturbances and endanger marine transportation. The typical Antarctic iceberg rises to around one hundred feet above the water surface; up to ten times that amount remains submerged.[72] From a ship one to two hundred miles off the Antarctic coastline between fifty and two hundred icebergs are likely visible at any one time, and close inshore a ship may have to maneuver through closer assemblages.[73] Some parts of the coast, such as the western margin of the Weddell Sea, are inaccessible due to the persistence of heavy ice near the coast.[74] Until 1960, no landing had been successfully attempted on the forbidding coast of Marie Byrd Land. Two U.S. icebreakers finally accomplished the task, but only after laboring through nearly one thousand miles of ice-encrusted sea.[75] Winter access is difficult everywhere except for the sub-Antarctic islands.[76]

Reaching Antarctica by sea presents one other formidable challenge, the Southern Ocean, which offers the worst sailing conditions on earth. High winds and huge waves make navigation in this part of the world especially dangerous. In fact, the Southern Ocean between 40° south latitude and the Antarctic circle (66° south latitude) is home to the strongest sustained winds on earth. Quite aptly, the winds around the mid-latitudes of the Southern Ocean are known to sailors as the "roaring forties," "furious fifties," and "screaming sixties."[77] Describing the ice-infested sea blocking the entrance to Antarctica's Bay of Whales, Admiral Byrd reported in 1934:

It was forty feet from the crests of some of the waves to the troughs. If that were all, the situation might not have been half bad. But the tides and the currents were ceaselessly working on the bottom of the ice. You could hear it groaning and heaving in a dozen different places; and a spot that would offer safe transit one day would be a gaping crevasse by the next.[78]

The Antarctic Environment and U.S. National Interests

Environmental factors present broad implications for political behavior. Two stand out in the Antarctic context: (1) the importance of the state (defined in terms of central decision-making institutions and roles) as the locus of decision making; and, (2) the influence of environmental conditions on the selection and promotion of U.S. national interests in the region. The remoteness and special geographical character of Antarctica necessitate substantial logistical support, and cost for this support requires substantial degrees of governmental involvement.[79] The expense of these operations is just too great for institutions other than national governments to bear.[80] In establishing a wintering station in Antarctica, every item for housing, food, fuel, and equipment must be transported across the sea to the building site. The only thing available on location is snow for water. Due to these conditions, the cost of transportation to establish and operate bases in the Antarctic remains the most expensive element in most national programs.[81] In the U.S. program, transportation expenses for air, ship, and surface carriers account for nearly half the total annual expenditures.[82] During the International Geophysical Year (IGY), for example, the United States mounted a highly ambitious program, an effort costing about $250 million.[83] Only 2 percent of that sum was required for the scientific studies contemplated. As Admiral Dufek remarked, "We had to spend some $245 million just to set science up in business on the inhospitable ice."[84]

Environmental characteristics of the Antarctic play a critical role in shaping national policy in the area. The position of U.S. economic interests in the Antarctic illustrates this point. Significant in the U.S. decision over the last decade to subordinate economic interests to environmental interests was increased knowledge about likely dangers to the marine environment from oil pollution. The potential threat to the region's ecosystem from mineral exploration and exploitation prompted the country during the 1980s at first to support a minerals agreement severely limiting such activities, and then to abandon it in favor of an environmental protocol banning mineral exploitation in the Antarctic for fifty years. Increased scientific understanding about Antarctica's unique environment clearly affected U.S. policymakers' perceptions on the position that would best serve American national interests. Importantly, U.S. policy priorities were altered accordingly.

3 / The Antarctic Treaty

INTERNATIONAL agreements are widely recognized as the primary source of international law. Some commentators even contend that, because states must agree to treaty terms before they enter into effect, treaties furnish the only source of international law to which governments are truly bound. Treaties, therefore, contain substantial political content. They merge a state's national interest, as defined by its central policymakers, with considerations of international law. In that way treaties demonstrate the symbiotic relationship between law and politics.

In evaluating the Antarctic Treaty System from the perspective of U.S. national interests, it thus becomes necessary to examine major features of the political/legal regime that governs the Antarctic. This chapter serves that purpose. It first looks at the origins of the Antarctic Treaty, from the U.S. vantage point during its creation. The analysis then examines the treaty in terms of its main provisions, the issue of territorial claims, its decision-making structure, and its treatment of institutional development. The study underscores the role of the United States in the formulation and evolution of the treaty and scrutinizes the government's official position concerning national claims to Antarctic territory.

The Early Years 1821–1959

Contemporary U.S. policy regarding the Antarctic Treaty is best understood by appreciating the evolution of U.S. interests in Antarctica. Such a survey assumes particular relevance since it was prior to the Antarctic

Treaty, especially in the 1940s and 1950s, that initiatives emerged that later crystallized into official policy statements defining U.S. national interests in Antarctica. In the formative period, official and private American expeditions to Antarctica acquired a distinct political role. U.S. Antarctic interests had evolved from discovery through exploration to scientific investigation by the time of the IGY of 1957–59.[1] Apart from the prestige derived from scientific achievements, these American expeditions stimulated public interest in Antarctica and provided an impetus for fashioning U.S. policy toward the region. A survey of major U.S. Antarctic expeditions prior to the 1959 Antarctic Treaty highlights their importance in the formulation of U.S. Antarctic policy and their relevance in the evolution of organizational arrangements for carrying out this policy.

The United States Exploring Expedition

United States interest in the Antarctic is legendary. More than a century before the signing of the Antarctic Treaty, U.S. leadership constituted the backbone of activities in the region. Americans have been active in Antarctica since the early nineteenth century when the American sealer, Nathaniel Palmer, made the first known landing on February 2, 1821.[2] During the 1830s, private lobbying of Congress to authorize an expedition to Antarctica bore fruit.[3] On May 18, 1836, Congress amended the Naval Appropriation Bill, authorizing the president to send an exploring expedition to the Pacific Ocean and the South Seas and appropriating $300,000 for that purpose.[4] In 1840 Congress authorized financing for the expedition, which, though it surveyed part of Antarctica's coast, did not produce a U.S. claim to territory.[5] The accomplishments of the U.S. Exploring Expedition of 1838–42, commonly called the Wilkes Expedition after its commander, Lieutenant Charles Wilkes, were many. It remains a milestone in American science, though science was not the expedition's only accomplishment. Because he sighted several points of land while cruising some 1,500 miles along the edge of the Antarctic coast,[6] Wilkes became the first person to recognize and provide proof of the existence of an Antarctic continent.[7]

The Wilkes Expedition produced intense American interest in the south polar region at a time when two prominent nations, France and the United Kingdom, also had expeditions in Antarctica.[8] President Andrew Jackson, as indicated in letters to the secretary of the navy, personally took an active interest in the undertaking.[9] Although it would be a hundred years before the United States sponsored and funded its next expedition to the Antarctic, the Wilkes Expedition established a notable precedent for American involvement in the southernmost continent.

Private Expeditions

The 1920s and 1930s witnessed an escalation of American interest in Antarctica, though it was primarily private individuals who seized the leadership. Prior to World War II, the achievements of Richard E. Byrd, Lincoln Ellsworth, Finn Ronne, and others kept American involvement and interest in Antarctica alive. The man who dominated Antarctic activities during the interwar period was Admiral Richard E. Byrd, who led five successive Antarctic expeditions, beginning with that of 1928–30.[10] His expeditions were largely responsible for spurring the U.S. government to become actively involved in Antarctic politics. They also demonstrated the feasibility of sustained operations on the continent and ushered in the concepts of modern exploration that have led Americans to "nearly every part of the continent."[11]

Unable to obtain financial support from the government, Byrd sought private backing for his 1928–30 Antarctic expedition.[12] He set out for Antarctica with forty-two men, two ships, and three aircraft in the late summer of 1928 at a cost of approximately $800,000.[13] His Little America was the most complex base yet established in Antarctica, and the group was the largest to winter over on the continent up to that time.[14] The expedition, insofar as it made use of the airplane, the radio,[15] and the snowmobile, may be considered the first full realization of the mechanical age of exploration in Antarctica.[16] The highlight of Byrd's expedition was his sixteen-hour flight to the South Pole and back on November 28–29, 1929.[17]

The achievements of the first Byrd Antarctic expedition were impressive. The expedition enhanced the prestige of the United States abroad. More importantly, Byrd's unofficial claim to Antarctic territory east of 150° west for the United States, naming it Marie Byrd Land for his wife,[18] presented the Department of State with problems and opportunities that it could not ignore.

The exploits of the Byrd expedition also boosted congressional interest in Antarctica. Some members of Congress reacted by urging the Department of State to recognize the strategic and economic potential of Antarctica and to act to secure the results of Byrd's expedition.[19] These concerns were reflected in a congressional resolution introduced by Senator Millard Tydings on July 1, 1930, which directed the president "to lay claim to all areas in the Antarctic which have been discovered or explored by American citizens."[20] Although this nonbinding resolution did not pass, its introduction generated a response from Department of State officials.[21] In a memorandum dated July 22, 1930, the department's geographer urged conducting a study to determine which states were entitled to claim which areas of Antarctica and to reevaluate the U.S. political stance. This memorandum appears to be one of the earliest statements on the Antarctic political situation prepared by the U.S. government.[22]

The second Byrd expedition to Antarctica was again privately funded. Although the expedition was not officially sponsored by the U.S. government, President Roosevelt wrote to Admiral Byrd that he "had the full support of the United States Government" and that Byrd could "call on the Government in case of need or emergency."[23] Returning to Antarctica in 1934, Admiral Byrd conducted large-scale exploratory field trips into uncharted parts of Antarctica, as well as extensive aerial surveys over new territories. He also wintered alone at a weather station built 123 miles inland from the main base, from which weather observations could be made throughout the four-month winter night.[24] This outpost was the first scientific station established in the interior of the continent. While Byrd's expedition was exploring the eastern region of the unclaimed sector, Lincoln Ellsworth was carrying out flight operations to the west. In November 1935 Ellsworth made unofficial U.S. claims to the entire region between the British and New Zealand sectors.[25] Ellsworth also completed the first transantarctic flight in November 1936.[26]

The United States Antarctic Service

The widespread popular interest generated by Byrd and Ellsworth had sufficed to fund their expeditions. This support had come entirely from private or institutional sources. In the late 1930s, the success and achievements of the expeditions provided government officials with a strong stimulus to begin serious thinking about U.S. interests in the Antarctic. Until 1939 government-funded activities in Antarctica had been quite limited. This history changed in 1939 with the first Antarctic expedition organized by the United States in this century. Politics as well as scientific exploration had come to the frozen continent, and several states had begun to claim sovereignty over Antarctic territory. These events, together with the landing of explorers from Nazi Germany on Antarctica and Byrd's plans for a third expedition, spurred the Roosevelt administration to implement a more aggressive policy regarding the Antarctic. Acting on a Department of State study that recommended a more active American role in Antarctic affairs, the president authorized an official expedition to Antarctica in 1939.[27] A century after the Wilkes expedition in 1839, therefore, the U.S. government again sponsored an Antarctic expedition. This time, however, Americans, under their government's auspices, would remain in the Antarctic.[28]

American activities in Antarctica became known as the United States Antarctic Service (USAS).[29] President Roosevelt preferred not to use the word "expedition" since it conveyed a temporary American presence in the Antarctic.[30] Although the USAS was established to support one expedition, the president apparently intended to make the American presence in Antarctica a continuing one, with bases occupied by rotating personnel.[31] He took an active interest in the USAS from its inception.[32] This support enabled the USAS to be the largest Antarctic expedition up to that time. Each of its three fields of operations matched the entire program of earlier expeditions. Other than the cooperation of several learned societies in the design of a scientific program and the involvement of civilian scientists, the Wilkes Expedition had been carried out entirely by the navy. In contrast, the USAS was organized as a civilian service. On November 25, 1939, President Roosevelt signed the organic directive that created the USAS as a government agency.[33] The USAS was placed within the Department of Interior, in the Division of Territories and Island Possessions.[34] In establishing the agency, the president set up an executive committee comprised of representatives from the Departments of State, Interior, Treasury, War, and Navy. The committee was to manage the Antarctic program, with principal responsibility for the program's execution resting with the Departments of Interior and Navy.[35] The president designated Admiral Byrd as the commanding officer of the USAS and ex-officio member of the committee. Representation from the five departments continued throughout the life of the organization.

The directive establishing the USAS made it responsible for investigating and surveying the natural resources of the continental and circumpolar seas and for conducting a comprehensive scientific program outlined by the National Research Council of the National Academy of Sciences.[36] More detailed guidelines were spelled out in a letter written in

July 1939 from the president to Admiral Byrd. Of particular significance here was the similarity between objectives of the USAS and those of future U.S. Antarctic programs. President Roosevelt wrote:

> The most important thing is to prove (a) that human beings can permanently occupy a portion of the Continent winter and summer (b) that it is well worth a small annual appropriation to maintain such permanent bases because of their growing value for four purposes—national defense of the Western Hemisphere, radio, meteorology and minerals. Each of these four is of approximately equal importance as far as we now know.[37]

The demands of World War II led to abandonment of the USAS in 1941. Although the USAS had achieved remarkable successes, increased apprehension over U.S. involvement in the war and pressures on the national budget forced Congress to rein in American ventures in Antarctica.[38] The executive committee of the USAS was aware as early as January 1940 that the Antarctic Service's appropriation for fiscal year 1941 would face serious difficulties in Congress.[39] Contrary to the wishes of the president, who wanted to continue the work of the Antarctic Service, the Congress allocated only $171,000 to it.[40] The two American bases were evacuated, and the expedition returned home in May 1941.[41]

The Post-War Years and the IGY

After World War II, a dramatic surge occurred in Antarctic exploration and research, and the United States was at the forefront. In 1946, the country reasserted its interest in the region by launching an expedition that still ranks as the largest ever sent to the continent: the United States Development Project of 1946–1947, better known as Operation Highjump.[42] Though the Department of the Navy was responsible for Operation Highjump, other federal agencies collaborated in scientific research in non-military fields.[43] Admiral Byrd was designated officer-in-charge and given technical control of the field programs.[44] The stated chief objectives of Operation Highjump were to test equipment and train personnel in polar operations and to continue the work of the USAS.[45] Less publicized but also important was the desire to consolidate and extend the basis for U.S. claims in Antarctica, should such claims ever be made.[46] After all, a recurrent presence would reinforce any future territorial claims, while scientific investigation remained a worthy objective in its own right.[47] The important contributions that Operation Highjump made to the knowledge of Antarctica were, therefore, not devoid of political content. As Paul Siple noted, "Mapping all of the Antarctic and being the first to sight the still unseen two-thirds of the continent" placed the United States "in a unique position to claim Antarctica for its own, should it so desire."[48]

U.S. Antarctic policy during this period was spelled out in "Antarctica," a paper prepared by the Policy Planning Staff of the Department of State on June 9, 1948. Designated PPS-31, the document's core elements contained much of the substance that later came to comprise the Antarctic Treaty.[49] Three issues were central to PPS-31. First, concerning the territorial question, the paper espoused the "no claims" principle; that is, the United States would recognize no claim and would make none, but it reserved the right to do so at a later date. On the second issue—the economic potential of Antarctica—PPS-31 concluded that, even if available, the prospects for resource exploitation were bleak due to the severe problems posed by the frigid climate and the hostile environment. Third, PPS-31 addressed strategic concerns. It recognized that, although Antarctica was not critical to immediate U.S. strategic security interests, denying access to the continent and circumpolar islands to adversaries, particularly the Soviets, was important to the country. A real concern was the possibility that the Soviet Union might set up bases in the unclaimed portion of Antarctica, where the United States had established its firmest legal grounds for a future claim.

The National Security Council (NSC) also issued in 1948 the first comprehensive statement of U.S. interests in Antarctica.[50] This formulation strongly indicated that, though U.S. Antarctic interests had evolved since 1948, they remained substantively unchanged. As elaborated by the NSC, American interests in Antarctica involved: (1) using the Antarctic for peaceful purposes only and preventing it from becoming a source of international discord; (2) protecting U.S. rights and interests; (3) guaranteeing freedom of exploration and scientific research; (4) providing free access to develop natural resources; (5) devising and instituting nonpreferential rules to guide activities in the region; and, (6) providing for orderly administration of the area.[51]

Emerging Organizational Framework

Early in 1954, President Eisenhower endorsed American participation in an IGY. That year, he also established an organizational structure for Antarctic policy and programs. He designated the Operations Coordinating Board (OCB) of the National Security Council as the agency responsible for overseeing and coordinating broad U.S. plans and policies for Antarctica.[52] To assist in this task, the OCB created an Antarctic Working Group composed of representatives from the executive branch agencies most involved in Antarctic matters—the Department of State, the National Science Foundation (NSF), and the Department of the Navy.[53] The Working Group was responsible for developing general policies for consideration and approval by the board.[54] The Department of State continued to be responsible for the political aspects of the U.S. Antarctic program, which was not surprising; State Department officials occupied an influential position within the OCB. For most of its tenure, the OCB was chaired by the undersecretary (since then designated as deputy secretary) of state,[55] and, at one time, State Department officials chaired thirty-three of the forty-two OCB working groups, including the Working Group on Antarctica.[56] The NSF was made responsible for coordinating the planning and management of the U.S. scientific program in Antarctica.[57] The navy, with its long history of involvement in Antarctica, was to provide logistical support, including the establishment and maintenance of facilities.[58] For this purpose, on February 1, 1955, the navy established the U.S. Naval Support Force, Antarctica.[59] The president appointed Admiral Byrd as officer-in-charge, responsible for the entire U.S. Antarctic program. Although this position was discontinued after Admiral Byrd's death on March 11, 1957, Captain George Dufek, as commander of the support force, inherited essentially the same responsibilities.

Along with these developments, in the late 1950's a movement arose within the U.S. Congress to create an independent agency for managing the U.S. Antarctic program. On May 31, 1957, Senator Alexander Wiley of Wisconsin introduced a bill that would have created an Antarctic Commission.[60] An identical bill was introduced in the House on June 30 by Representative Clement J. Zablocki, also of Wisconsin. The bills were not passed by Congress, largely because of strong opposition by the executive branch and lack of significant congressional interest.[61] The executive branch opposed the legislation on the grounds that existing arrangements for the coordination of Antarctic activities functioned satisfactorily, making an independent commission unnecessary and possibly leading to unnecessary duplication of effort.[62] Preempting further congressional action, President Eisenhower directed the Bureau of the Budget to study the organizational framework for the U.S. Antarctic program.[63]

The turning point regarding official U.S. policy toward claims to Antarctic territory came in 1957–58. Two factors account for this attitudinal change. One was the universal agreement to set aside overt political considerations about claims during the IGY. The other was the perception by key American policymakers that the United States stood to gain more from having access to the entire continent than by staking a claim to only a part of it. This policy revision led to the formal removal of the claims objectives from the orders of Operation Deep Freeze, the support force for U.S. IGY stations. Nonetheless, commitment to the IGY, motivated somewhat by political considerations, resulted in a dramatic increase in American activities in the Antarctic. One reason for this enhanced commitment was the NCS's decision in 1957 that "the United States would be second to none in Antarctica."[64]

Preparations for the IGY involved two expeditions: Operation Deep Freeze I (1955–56) and Operation Deep Freeze II (1956–57). These expeditions were charged with building and supplying seven American bases, including the base at the geographic South Pole.[65] The National Science Foundation assumed the major task of representing the needs of the IGY before the U.S. Congress. The National Academy of Sciences established a Committee on Polar Research to advise the NSF on possible projects and long-range scientific planning.[66] The State Department, through its Office of the Science Adviser, contributed its diplomatic experience in bilateral negotiations with foreign governments.

The first indication of a continued U.S. research program following the IGY surfaced in January 1958, and in 1959 the decision was made and supported by the Congress that the NSF should "bear full responsibility for the formulation and coordination of the total U.S. Antarctic program with the DOD supplying logistics and having primary responsibility for carrying out the Antarctic operation as planned."[67] Between 1953 and 1958, the U.S. National Committee for the IGY, a National Academy of Science committee, had outlined research programs and selected scientists who were funded by the NSF, although there was no Antarctic staff per se at the NSF during this period. In 1958, when the decision was made to continue operations in the Antarctic, T. O. Jones was appointed Antarctic program director under the Office of Special International Programs in the NSF.[68]

Establishing a dominant management role for the NSF over Antarctic research was achieved only with strong backing of the Bureau of the Budget.[69] Following the decision in 1958 to continue a permanent, extensive research program, an interagency power struggle emerged over how the effort should be organized and who should succeed to authority positions after the U.S. IGY program.[70] The Department of Defense challenged the proposal to assign the dominant role to the NSF. In a report before his retirement in 1959, Admiral Dufek wrote that making the NSF responsible for the U.S. Antarctic program "would prove fatal to our national interests," because the NSF's "preoccupation with science would lead to the neglect of our political, economic, and strategic interests."[71] Admiral Arleigh Burke, chief of naval operations, also objected on the grounds that "neither by law nor experience does the National Science Foundation seem equipped to manage a program with so many implications outside its usual field of endeavor."[72] Likewise, the Committee on Polar Research (subsequently the Polar Research Board) was concerned that its prerogatives might be compromised by having the NSF exercise ultimate authority over decisions about what research projects would be funded. Alan T. Waterman, then director of the NSF, defused much of this criticism and eased political pressure by creating an Interdepartmental Committee on Antarctic Research, upon which any federal agency could sit and thus communicate its views to the NSF.[73]

Treaty Origins and the U.S. Role

The United States played a central role in the development and evolution of the 1959 Antarctic Treaty, the core politico-legal arrangement for the management of Antarctica. The United States was not only the chief architect of this agreement but was its foremost supporter as well. The Antarctic Treaty, which has enabled otherwise politically antagonistic states to coexist and work together harmoniously, has served U.S. interests appreciably. It was this perception of the national interest that prompted the Eisenhower administration to negotiate the agreement. Four American goals for the region articulated in a March 1958 policy statement are embodied in the treaty: (1) to prevent the use of Antarctica for military purposes; (2) to provide for freedom of scientific investigation; (3) to establish an orderly joint administration of Antarctica by the countries directly concerned; and, (4) to preserve Antarctica for peaceful purposes only.[74] These four goals still express the core national interests of the United States in Antarctica today.

The direct origin of the Antarctic Treaty is generally identified with the International Geophysical Year (IGY) of 1957–58, a multinational effort designed to foster scientific cooperation in a number of fields, with Antarctic research figuring prominently among them.[75] Beyond the stimulus provided by the IGY, the origins of the treaty can be traced directly to American efforts.[76] It was U.S. perseverance in securing an agreement that would continue the cooperation that prevailed during the IGY and preserve the continent as a zone of peace.

Although American efforts to secure an international agreement for Antarctica began much earlier, officials did not succeed until 1959. In the wake of growing conflict among Argentina, Chile, and the United Kingdom—the three states with overlapping sovereignty claims in Antarctica—the Department of State during late 1947 and early 1948 began considering possible political arrangements for administering the region. This perception of national interest led the United States in 1948 to approach the states that had a direct interest in Antarctica "with the suggestion that the promotion of scientific investigation in Antarctica and the solution of the problem of conflicting claims might be accomplished through some form of internationalization."[77] The "suggestion" included a concrete proposal for regulating activities on the continent through some form of international regime.[78]

Discussions at that time centered on two types of arrangements for managing Antarctica. First, an international trusteeship comprised of the states with interests in Antarctica—which entailed surrendering all national claims of sovereignty—would maintain administrative authority for governance of the continent.[79] Second, a multiple condominium (excluding the Soviet Union) would "pool" all legal rights and interests in favor of a governing body invested with comprehensive powers over administrative, defense, and legal matters.[80] The United Kingdom was receptive to the idea of internationalization.[81] Other claimant states, however, overwhelmingly rejected proposals that involved relinquishing sovereignty claims in favor of internationalization, in particular Chile and Argentina.[82] Although not a claimant, the Soviet Union declared in 1950 in a strongly worded diplomatic note that any decision about the future of Antarctica would have to involve its participation.[83] The United States continued the discussion with modified versions of the proposals, but agreement could not be reached. A period of international diplomatic stalemate ensued into the 1950s, although at the 1956 session of the United Nations General Assembly the Indian delegation briefly sought to place the question of Antarctica on the agenda.[84]

American efforts to reach an agreement on managing Antarctica receded as planning for the International Geophysical Year began in earnest in early 1954; still, "the very fact of the intensification of Antarctic activity during the IGY revived interest in the idea of establishing an international regime for the area."[85] By promoting research on meteorology, the upper atmosphere, cosmic rays, and other areas, the IGY inaugurated a unique scientific effort; nowhere was this cooperative effort more concentrated than in the polar regions. Prior to the IGY, scientific research had been conducted in the Antarctic by privately or government-sponsored expeditions. These efforts were eclipsed by the scope of the IGY, with twelve nations operating sixty-six stations in Antarctica and with the United States at the forefront.[86] The success of the IGY became the catalyst for a renewed U.S. effort to secure a formal agreement.

The scientific results obtained during the IGY mark the project as an unqualified success. The IGY was a success in another way as well: It set the stage for the political compromise on territorial claims that underlies the Antarctic Treaty. The understanding that all IGY activities would be nonpolitical and that none would serve as the basis for territorial claims clearly influenced the decision of claimant states to "freeze" their claims to Antarctic territory.[87]

Prior to the first IGY Antarctic planning conference held in June 1955, concern was prevalent that claimant states would object to the establishment of scientific stations on "their" Antarctic territory. These anxieties were dispelled when the claimant states decided to welcome the presence of other national stations on all Antarctic territory for the duration of the IGY. It was more or less tacitly agreed among countries interested in Antarctica that political problems regarding the region would be shelved for the duration of the IGY. By the time advance parties set out in late 1955 to build the new Antarctic stations, the stage was thus set for cooperation on a broad scale.[88]

Scientists participating in the planning conference for Antarctica played a major role in ensuring open access to the entire continent during the IGY. At the 1955 conference, the participating scientists passed a resolution that sought to avoid any political complications. The resolution recorded an understanding that IGY activities were temporary measures that "do not modify the existing status of the Antarctic regarding the relations of the participating countries."[89]

Disputes after World War II over the complex legal status of Antarctic claims had hampered scientific research and, occasionally, had threatened to provoke more serious conflict.[90] This tension and the success of the IGY intensified the desire among participants to continue some form of cooperative arrangement beyond the conclusion of the IGY.[91] The United States took the lead. Indeed, the Antarctic Treaty represents the culmination of a series of negotiations initiated by the U.S. government.[92] In 1958 President Eisenhower proposed in a letter addressed to the seven nations with claims to Antarctic territory and to five others that had scientific interests in the area that a conference be held to draw up a treaty concerning the future of Antarctica.[93] He publicly stated on May 3, 1958, that the purpose of the conference was to prevent Antarctica from becoming "an object of political conflict" and to keep the region "open to all nations to conduct scientific and other peaceful activities" there.[94] The Eisenhower note reiterated that the United States "has had and at the present time, continues to have, direct and substantial interests in Antarctica," and that the United States reserved "the right to assert a territorial claim or claims" on the continent.[95]

Negotiating the treaty was difficult.[96] Department of State officials suspected that the major obstacle to a successful completion of the conference could well be the positions of Chile and Argentina. Their suspicions were well founded. Public opinion in these two countries made it unlikely that both governments would accept any arrangement that gave even the appearance of diluting their respective claims to Antarctic territory.[97] It took several meetings to establish a common basis of agreement. The United States, which initiated the preparatory talks, exerted "a major influence on the course and outcome of the negotiations."[98] Especially prominent was Ambassador Paul C. Daniels, whose personal influence was pervasive in smoothing points of controversy and providing working papers that ensured a strong U.S. influence in the final treaty.[99]

The major objectives posited by the U.S. delegation to the conference were to: (1) preserve Antarctica for peaceful purposes, including the prohibition of military operations in the area; (2) preserve freedom of scientific investigation in the region; (3) reduce the political tensions arising out of territorial claims asserted in Antarctica; and, (4) establish a system of continuing consultation among participants to further the objectives of the treaty and "to provide machinery for dealing with problems and opportunities in the Antarctic that only time will disclose."[100] The conference worked toward those ends.

Following sixteen months of extensive preparatory talks, the Antarctic Treaty Conference began in Washington, D.C., on October 15, 1959.[101] Representatives from twelve countries met on December 1, 1959, in Washington, D.C., to sign the treaty.[102] The United States became the fifth state to ratify it.[103] The U.S. Senate gave its advice and consent to the treaty by a vote of 66 to 21 on August 10, 1960, but only after some lively and heated objections. The treaty was praised in the Senate as "containing all the provisions which the U.S. believed were required for the protection of its national interest"[104] and as setting a precedent in the field of disarmament, prohibition of nuclear explosions, and the law of space.[105] At the same time, it was criticized as "politically unsupportable," as "morally and constitutionally wrong," and as hindering "our well-earned economic expansion, thus chaining our future generations to debt."[106] Congressional opposition to the treaty was not directed at any of its specific provisions but rather at the American refusal to assert any claim to Antarctic territory.[107] The treaty entered into force on June 23, 1961, when Chile, Argentina, and Australia, the last three states to sign the treaty, deposited their instruments of ratification with the United States (the depository state).[108] Announcing the ratification, President Kennedy declared that he "earnestly believed that the Antarctic Treaty represented a positive step in the direction of world peace."[109] The jurisdictional ambit of the treaty applies to the area south of 60° south latitude, including all ice shelves.[110]

The Antarctic Treaty: Major Features and Objectives

The Antarctic Treaty supplies the key legal instrument for governing the Antarctic. Reflecting the U.S. influence in drafting the treaty, the Preamble emphasizes two cardinal objectives that the United States had championed: the peaceful use of Antarctica and scientific cooperation.[111] In a message transmitting the Antarctic Treaty to the Senate, President Dwight Eisenhower referred to it as "unique and historic" and as "an inspiring example of what can be accomplished by international cooperation in the field of science and in the pursuit of peace."[112] American foreign policy officials continue to regard the treaty as a forward-looking example of international cooperation and as a sound basis for promoting U.S. national interests in the region.[113]

The Antarctic Treaty broke new ground in several respects.[114] Consisting of a Preamble and fourteen articles, the treaty is unique in the history of international agreements in demilitarizing and denuclearizing an entire continent.[115] Article I reflects an important goal of the United States: demilitarization of the continent. This provision prohibits any military activities; in particular, the establishment of military bases and fortifications, the conduct of military maneuvers, and the testing of any type of weapons. Article V bans nuclear explosions and forbids disposal in the Antarctic of radioactive waste materials.[116] Article VI led the way for an open access inspection system designed to ensure compliance with the provisions of the treaty.[117] Other notable provisions stipulate freedom of scientific investigation and cooperation, including the free exchange of research data and results as well as exchange of personnel (Articles II and III). Article XI established mechanisms for the settlement of disputes. Special importance attaches to Article IV, which contains the contracting parties' approach to the delicate issue of territorial claims. Article IV is often considered the cornerstone of the agreement because its innovative diplomacy helped ensure ratification of the treaty. The article sidesteps the intractable issue of territorial claims by specifying that activities under the treaty shall not constitute a basis for asserting, supporting, or denying a claim to territory; nor does the treaty create any right of sovereignty for any of the states party to it.

Since the treaty did not create any formal administrative body, Article IX assumes particular importance. Article IX calls for a meeting of the contracting parties within two months of the entry into force of the treaty and for regular meetings thereafter. The stated purpose of these meetings is to address matters of common interest pertaining to Antarctica and to formulate and recommend to the contracting parties' governments measures promoting the principles and objectives of the treaty. The Antarctic Treaty owes its dynamic character and ability to deal with new problems to this provision, since it provides treaty members with a means for identifying and responding to difficult issues as they arise. This aspect of the agreement has been largely responsible for the evolution of a politico-legal framework for managing the Antarctic and has generated a multifaceted system consisting of the treaty itself and several auxiliary instruments concluded pursuant to it.

The 1959 treaty does not deal with Antarctic resources, for which it has been criticized;[118] no provisions deal with economic development,[119] a conscious omission.[120] Silence on the issue of resources can be explained in at least two ways. First, the treaty was negotiated at a time when the parties' central concern focused on containment of the Cold

War.[121] The primary impetus for the Antarctic Treaty was to allow scientific research to continue in the face of conflicting claims of territorial sovereignty. Given differing positions on territorial claims, proposals to address resource development and commercial activities in Antarctica would likely have met strong opposition during the treaty negotiations. Second, the contracting parties did not envisage discovery of exploitable reserves of minerals in the Antarctic in the near future. To be sure, at that time there was no prospect of resource development other than marine living resources, which could be exploited without raising conflicts about territorial claims. Hence, no urgency surfaced in 1959 to confront the possibility of mineral resource exploitation.

The treaty also rings silent on another important matter, namely legal jurisdiction over personal conduct on the continent. One common criticism of the Antarctic Treaty is that no clear legal jurisdiction in civil and criminal cases is provided. Under the treaty, contracting parties retain sole jurisdiction over their nationals operating in the continent.[122] In such cases, the jurisdictional principle of nationality takes precedence over that of territoriality. This situation could produce complicated jurisdictional problems in multinational scientific expeditions.

Like most agreements, the Antarctic Treaty specifies how states may join or depart from the membership. Article XIII stipulates that the treaty will enter into force when all twelve original parties have ratified it. The Treaty may be modified or amended at any time, but change requires a unanimous vote of those contracting parties entitled to consultative status. The Antarctic Treaty is of indefinite duration; however, paragraph 2(a) of Article XII provides the opportunity for a review conference thirty years after the treaty was entered into force (i.e., 1991), if any consultative party at that time or afterward desires it. At such a conference, the treaty can be amended by majority agreement of all contracting parties.[123] Any modification or amendment to the treaty will enter into force, however, only if there is unanimous consent among the consultative parties.[124] Interestingly, the thirty-year anniversary passed without any consultative party indicating an interest in renouncing membership in the Antarctic Treaty System, nor has there been any move since 1991 to call such a review conference.

Territorial Claims

Antarctica is geopolitically sliced up like a pie, with wedge-shaped national claims radiating outward from the South Pole to termination lines along the 60° south latitude mark. Sovereignty was subject to conflicting claims when the Antarctic Treaty was ratified in 1961.[125] Between 1908 and 1943 seven states made territorial claims in Antarctica.[126] These claims rely primarily on a combination of two legal principles as grounds for asserting sovereignty over Antarctic territory: (1) the classic principle of discovery and occupation;[127] and, (2) the so-called sector principle.[128] Chile and Argentina also invoke the concept of *uti possidetis* to bolster their claims to Antarctic territory.[129] Lastly, some claimants utilize acts of symbolic administration (e.g., issuing decrees and postage stamps, employing magistrates, and establishing family groups) to bolster their claims. Only five claimants—Australia, France, New Zealand, Norway, and the United Kingdom—mutually recognize each other's claims. All seven claims cover about 80 percent of the continent, leaving the remainder of Antarctica as the largest unclaimed piece of territory on earth.[130] The other original signatories have not asserted any claims and do not recognize those of other nations, albeit the United States and the former Soviet Union (now Russia) have reserved the right to assert claims at a later date.

The Antarctic Treaty remains a remarkable political accomplishment in that it preserves and protects the inherently contradictory positions of both claimant and nonclaimant states.[131] In principle the treaty does not compromise either the sovereignty claims or nonrecognition of those claims. Article IV(2) expressly preserves the position of claimants and nonclaimant states, serving the interests of both by banning new claims or expansion of existing ones. The treaty thus "froze" the legal status quo in Antarctica, suspending the struggle over claims and minimizing opportunities for disputes between claimants and nonclaimants. Agreement would not have been possible without this provision, and continued abeyance of the territorial claims issue remains essential for sustaining Antarctic cooperation.

This "freezing" of territorial claims carries important implications for promoting national interests and has proved critical for the treaty's further development. For one, claimants need not take action to protect their claims. They are assured that nonclaimants will not make claims and are guaranteed that foreign activities in claimed areas will not prejudice existing claims. In turn, Article IV assures nonclaimants access to Antarctica that is unhindered by sovereignty claims.[132] The ingenious solution of "freezing" claims to territory has served U.S. Antarctic interests well. By "removing sovereignty as an issue, the United States was relieved of its own sovereignty dilemma."[133] This situation also permitted the United States unrestricted access to the entire continent.

Significantly, the United States has not made any official claim to Antarctic territory despite having the most extensive history there;[134] nonetheless, the United States (as well as Russia) still reserves the right to make a claim.[135] As noted earlier, the U.S. position has not always been firm. During the early period of American interest in Antarctica, public pronouncements, in fact, cloaked changes in official attitudes toward claims. The documentary record uncovers four official U.S. attitudes regarding claims during this century: (1) prior to 1924, when the United States had no formulated policy; (2) from 1924 to the mid-1930s, when the United States nearly denied any possibility of states making claims to Antarctica; (3) from the mid-1930s to the beginning of the IGY in 1957–58, when the United States encouraged its nationals to claim territory on its behalf; and, (4) from 1958 to the present, when the country tempered its ambitions for making a claim and instead fully supported an international regime of cooperation, undergirded by a moratorium on claims and bases for claims.[136]

U.S. Antarctic expeditions, like those of other states with interests in the continent, have been motivated by a combination of factors, not least of which has been the political element. That the United States did not assert a claim before signing the Antarctic Treaty does not mean that it had not laid the foundation for a future claim. Several explorers advanced U.S. rights to Antarctic territory, among them Palmer, Wilkes, Byrd, Ellsworth, Dufek, Ketchum, Ronne, and members of their parties. The regions explored include, in addition to Marie Byrd Land and the area south of the Norwegian claim, the Palmer Peninsula, the Ross Ice Shelf, Wilkes Land, the American Highland, the South Pole, and various areas that have been flown over or mapped by U.S.- owned aircraft.[137] Nevertheless, in commenting officially on whether the United States should annex Wilkes Land, Secretary of State Charles E. Hughes asserted in 1924 that "the discovery of lands unknown to civilization, even when coupled with a formal taking of possession does not support a valid claim of sovereignty unless the discovery is followed by an actual settlement of the discovered territory."[138] This remains the U.S. position today.

By the late 1930s the increasing pace of U.S. activities on the continent spurred the Department of State to adopt a view more flexible than that espoused by Secretary Hughes. The private expeditions of Richard E. Byrd in 1929 and Lincoln Ellsworth in 1935–36 made extensive unofficial claims for the United States. Admiral Byrd claimed all the territory explored by him east of longitude 150°W (Marie Byrd Land).[139] During their transcontinental flight from the Antarctic Peninsula to the Ross Ice Shelf, Ellsworth and Hollick-Kenyon discovered the Sentinel Range and the Hollick-Kenyon Plateau. Ellsworth made territorial claims on behalf of the United States, but the U.S. government took no action to support them.[140] Even so, Congress awarded Ellsworth a medal for "claiming on behalf of the United States approximately 350,000 square miles of land in Antarctica."[141] Although President Roosevelt did not officially sanction this claim, at that time U.S. activities on Antarctica were motivated largely by sovereignty concerns. In fact, a major impetus for an American Antarctic came from President Roosevelt. In 1939 the United States decided to set up bases to further its claims, although the State Department was most concerned not to antagonize the Latin American nations. In an initiative that appears to have been the personal decision of President Roosevelt, Argentina and Chile were informed that U.S. activities in the South American sector would be carried out on behalf of "all the other American Republics."[142] A Department of State Policy and Information Statement dated January 27, 1947, nevertheless reveals that the U.S. Antarctic Service Expedition "was organized for the specific purpose of establishing and strengthening US claims within the sector previously explored by Admiral Byrd and Lincoln Ellsworth to the east of Little America."[143]

Members of the U.S. Antarctic Service were thus encouraged by President Roosevelt in 1939 to take and record any appropriate acts that might later assist the United States in supporting a sovereignty claim.[144] In fact, President Roosevelt's instructions to the U.S. Antarctic Service were that no member of the service "shall take any action or make any statements tending to compromise the non-recognition position," and that "members of the Service may take any appropriate steps such as dropping written claims from airplanes, depositing such writings in cairns, et cetera, which might assist in supporting a sovereignty claim by the United States Government." Public disclosure of such activity, however, required specific permission from the secretary of state.[145] Secret directives for three U.S. expeditions—Operation Highjump (1946), Windmill (1947), and Deep Freeze I (1955)—listed extension and consolidation of "United States sovereignty over the largest practicable area of the Antarctic continent" among their objectives.[146] As a result, members of the expeditions left markers and U.S. flags in many of the places surveyed.[147]

The Department of State also encouraged expedition personnel to record and document their observations should the United States eventually decide to stake a claim.[148] In December 1946 Acting Secretary of State Dean Acheson advocated a "policy of exploration and use of those Antarctic areas to which we already have a reasonable basis for a claim" so that the United States might gain "maximum advantage."[149] This policy changed when references to strengthening territorial claims were omitted from the specific objectives of Operation Deep Freeze in 1956–57, pursuant to a "gentlemen's agreement" that no activities undertaken during the IGY should serve political objectives.[150]

Some scholars hold that the United States lacked a meaningful policy on territorial claims, and, as a result, the official U.S. position on claims emerged by default. This view fails to justify how claims to any portion of the continent might be denied, and at the same time how undefined U.S. rights might lawfully be asserted. To F. M. Auburn the State Department's inability to make up its mind resulted in the decision to make claims but keep them secret.[151] Auburn maintains that "failure to take timely action in Antarctica may have well cost America a claim to a substantial part of the continent."[152]

Political pragmatism, however, provides a more compelling explanation for the persistent hesitation of the United States to assert a claim to Antarctic territory. The prevailing view among U.S. policymakers holds that assertion of territorial claims has not been necessary to achieve the basic objectives of U.S. policy and that such action might even be detrimental to these objectives.[153]

Several factors help explain the ambivalence of American policymakers on the issue of a U.S. claim to territory.[154] First, there is the problem of determining a suitable area to claim. The unclaimed sector, the region to which the country possesses the strongest bases for a claim, is among the least accessible and least hospitable areas of the continent, which is precisely why it has gone unclaimed by any government. One reason for its inaccessibility is that the waters along the coast of Marie Byrd Land are perpetually frozen.[155]

The unwillingness to antagonize friendly states also weighs in heavily for the "no-claim" decision. Although the United States could logically support tentative claims in most, if not all, sectors claimed by other states, making a claim might produce disagreeable consequences. Close relations with Great Britain and U.S. commitments in the Western Hemisphere would be threatened by U.S. claims. Equally clear is that by asserting a claim, the United States might present Russia (formerly the Soviet Union) with an excuse to follow suit. Most importantly, the United States could hardly lodge a claim without recognizing claims made by other governments, thereby endangering the principle of open access.[156] A claim could therefore diminish American freedom to move and establish bases anywhere on the continent.[157] A related concern is that, "if the US were to make specific claims throughout Antarctica, the result might be an apparent downgrading of US rights in areas not claimed."[158] The United States, then, "might be deemed to have less rights in other areas of Antarctica if it claimed superior rights in specific areas."[159]

Although the United States has not asserted a formal claim to Antarctic territory, there can be no doubt about the government's fundamental interests in the region and its determination to participate in any policy decisions concerning the future legal framework, economic utility, or environmental protection of the area. As Kenneth Bertrand correctly observes, American interest in the Antarctic "is as old as the nation itself. . . . There has been no decade when at least some Americans have not been in the Antarctic as seal hunters, whalers or explorers."[160] Extensive U.S. activity in Antarctica has created widespread appreciation of special U.S. interests, including de facto recognition by other concerned governments that the American position is preeminent in the vast, unclaimed portion of the continent.[161]

Under the treaty, the United States was careful not to surrender any of its rights to Antarctic territory, which have been symbolically upheld through continuous occupation of a station at the geographic South Pole. Indeed, as early as 1961 one observer opined: "It should not be assumed that the United States, which rejects all other claims and has generally insisted that occupation should be effective in the more classic sense, has refrained from laying and carefully preserving at least the foundation for claims on its own behalf."[162] Viewed in historic and scientific perspective, the United States has sought to do precisely that. At the same time, it is well to recognize that whatever may be the "merits of American historic rights, it must be conceded that other countries have interests of equivalent validity in certain parts of the continent."[163]

Decision Making

The Antarctic Treaty has been criticized because it fails to establish any formal governmental mechanism. There exists neither a secretariat nor a permanent international headquarters.[164] The only organizational structure created by the treaty is the periodic meetings of the consultative parties. The United States currently supports creation of a secretariat, though it did not always hold that view.[165] During early meetings of the consultative parties, a highly controversial issue arose over Australia's desire to establish a secretariat in Canberra. U.S. concern with excessive bureaucratization prompted the United States to be a leading opponent against creating a secretariat. Privately, the United States opposed Australia's recommendation for at least two other reasons. First, a secretariat would include Soviet representation. Second, the institution might invite unwanted United Nations' interference in Antarctic affairs. In any event, Australia's recommendation was defeated without the United States having to speak against it.[166]

Though the Antarctic Treaty did not create what might have become an unwieldy bureaucracy, it did establish machinery to deal with problems of mutual concern. Antarctic policy is made by the consultative parties in regular meetings.[167] Representatives of the member states meet to exchange information, to consult on matters of mutual interest pertaining to Antarctica, to search for solutions to common problems, and to ensure implementation of treaty principles and objectives. Since 1961 there have been eighteen consultative meetings held in rotation in the capitals of member states, although the 1994 meeting convened during April 11–22, 1994, in Kyoto, Japan.[168] Consultative meetings are supplemented by other formal and informal contacts, such as special consultative meetings and "meetings of experts." Both regular and special consultative meetings have created a framework for member states to carry on many activities, ranging from the exchange of views to the negotiation of binding legal instruments. These meetings have been critical for reaching agreements designed to administrate and regulate activities in the south polar region.

Criticism has been leveled at the general policy of secrecy during consultative and other meetings, which officially prevent public access to documents, including reports of meetings and working papers. This situation affords scholars and the general public only a limited view of the real policy options under discussion.[169] The secrecy with which the Antarctic Treaty Consultative Parties (ATCPs) have conducted their deliberations has prompted several groups, particularly environmental organizations, to call for wider circulation of documentary material and more detail in reports of meetings. There has been some progress in this direction, but it has been slow. Most commentators agree that whenever publicity would affect negotiations, confidentiality should be maintained. Secrecy under such circumstances allows acceptable compromises that are less inhibited by political considerations. After decisions are reached, however secrecy is harder to justify. Outside commentators often regard secrecy as a screen behind which the Consultative Parties can hide various misdeeds.[170] Despite changes to provide greater dissemination of information, deliberations in consultative meetings and related sessions still tend to be more private than in most international meetings.[171]

The frequent assertion that the consultative parties embody an exclusive club that encumbers other states from participating in decision making entails more ideological rhetoric than actual substance. In addition to the original parties, any state signing or acceding to the treaty can obtain consultative status, if it demonstrates interest in Antarctica by conducting "substantial scientific research" in the region.[172] This requirement is not unreasonable given the many obligations that maintenance of the Antarctic regime imposes on treaty members and the high costs involved in hosting consultative party meetings. As of 1996, the twenty-six consultative parties consist of the original twelve signatories to the treaty, plus fourteen states that have gained consultative status since the treaty was signed in 1959.[173] Another seventeen states have acceded to the treaty. Acceding or non-voting parties accept all obligations under the treaty but do not participate in its operation.[174]

The first consultative meeting convened in Canberra on July 10, 1961, three weeks after the agreement entered into force. At this gathering, delegates recommended administrative arrangements for future meetings, including the preparation and dissemination of the final report.[175] The principal products of consultative meetings since then have been "recommendations" adopted by the ATCPs and submitted to their governments for approval.[176] These recommendations have supplied the main channels through which the consultative parties have sought to develop the Antarctic regime and further the treaty's objectives and principles.[177] Recommendations have covered a wide range of subjects, including special uses of the Antarctic; preservation and conservation of wildlife and living resources; facilitation of scientific research and international scientific cooperation; implementation of exchanges of information; and safety of operations and logistics on the continent.[178]

Decisionmaking is done by consensus,[179] which became ensconced at the first ATCP meeting by the requirement that recommendations be approved by all representatives present.[180] As of 1996, at least 209 recommendations have been adopted.[181] Recommendations, which have the status of executive agreements in U.S. law, become effective when approved by all the consultative parties[182] and must be written into each meeting's final report.[183]

Since 1977 the ATCPs have also held special consultative meetings to address specific issues, usually regarding either applications for admission to consultative party status or resource-related questions.[184] Other mechanisms have evolved for dealing with matters requiring common action, expert guidance, or links with outside bodies. To illustrate, meetings of experts have been held on Antarctic communications in 1962 and Antarctic minerals in 1973 and 1979.[185] The Scientific Committee on Antarctic Research (SCAR)[186] serves as a nongovernmental scientific advisory body to the member states. The ATCPs seek scientific advice from SCAR when needed to facilitate the management of the south polar region. For instance, SCAR has proffered advice on the creation of Specially Protected Areas and Sites of Special Scientific Interest, as well as on the regulation of pelagic sealing.[187]

Institution Building

One hallmark of the Antarctic Treaty remains its dynamic quality. Member governments point with undisguised pride to the treaty's capacity to expand as one of its most distinctive features. The Antarctic agreement has consistently sought to explore issues of concern and then create new institutional structures. The increase in human activities in Antarctica has brought greater understanding of the geophysical characteristics of the region and realizations about the need to preserve its pristine nature. As a result, the ATCPs have responded by producing additional measures and instruments to regulate new activities. While discussing Antarctica, the term "Antarctic Treaty System" often arises as a convenient label for subsuming those agreements and legal instruments appended to the treaty of 1959.

Throughout the life of the treaty, activities of the ATCPs have placed considerable emphasis on conservation and environmental obligations. Recommendations have been supplemented by formal legal agreements. The first important addition came in 1964 as the Agreed Measures for the Conservation of Antarctic Flora and Fauna,[188] followed in 1972 by the Convention for the Conservation of Antarctic Seals, which entered into force in 1978.[189] Two associated conventions also hold importance: the 1980 Convention on the Conservation of Marine Living Resources, which entered into force in 1982,[190] and the Antarctic Minerals Convention, adopted in May 1988 and opened for signature later that year.[191] The Environmental Protocol adopted by the ATCPs in October 1991 is the most recent legal appendage to the Antarctic Treaty System.

Conclusion

There is little doubt that the maintenance and continued evolution of the Antarctic Treaty System remains a prominent U.S. national interest; after all, the Antarctic Treaty developed from an American initiative. The United States has maintained a leading position in Antarctic policymaking and in fashioning and preserving the politico-legal order for managing the Antarctic region. The Antarctic Treaty System (ATS) has furthered the interests of the United States by preventing use of the Antarctic that would be detrimental to American scientific, environmental, security, and economic interests in the region. It seems obvious given the history of international claims and physical challenges in the Antarctic that the maintenance of the ATS remains essential for the effective pursuit of U.S. national interests in the region.

Since the Antarctic Treaty came into force, U.S. foreign policy officials have been resolute and persistent in their support for it and its subsequent development. The record shows that the United States has periodically reaffirmed the importance of the provisions that ensure the region remains free from military conflict and political contention.

4 / Making U.S. Antarctic Policy

SINCE THE Antarctic Treaty entered into force on June 23, 1961, United States Antarctic policy has been driven by two major objectives: To maintain and strengthen the treaty system and to maintain an active and influential American presence in the region.[1] Successful pursuit of U.S. interests in the Antarctic depends on the preservation of the Antarctic Treaty System (ATS).[2] Appreciation of U.S. Antarctic policy and programs, and the national interests that they promote, requires an understanding of the organizational structure for formulating and implementing goals and objectives. It is also important to identify the most influential actors shaping policies and programs, and the institutional mechanisms in place, as well as the outside forces, whether domestic or international, that bear on decision making. This chapter undertakes this task. In doing so, it emphasizes that, in the most fundamental sense, the executive branch determines the U.S. Antarctic policy agenda. The most influential actors—the central or chief policymakers—are the president, top advisors at the National Security Council, the secretary of state, the director of the National Science Foundation, and, to a lesser degree, the secretary of defense. The most important institutions are the White House, the Department of State, and the National Science Foundation.

Actors and Institutions in Antarctic Policymaking

The Executive Branch

A striking feature of U.S. Antarctic policy, particularly since the treaty's inception, is the remarkable consistency in the articulation of American interests. Aside from more explicit emphasis on environmental concerns and resource issues, American goals pursued in Antarctica today differ little from those pursued in the period immediately preceding the treaty. These goals have been supported by every presidential administration since 1959, which conducted major reviews of Antarctic policy and arrived at similar conclusions.[3] American presidents and administration officials have consistently affirmed their commitment to the pursuit of keeping the region free of conflict, promoting scientific research, protecting the environment, and affording nondiscriminatory access to Antarctic resources whenever consistent with rigid environmental standards. Chief policymakers have repeatedly expressed support for the Antarctic Treaty as the governing regime that best protects U.S. interests in the region, including preservation of U.S. historic rights and protection of future national interests in Antarctica. In 1965 Harland Cleveland, then assistant secretary of state for International Organization Affairs, articulated U.S. policy goals for Antarctica, which nearly mirrored those formulated in 1948. In summary, Ambassador Cleveland stated that the U.S. goals in Antarctica were: (1) to ensure that Antarctica is only used for peaceful purposes; (2) to foster international cooperation among nations active in Antarctica; (3) to support scientific research; (4) to support exploration; (5) to acknowledge the potential discovery of resources; and, (6) to preserve Antarctic animal and plant life.[4]

A decade later, when describing U.S. national interests in Antarctica, Dixy Lee Ray, assistant secretary of state for Oceans and International Environmental and Scientific Affairs, remarked that throughout the history of U.S. involvement in Antarctica these national interests had remained fairly consistent. Dr. Ray stressed that the United States has consistently held that Antarctica be used for peaceful purposes only and not constitute a source of international discord. American rights and interests must be protected; freedom of exploration and scientific research should be guaranteed; activities in Antarctica should be guided by established, nonpreferential rules; orderly administration of the area should be established; and uniform and nonpreferential rules applicable to all countries and nations for any development of resources in the future should be clear.[5]

The goals of U.S. Antarctic policy delineated four years later by Richard Atkinson, then director of the National Science Foundation, remained essentially the same: to ensure that Antarctica is only used for peaceful purposes; to prevent Antarctica from becoming an object of international discord; to foster cooperative scientific research; to promote the equitable and wise use of Antarctic living and nonliving resources; to protect the Antarctic environment; and to preserve U.S. historic rights and protect future national interests in Antarctica.[6] The same policy is evident today. The consistent commitment by chief U.S. policymakers to varied but interrelated national interests helps explain the coherence in Antarctic policy and provides direction and a unifying framework for dealing with Antarctic matters.

Even prior to signing the Antarctic Treaty, the executive branch occupied a preeminent position in Antarctic policymaking. More specifically, the most influential actors in Antarctic policy and programs have been officials within the White House, the Department of State, the National Science Foundation, and to a lesser degree, the Department of Defense. As with expressions of U.S. national interests, the basic structure of decision making for U.S. Antarctic policy has not changed significantly since the early 1960s, especially concerning responsibilities for policy and programs.[7] Upon conclusion of the Antarctic Treaty, the need arose to formalize the basic organization for conducting U.S. Antarctic activities. In early 1960 the president instructed the Bureau of the Budget to review the organizational framework for U.S. Antarctic policies and programs. Upon completion of the study, and following President Eisenhower's approval of its recommendations, the bureau on August 3, 1960, issued Circular A-51, entitled "Planning and Conduct of the United States Program in Antarctica."[8] The assignments of responsibility contained in the circular essentially confirmed existing arrangements.[9] The National Science Foundation would continue to coordinate and manage the U.S. Antarctic research program, including planning and funding requests for scientific research programs but excluding logistics support.[10] The NSF was also to serve as the clearinghouse and source of information on Antarctic records. In turn, the Department of Defense would continue to provide logistic support for U.S. activities on the continent.[11] Responsibility for U.S. Antarctic policy and programs remained with the National Security Council's (NSC) Operations Coordinating Board (OCB).

In early 1961, executive branch officials modified the policy framework for Antarctic affairs. President Kennedy believed that the Eisenhower National Security Council machinery, including the OCB, had sapped initiative from the Department of State.[12] On February 18, 1961, after only one month in office, he abolished the OCB. In doing away with the OCB (and some fifty interdepartmental committees that died with it), President Kennedy stated that "we will center responsibility for much of the Board's work in the Secretary of State."[13] With this shift of responsibility, the Department of State emerged as the locus "not only for providing policy guidance, but for ensuring coordination of all U.S. activities in Antarctica."[14] Shortly thereafter, in March 1961, the department's Antarctic staff in the Bureau of International Organization Affairs established the Interagency Committee on Antarctica.[15] The committee was not "a policy or decision-making body," but its work was critical for identifying areas where policy changes might be needed or where new policy should be developed.[16] Attendance at these meetings included representatives from the National Science Foundation, the Departments of Defense, Commerce, and Interior, the U.S. Information Agency, and other interested agencies.[17] The NSF, specifically its Office of Antarctic Programs, continued to coordinate and manage Antarctic scientific research.[18] Likewise, the Department of Defense remained responsible for planning and carrying out operations in support of scientific and other programs in Antarctica.[19]

The decision-making structure established after the abolition of the OCB was too informal for addressing a subject as multifaceted as Antarctica. This deficiency prompted the Johnson administration in 1965 to propose a more systematic arrangement for carrying out U.S. Antarctic policy. President Johnson formalized this structure on April 10, 1965, when he established the Antarctic Policy Group (APG).[20] The APG was to be directly responsible for "guiding our Antarctic policy and helping develop the U.S. program in that region."[21] The presidential imprimatur was placed on the APG shortly after its inception when, on May 1, 1965, President Johnson announced receipt of the first report from "his" Antarctic Policy Group established "at his request." Later, on May 20, 1965, the president reaffirmed United States goals in Antarctica as follows:

We stand behind the Antarctica Treaty and will do everything in our power to ensure that the Antarctica region will be a place of peace rather than a place of hostile international rivalries; we strongly favor international cooperation among the nations that are active in Antarctica; we support with all of our resources, scientific research in Antarctica, further exploration and charting of Antarctica, the development of new methods of transport and logistics in that vast region, and the preservation of unique plant and animal life. Finally, we earnestly hope that these projects of peaceful cooperation will yield resources which every nation needs and every nation can use.[22]

In a letter to President Johnson following the establishment of the APG, Acting Secretary of State Ball asserted that "the APG will take appropriate steps to insure that the views of all U.S. agencies having an interest in Antarctic affairs are adequately reflected in the work of the group,"[23] a practice that remains in place to this day.[24] Referring to the workings of the APG, Assistant Secretary Cleveland indicated as early as 1965 that the decision-making arrangement was both practical and comprehensive. "There is joint participation by the agencies concerned, beginning with policy and extending through programming to operation," and "the relations among the agencies involved are good; the lines of responsibility are clear;" and "the leadership of both the support force and the scientific program is in excellent hands."[25]

Responsibility for Antarctic affairs in the Department of State resided in the Bureau of International Organization Affairs (IO) until 1970. A departmental reorganization effective April 7, 1971, transferred responsibility for Antarctic affairs to the Bureau of International Scientific and Technological Affairs.[26] In conjunction with this realignment, the secretary designated the assistant secretary for scientific and technological affairs to be his representative and chairman of the APG. In January 1975, the Office of International Scientific and Technological Affairs was reorganized into the Bureau for Oceans and International Environment and Scientific Affairs (OES), and the assistant secretary for OES became the chairperson of the APG.[27] The NSF was represented by the assistant director for geosciences, and The Department of Defense was represented by the assistant secretary of defense for international security policy.

With the increasing importance of scientific activity in Antarctica, a National Security Council study in 1970 recommended, and the president agreed, that program funding and management responsibility for U.S. Antarctic research be consolidated within the NSF.[28] While announcing this change, President Nixon used the opportunity to reaffirm U.S. interests in Antarctica. In so doing, he noted that U.S. policy in Antarctica had three overarching objectives: (1) the maintenance of the Antarctic Treaty and the continued use of the continent for peaceful purposes, including preventing international discord; (2) the promotion of cooperative scientific research for the solution of worldwide and regional problems, including environmental monitoring and prediction and assessment of resources; and, (3) the protection of the Antarctic environment and the development of appropriate measures to ensure the equitable and wise use of living and nonliving resources.[29] These objectives remain the essential framework for U.S. national interests in the Antarctic today.

President Nixon's decision was promulgated in National Security Decision Memorandum (NSDM) No. 71, dated July 10, 1970, which directed, inter alia, the orderly transfer of the U.S. Antarctic Program to the National Science Foundation.[30] Reflecting the president's decision, the Office of Management and Budget (the former Bureau of the Budget) on August 4, 1971, replaced Circular No. A-51 with Circular No. A-51 Revised. Circular A-51 Revised provided the basis for consolidated planning, funding, management, and conduct of the U.S. Antarctic program. Concerning assignment of responsibility, the circular stated that the Antarctic Policy Group (APG) was to continue to guide U.S. Antarctic policy and instructed it to conduct annual reviews, in coordination with the budget cycle, of operations in Antarctica. The membership of the APG, according to the directive, consisted of the secretary of state (chairman), the director of the National Science Foundation, and the secretary of defense, or their designees. The Chairman had the authority to invite representatives of other agencies to participate on an *ad hoc* basis. The Interagency Antarctic Committee was to provide a coordinating function for the APG. Representation on the committee, as stipulated in the directive, included individuals from all agencies having significant interests or program activities in Antarctica as determined by the APG. The Directive also gave the APG authority to establish subsidiary committees to facilitate its work.[31]

Key members of the APG considered consolidation a welcome development. According to Richard Atkinson, then director of the NSF, centralizing responsibility would permit more accurate estimates of total program cost, make program planning more efficient, and facilitate the provision of operational and logistics support in a way more directly responsive to research program needs.[32] R. Tucker Scully, director of the State Department's Office of Oceans and Polar Affairs, similarly opined in 1979 that requiring the program to be managed as a single package "significantly added to the effectiveness and efficiency of the operation of the U.S. Antarctic research program and thus to the general satisfaction of our objectives in Antarctica."[33] Rear Admiral Travers of the Department of the Navy also regarded the transformation as enabling "the realization of a considerably more cohesive, effective and efficient program," providing "more effective program control and coordination," and improving "flexibility and efficiency in allocating program resources."[34]

Since the mid-1970s, all funding for the Antarctic research program has gone directly to the NSF.[35] As a result, the program's budget requests are made and defended by the foundation. The Department of the Navy continues to provide logistic support on a reimbursable basis.[36] As a reflection of the diminished role of the navy in Antarctic operations, the undersecretary of the navy on March 20, 1971, designated the navy's assistant secretary for research and development as the representative for Antarctic matters,[37] replacing the Office of the Chief of Naval Operations. As for all logistic support, the NSF pays for Coast Guard icebreaking operations in the Antarctic. The NSF employs a private contractor, Antarctic Services, Inc., to provide procurement and supply services, to manage seasonal construction activities, and to provide maintenance and operational support at Palmer, South Pole, Siple, and McMurdo stations. It also operates the program's only research vessel, *Hero*.[38]

It falls upon the NSF to develop goals for scientific research and to fund university, other nongovernment, and all government scientific research programs in the region. The Division of Polar Programs administers the U.S. Antarctic research program.[39] The division conducts no research itself but administers grants to scientists. Through its senior United States representative in Antarctica, the NSF provides onsite management for field programs. The NSF, in conjunction with the State Department, also arranges for cooperative research with other Antarctic Treaty states.[40] Although other executive branch agencies have interests and responsibilities in Antarctic research (e.g., the National Oceanic and Atmospheric Administration, the Department of the Interior, and especially the Geological Survey), none of these agencies has funds specifically for Antarctic research, and each can make a contribution only as a spinoff of its respective mission. The NSF can seek counsel of these agencies or provide funding for special projects.[41] As a practical matter, however, American scientists cannot conduct research in the Antarctic without approval and support of the NSF.[42]

In 1976, President Ford reviewed the funding levels and management arrangements for the Antarctic program.[43] As a result, he reaffirmed, with minor refinements, the principles contained in NSDM 71.[44] This decision was set forth in National Security Decision Memorandum 318 (NSDM 318), "U.S. Policy for Antarctica."[45] NSDM 318 clarified that funding was the NSF's responsibility, that the program would be managed as a single package, that civilian contractors should be used whenever cost effective, and that the Department of Defense and the Coast Guard were to maintain the necessary support capability.[46] NSDM 318 underscored the importance of maintaining an active and influential United States presence in the Antarctic. The president also noted that U.S. interests in Antarctica extended well beyond the normal range of responsibilities of the NSF and directed that the funds available to the foundation for the Antarctic are not to be used for other purposes.

President Reagan in early 1982 provided unmistakable evidence of the importance attached to the preservation and furtherance of American interests in Antarctica. In White House Memorandum No. 6646 of February 5, 1982,[47] the president reaffirmed the U.S. policy of maintaining an active and influential presence in Antarctica, asserting that such a presence is "designed to support the range of U.S. interests."[48] The policy stated that this presence was to include scientific activities, year-round occupation of the South Pole and two coastal stations, and availability of related, necessary logistic support. The essence of U.S. national interests in Antarctica was expressed again in 1984 by R. Tucker Scully, a senior official of the Department of State with extensive involvement in Antarctic policy. In testimony before Congress, Mr. Scully reiterated U.S. interests in the region: to support the Antarctic Treaty; to reserve Antarctica for peaceful purposes only; to prevent the region from becoming the object of international conflict; to ensure the protection of the Antarctic environment; to maintain Antarctica as a laboratory for scientific research; and to ensure that if resource activities become environmentally possible, the "United States will have a reasonable shot at participating in those activities on reasonable terms and conditions."[49]

The Bush administration remained firm on the Antarctic Treaty as the linchpin of U.S. Antarctic policy and on the steady pursuit of the full range of American interests in the region.[50] While hardly altering the NSC machinery, the President transformed the formal structure for consideration and development of U.S. policy. These changes were reflected in National Security Decision (NSD) 1, dated February 4, 1989. NSD-1 established three NSC subgroups: the Principals Committee, the Deputies Committee, and the Policy Coordinating Committee.[51] As a result, the Antarctic Policy Group became rechristened the Policy Coordinating Committee for Oceans, Environment, and Science, which officially convenes at the assistant secretary level.[52] Despite this change, the basic process for formulating Antarctic policy within the executive branch remains basically the same as it has been since the mid-1960s.[53] The State Department representative still chairs the group;[54] the lead agencies continue to be the Department of State, the National Science Foundation, and the Department of Defense. At the working level, the Antarctic Policy Working Group continues to function under the leadership of the director of the Office of Oceans and Polar Affairs, and the "action officers" are essentially the same people.[55]

Another notable development was the accompanying statement made by President Bush on November 16, 1990, while signing the "Antarctic Protection Act of 1990" (H.R. 3977). The most striking feature of the president's statement is that it is virtually identical to the formulations of U.S. Antarctic policy articulated earlier. President Bush emphasized that any new agreement among the ATCPs must reinforce the essential elements of U.S. Antarctic policy, namely: (1) maintenance of Antarctica as a zone of peace; (2) comprehensive protection of the region's environment; (3) preservation of Antarctica as a place for conducting scientific research essential to understanding the dynamics of the planet's natural systems; and, (4) maintenance of the Antarctic Treaty System as the framework for pursuing these goals.[56] These interests mirror the historical objectives of U.S. foreign policy in the Antarctic since the treaty's inception in 1959.

The Decision-making Process

From its beginnings formal representation on the Antarctic Policy Group consisted of the heads of the three lead agencies: the Department of State (acting as chair), the National Science Foundation, and the Department of Defense. Today, the APG reports on Antarctic matters as a body to the secretary of state and, through him, to the president. Some thirteen other departments and agencies, including Commerce, Interior, the

National Oceanic and Atmospheric Administration,[57] the Environmental Protection Agency, Transportation, and the Marine Mammal Commission, are ad hoc members, especially as working groups of the APG that focus on particular issues. The Department of State, however, retains primary responsibility. As chairman of the APG and lead negotiator on Antarctic policy issues, the State Department represents the United States at all Antarctic Treaty Consultative Meetings (ATCMs) and maintains requisite diplomatic relations with other participating governments. These duties include responsibility for legal matters related to interpretation and implementation of the Antarctic Treaty and related instruments. The Department of State also performs duties attendant to the special status of the United States as the depository government for the Antarctic Treaty, such as responding to queries from other governments about requirements for accession.

Each member of the APG can and usually does delegate participation to a lower ranking departmental official. When the APG was established in 1965, Acting Secretary of State George Ball designated the assistant secretary of state for international organization affairs to represent the department and to chair the APG.[58] The APG has convened at the assistant secretary level for almost its entire history.

The APG reviews U.S. activities in Antarctica each year as well as proposals for future U.S. Antarctic activities on a schedule coordinated with budget decisions.[59]

The APG is also responsible for dealing with ATCP meeting recommendations and decision making. The United States formally regards ATCM recommendations and decisions as binding in principle.[60] It also takes the view that these decisions apply to U.S. policy in the interim before they enter into force under the Antarctic Treaty.[61] The secretary of state approves consultative party meeting recommendations, following their review by the APG.[62] The APG meets as conditions require it, at the call of the chairman, or at the request of a member.[63]

Since the APG convenes infrequently,[64] a coordinating committee with representation from various interested agencies functions as a working group.[65] Again, the Antarctic Policy Working Group is chaired by the Department of State's director of the Office of Oceans and Polar Affairs.[66] The working group normally meets once a month but may hold more frequent meetings when necessary.[67] For example, during recent negotiations on an environmental protocol and the regulation of mineral activities in Antarctica, the group met at least once, and sometimes twice, a week.[68] The Antarctic Policy Working Group devotes considerable attention to developing and coordinating positions for use by U.S. delegates to regular and special consultative party meetings. The group also spends time on other negotiations under the treaty, such as the Commission for the Convention for the Conservation of Antarctic Marine and Living Resources.[69]

The working group performs a function similar to the former Interagency Committee by providing a mechanism for ferreting out important issues and presenting them to chief decision makers. These issues typically deal with areas where policy needs to be clarified or new policies adopted. The group offers policy options and recommendations for action.[70] Antarctic issues are mainly decided at the APG and working group levels; they are usually not so contentious that they must be resolved at the level of the NSC.[71] Also, executive branch officials do not operate in a vacuum in reaching decisions on Antarctic matters,[72] since an underlying consensus seems to prevail among those participating in U.S. Antarctic policy. Policy decisions made at the upper levels of the U.S. government thus frequently depend substantially on the substance (and slant) of information supplied from below. Accordingly, information furnished by the Antarctic Policy Working Group can be mainly responsible for shaping policy considerations and various perceptions about policy outcomes. Highly salient Antarctic issues that can not be resolved at the working group or Policy Coordinating Committee (PCC) level are usually referred through the National Security Council to the president for decision, though this occurence is rare.[73]

U.S. Antarctic Policymaking: The Minerals Issue

The decision in mid-1991 regarding the U.S. position on the Madrid environmental protocol reveals the difficulties that Antarctic questions can pose for American foreign policymakers. For that reason, the episode is worth closer examination. In 1990–1991, as Antarctic Treaty member states shifted national positions on the minerals treaty, it became clearer that U.S. efforts to get the Convention on the Regulation of Antarctic Mineral Resource Activities (CRAMRA) approved were failing. As governments declared neutrality or moved away from CRAMRA, the United States began to alter its own internal position. American policymakers became hamstrung by the lack of a decision. To set the stage for

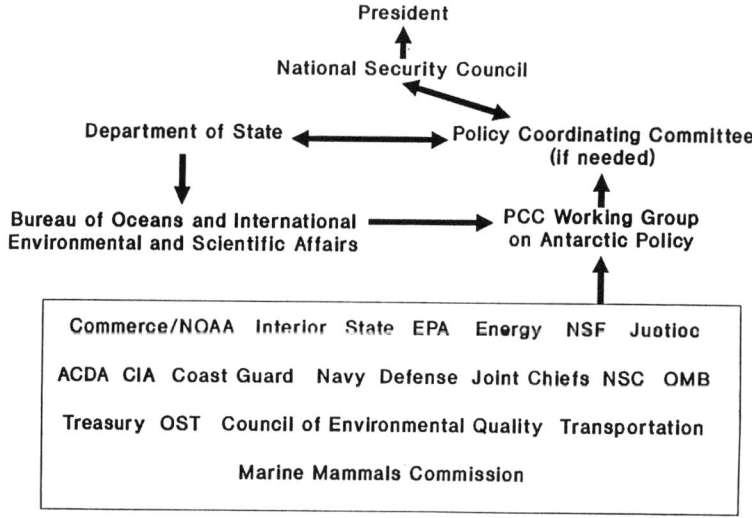

this situation, one has to appreciate the international political context in which it occurred.

The Madrid protocol emerged out of the diplomatic wreckage left by the failed promulgation of the Antarctic minerals convention. Between 1982 and 1988 the ATCPs negotiated a special regime for the regulation of prospecting, exploration, and development of mineral resources in the Antarctic. Agreement came in June 1988 on the text for a new minerals agreement, and in November 1988 CRAMRA was opened for signature in Wellington, New Zealand.[74]

Notwithstanding strong support for its proactive attributes by the United States, this Wellington minerals convention was stillborn. What had turned this formal legal agreement into such a disagreeable diplomatic option? Critical concerns were not addressed by the proposed minerals regime, and these difficiencies were translated by environmentalists in the United States and elsewhere into potent political weapons.

Environmentalists were seriously disturbed that the Wellington convention was a slippery slope leading inevitably to exploitation and development of mineral resources in and around Antarctica. It was feared that the agreement would provide incentives for commercial mineral activities. Its entry into force would have lifted the policy of voluntary restraint and effectively made commercial mining lawful. These developments could have increased the chances for discovering commercially exploitable deposits. Clearing the way for legal mining activity would promote prospecting, which could lead to mineral discoveries and exploitation. That pattern would inevitably produce environmental degradation.[75]

By the summer of 1989, political circumstances had scuttled the prospects for the Wellington convention. Two states essential for implementing the instrument—Australia and France—announced that they would not ratify the agreement. Any possibility of the mineral convention's entry into legal force was thus effectively preempted.[76] A sense of betrayal overcame those government officials in the United States, United Kingdom, and New Zealand who had worked painstakingly for seven years to diplomatically craft this minerals regime, only to have it torpedoed in the home stretch by Australia and France.

Aside from the political fallout among the ATCPs from the Australian-French action, two salient points deserve particular mention here. First, environmental considerations clearly motivated the Australian (as well as the subsequent French) decision to switch course on a minerals treaty for the Antarctic. Four publicized environmental disasters in polar waters early in 1989 no doubt seized the Hawke government's attention. On January 28, the Argentine supply ship *Bahia Paraiso* hit rocks offshore the U.S. Palmer Research Station on the Antarctic Peninsula, spilling some 250,000 gallons of diesel fuel into the sea; in the process thousands of krill and scores of penguins and other sea birds were killed, and a number of scientific projects along the coast were ruined. On February 7, the British resupply ship *HMS Endurance* hit an iceberg near Deception Island, reportedly creating an oil spill in Esperanza Bay. On February 28, the Peruvian research vessel *BIC Humboldt* ran aground and leaked oil into Fildes Bay off King George Island. But these three Antarctic episodes paled in comparison to the Arctic disaster on March 24 when the tanker *Exxon Valdez* struck a reef off Prince William Sound, Alaska, spilling some 11 million barrels of crude oil in the frigid waters. Thousands of otter, birds, and fish were killed as the oil washed ashore along a 45-mile pollution zone.[77]

The *Exxon Valdez* tragedy in the Arctic paradoxically became a media godsend for Antarctic environmentalists. It dramatically demonstrated the severe costs and real risks of transporting crude oil in frigid waters.

Genuine concern also arose that future minerals activities might exacerbate ozone depletion and thereby contribute to increased skin cancer rates, which already afflicted Australians in record numbers. Domestic political considerations over the rise of local Green Party power in Tasmania and the desire to retain Australia's sovereign influence over mining royalties in its claimed Antarctic sector also remained prominent motivations behind Australian Prime Minister Robert Hawke's decision. Still, the pervasive international publicity generated by this eco-catastrophe clearly bolstered the decision by the Australian government to oppose the minerals treaty.

The demise of the Wellington minerals convention opened the diplomatic door for considering and negotiating a potentially more far-reaching instrument for comprehensive environmental protection of the Antarctic environment. Ill feelings among ATCPs over the Australian-French policy reversal gave way to new aspirations for environmental protection. Prior to the October 1989 fifteenth Antarctic Treaty Consultative Party Meeting in Paris, Australia and France circulated a joint proposal calling for comprehensive measures to protect the Antarctic environment and its dependent and associated ecosystems.[78] Stimulated by this development, four other states—the United States, Chile, New Zealand, and Sweden—submitted draft proposals for comprehensive protection measures at the Paris ATCP meeting.[79] From these events came the decision to convene a special ATCM in Viña del Mar, Chile, which met from November 19 to December 6, 1990.

The stated purpose of the meeting in Viña del Mar was to discuss the feasibility of developing a comprehensive regime for protecting the Antarctic environment, but impasse over the minerals issue largely dominated the discussions. In the session's closing hours, however, a draft protocol on the Antarctic environment was personally submitted by Norway's Rolf Trolle Andersen. This so-called Andersen draft succeeded in supplying compromise provisions for a broad, generalized environmental regime for the Antarctic and became adopted as an unofficial working draft for subsequent sessions of the eleventh special consultative meeting.[80]

Following the November-December 1990 special ATCMs in Viña del Mar, the APG convened in January. During that meeting it was decided that any discussion concerning altering the U.S. position on the minerals treaty would not be permitted before the April 1991 special meeting in Madrid. Members on the APG were instructed to maintain confidentiality and not to discuss issues or exchange views on mining in Antarctica as they related to the protocol. Consequently, over the next three months the APG did not actively solicit alternative proposals on how the mining ban might be dealt with. The reason ostensibly given for this strategy was that the State Department was unable to reach internal agreement over the issue.

Just prior to the April 1991 special consultative meeting in Madrid, the APG reconvened. No new position was adopted at that time by the APG. Instead the decision was made to stay with the old position on CRAMRA. At Madrid, negotiations over the mining ban went forward, but agreement over the nature of the ban was elusive. The center of tension and difficulty lay not so much in international negotiations as between the U.S. delegation in Madrid and the instructions being sent from Washington, D.C. It became apparent that another negotiating session, set for June 1991 on the occasion of the Antarctic Treaty's thirtieth anniversary, would be necessary to reconcile differences not only between the United States and other ATCP positions but also within the U.S. position itself.

Between April and June 1991, the issue of Antarctic minerals and their potential strategic value for the United States became a highly charged political issue, generating intimate involvement at high levels of the U.S. government. The issue was vetted through the White House staff (in particular, the Domestic Policy Council), the Office of Management and Budget, and the National Security Council. The White House Chief of Staff John Sununu, who was known for his rather critical views on U.S. environmental policy, helped frame the discussion. An ad hoc "Environmental Policy Review Group" in the Office of Management and Budget (OMB) also figured prominently in the internal debate, in which Robert E. Grady, assistant director for national resources at OMB, and Roger Porter, assistant to the president for economic and domestic policy, played key parts. Exchanges also took place at various assistant and undersecretary levels about the Antarctic minerals question and the criticality of those resources to U.S. national interests in the coming fifty years. Undersecretary for Economic Affairs Robert Zoellick strongly endorsed a policy position that would preserve future U.S. access to Antarctic minerals. The assistant secretary of state for OES, Ambassador E. U. Curtis Bohlen, argued for a position more closely attuned to the draft environmental protocol. Just before the Madrid meeting in June 1991, a new American position was formulated, one that advocated support for an indefinite moratorium as opposed to a permanent ban on mining.

The APG process was unable to reach consensus on whether to sign the environmental protocol in Madrid in June. Reservations were expressed by Interior and Treasury over the length of the mining moratorium, especially over whether the ban was to be "permanent" or "indefinite."[81] The assistant secretary of OES communicated from Madrid with the White House. After consultations in the National Security Council and with Secretary of State Baker, the undersecretary for economic affairs cabled the U.S. delegation in Madrid to delay signing the protocol until closer consideration of its implications could be given by administration officials.[82] Those instructions meant that, at best, another negotiating session would be required.

Later in June, the question of the protocol and its mining ban went to the Policy Coordinating Committee (PCC) and subsequently to the White House, where it was discussed in the Domestic Policy Council. Ultimately President Bush himself made the decision, and on July 3 the White House announced that the United States would sign the protocol.[83]

Though not publicly revealed, it is interesting to speculate about the motivations behind President Bush's agreement to the protocol. Very likely he did not wish to evoke a negative issue during the 1992 presidential re-election campaign. It would certainly appear incongruous for the proclaimed "environmental president" not to support an obviously popular environmental position in the United States, as well as among the Antarctic Treaty governments. Similarly, the president must have viewed Antarctica as not being a major national security asset. If mineral resources were there, they would be inaccessible to anyone for the foreseeable future. United States opposition to the protocol thus hardly seemed worth the price of opening the issue to public debate and creating a critically unpopular policy position. For President Bush, the domestic gains of signing the Madrid protocol must have been weighed against the potential political costs of opposing it.

Decision making over the issue of banning minerals development in the environmental protocol negotiations reveals salient points about the U.S. policymaking process. For one, U.S. Antarctic policy appears to be generally formulated in an adaptive fashion, even though the essential elements of that policy have been laid out, agreed upon, and constant for more than thirty years.[84] To the extent that incremental changes have occurred, they have come from the extension of that Antarctic policy abroad. United States Antarctic policy is well considered, debated, and discussed; moreover, it is constructed upon decisions taken or discussed in previous meetings. U.S. Antarctic policy may thus be modified, but these changes come out of an interagency process of consensus building.[85]

During the decision-making process, turf battles over interagency interests naturally occur, though they are usually confined to the APG. Personalities also become intertwined in issue positions, as might be expected in an interagency process of debate and deliberation; but issues rarely become acutely emotional or made public. People involved in the APG who work on Antarctic questions are genuinely interested in the subject area. This commitment generally outweighs bureaucratic concerns and tends to mitigate turf battles. Antarctic issues thus can be dealt with in a reasonable way, without becoming highly charged with bureaucratic infighting or from conflicting interagency political loyalties.[86]

The executive branch is not alone in effecting U.S. policy for the Antarctic. Policy reflecting national interests is also rooted in broad patterns of congressional and interest group behavior on Antarctic issues.

Congress

Congress is obviously an important actor in U.S. Antarctic policy formulation. Congress authorizes appropriations and passes legislation to effect policy and implement programs. The Senate Foreign Relations Committee supplies advice and consent for treaties that have been signed by the executive. Moreover, congressional committees hold hearings on the oversight of Antarctic matters. These formal gatherings serve as occasional forums for public discussion of Antarctic issues and at times play an important role in Antarctic affairs. On balance, however, the impact of Congress on the formulation of U.S. Antarctic policy is relatively modest. Antarctic issues are not salient to most congressmen, much less the American voter. As a consequence chief U.S. decision makers on Antarctic questions retain a reasonably insulated position in designing and setting policy.

Congress is not expected to assume the lead in developing U.S. Antarctic policy. The congressional approach to Antarctic affairs is strongly influenced by personal, local, or special interest considerations, whereas attempts to deal with the broad range of U.S. Antarctic interests plays a much less important role. More specifically, two factors seem particularly decisive for understanding the lack of sustained influence on the part of Congress toward Antarctic policymaking.

One factor is the episodic nature of congressional interest in Antarctic matters. In general, congressional involvement in Antarctic policy is cyclical, shortlived, and influenced by the level of public or group interest in the issue at hand. This behavior brings to mind Richard Mayhew's model of members of Congress as "single-minded seekers of reelection" who engage in those activities most likely to enhance their political prospects. [87] Senator Ernest F. Hollings echoed this view when he remarked that "Antarctica has not been a hot issue in Congress, Antarctica is a very long way away, and there are not many voters there."[88]

As later chapters reveal, the salience of an Antarctic issue and the intensity of support for a solution among influential interest groups are often important factors in the prominence of an issue on the congressional agenda. Periodic interest in Antarctic affairs and actions designed to placate interest groups and achieve short-term gains, however, mitigate against an influential role for Congress in sustaining a coherent and comprehensive long-range Antarctic policy.

The second limitation is the decentralized nature of the U.S. Congress. The committee system, made more fragmented by increasing proliferation of subcommittees, produces dispersion of responsibility and jurisdictional overlap. Many different subjects can cut across the jurisdiction of several existing congressional committees, and no one subcommittee focuses in on comprehensive consideration of all Antarctic-related matters. This lack of central responsibility is clear. As Representative Leo W. O'Brien remarked in 1965 when considering the need for an independent Antarctic commission,

There is no place in the Congress of the United States where all these threads are gathered together and where there is a direct and rather full responsibility for activities in the Antarctic. That in and of itself indicates a fundamental weakness, to me. The Appropriations Committee, yes, that is fine. But its members are never in contact with us or we with them on it. We get involved with a dozen different committees and you have a feeling that whenever you approach this subject, it is like trying to pick a bowling ball without holes in it. No one has the responsibility.[89]

The House authorizing committee for the NSF is the Committee on Science (formerly Science, Space, and Technology). In the Senate, two different committees share jurisdiction over appropriations for NSF—

the Committee on Labor and Human Resources and the Committee on Commerce, Science, and Transportation. Nonetheless, as Appendix F shows, the number of congressional committees and subcommittees that have been at least indirectly involved in Antarctic affairs is substantial. The consequences of this fragmentation of responsibility include piecemeal consideration of related policies and significant overlap, thus hampering formulation of U.S. national interests in toto. This fragmentation places severe limits on Congress's ability to define a set of national priorities and to offer a unified U.S. position for international negotiations.

The political influence of Congress on U.S. Antarctic policy was demonstrated during 1990 by its handling of the Antarctic minerals issue. Wielding its instruments of institutional influence—joint resolutions, public hearings, public statements by congressmen, and adopted legislation—the 101st Congress became a salient political force in shaping the course of U.S. foreign policy on the Antarctic minerals question.

The Congress assumed an unusually active role in exerting its authority to effect Antarctic policy.[90] During 1989–1990, Congress took decisive action to protect the continent from the potentially adverse effects of mining and began to examine the environmental practices of federal agencies in Antarctica. Unlike the previous Congress when no bills pertaining to Antarctica were offered, during the 101st Congress five bills and four resolutions were introduced on the subject of Antarctica. Why this sudden surge of congressional interest in the remote, desolate frozen south polar region? Several answers are apparent.

First, public attention in the United States had been focused on Antarctica since the early 1980s. No doubt one reason for Antarctica's public prominence was the detection in 1985 of an "ozone hole" in the atmosphere above the continent, a phenomenon that has recurred every successive spring. Clearly another reason for public interest in Antarctica is the growing awareness of environmental interdependence throughout the world. In the United States during 1990, several prominent periodicals featured cover stories on Antarctica, focusing public attention on environmental problems affecting the continent. Highlighted prominently was the prospect that mining for Antarctic minerals and hydrocarbons might become a serious future possibility. This notion was propelled by the successful negotiation among the ATCPs in 1988 of CRAMRA.

The critical decisions by Australia and France in May and June 1989 not to support CRAMRA triggered action in the U.S. Congress. On September 26, 1989, Senator Albert Gore introduced Senate Resolution

206, which called for the United States to encourage immediate negotiations for a new agreement among ATCPs for the full protection of Antarctica as a global ecological commons.[91] The resolution called upon the president to negotiate a new agreement that would close Antarctica for an indefinite period to commercial minerals development and related activities. On October 5, 1989, House Joint Resolution 418, a companion to the Gore resolution, was introduced in the House of Representatives by Congressman Wayne Owens.[92] Like the Gore motion, the Owens resolution recommended that the president not present the Wellington treaty to the Senate for confirmation, pending negotiation of a new agreement among the ATCPs that would ensure the full protection of Antarctica as a global ecological commons.

Introduction of joint resolutions concerning a pending international question is a device that has been used before by Congress to influence American policy. The U.S. Constitution authorizes the president to negotiate treaties and later to transmit them to the Senate for advice and consent;[93] nevertheless, members of Congress often furnish advice to the president during the negotiation process by issuing statements, introducing bills and resolutions, and actually participating as members of delegations.[94] A joint resolution by the Senate Foreign Relations Committee that condemns a treaty negotiated by the State Department should give pause to the president, which was precisely the political signal sent about the CRAMRA agreement with the Gore-Owens resolution in late 1989.

To add legal credibility to the resolutions, special bills were introduced in both houses of Congress that would make the policies of these resolutions binding upon American citizens. On February 7, 1990, Representative Silvio Conte introduced H.R. 3977, the "Antarctic Protection and Conservation Act of 1990."[95] This bill prohibited any U.S. citizen from engaging in, financing, or knowingly providing assistance to any mineral resource activity in Antarctica. The undersecretary of commerce for oceans and atmosphere (i.e., the administrator of the National Oceanic and Atmospheric Administration) was authorized to assess civil and criminal penalties to enforce these prohibitions.[96] On the Senate side a companion bill was introduced by Senator John Kerry on May 3, 1990.[97] The Kerry bill, entitled the "Antarctica Protection Act of 1990," prohibited U.S. citizens from mining in Antarctica and called on the secretary of state to negotiate with other ATCPs a new international agreement that would ban mineral resource activities by all countries.[98]

Significantly, the introduction of these measures touched off an extensive debate between Congress and the Bush administration over the merits of the Wellington treaty and whether mining should or should not go forward in Antarctica. The political issues and environmental stakes became more clearly spelled out in a series of public hearings held by the Congress during 1990. The diversity in perceived U.S. interests between the Congress and the Department of State was revealed during a March 14, 1990, congressional oversight hearing on the convention before the House Committee on Merchant Marine and Fisheries.[99] During official testimony, the deputy assistant secretary for oceans and international environmental and scientific affairs stated that the CRAMRA agreement served three fundamental U.S. interests: (1) in the political context, it would preserve the principles and purposes of the Antarctic Treaty; (2) in the environmental context, it would protect the Antarctic environment; and, (3) in the resource management context, it would ensure that U.S. operators would have access to Antarctic minerals on a fair and nondiscriminatory basis.[100] The overriding view in Congress, on the other hand, was that CRAMRA would provide a mechanism for mineral exploitation and development of Antarctica, notwithstanding its acknowledged environmental safeguards.[101]

In May 1990 the Committee on Merchant Marine and Fisheries hosted a briefing by Jacques-Yves Cousteau on Antarctica.[102] This session apparently left a notable impression on members of Congress as Cousteau dramatically appealed for their help in preventing Antarctica from being subjected to commercial minerals exploitation. During a July 27, 1990, hearing before the Senate Foreign Relations Committee, a shift in the administration's attitude towards the mineral agreement was indicated by the new assistant secretary of state for OES, Ambassador Curtis E. Bohlen. In his statement, Ambassador Bohlen testified that the "administration's position [on the Wellington minerals convention] is evolving." He further asserted that the administration recognized that there existed scant domestic support for ratification of that treaty, that international consensus was still wanting, and that he could support a moratorium on all minerals activities for an indefinite period.[103]

On September 13, 1990, the House Foreign Affairs Committee submitted the Owens resolution to the full House, and four days later, the Senate Foreign Relations Committee did the same with the Gore resolution. In an unmistakably strong indication of their position against mining in Antarctica, the House of Representatives on October 1, 1990 passed the Owens resolution by a vote of 398 to 11. The resolution called on the president not to submit the Wellington convention to the Senate until new provisions could be negotiated that would ensure the full protection of Antarctica as a global ecological commons and closed Antarctica to minerals development for an "indefinite period."

On October 4, the Senate passed the Gore resolution. Over the next three weeks negotiations among the House, Senate, and the Department of State produced an agreement on the text of a single resolution, which was passed by the House on October 23 and by the Senate on October 24. The Congress then sent the resolution to the president for signature. The Gore-Owens resolution was signed into law by President Bush on November 16, 1990.[104]

Negotiations over the Conte and Kerry bills were more protracted. The bills passed on October 15 and 16, 1990, in the House and Senate respectively, but substantial differences in their provisions still had to be reconciled. The Conte version in the House contained a provision that would apply the National Environmental Policy Act of 1969 (NEPA)[105] to federal actions in Antarctica. The Kerry bill recommended that the secretary of state negotiate a new agreement that would "prohibit or indefinitely ban" Antarctic mineral resource activities. The administration strongly opposed extension of NEPA beyond the boundaries of the United States. It also wanted to retain flexibility in subsequent international negotiations by not being bound to both a permanent moratorium or an indefinite ban on mining. Consequently, the NEPA provisions in the Conte bill were dropped from the final version of the text, and Congress was persuaded to compromise by urging the president to negotiate one or more international agreements that would "prohibit *or* ban indefinitely" Antarctic mineral resource activities. The compromise Conte bill was passed by the Senate on October 24, 1990, and by the House two days later. On November 16, 1990, President Bush signed into U.S. law the Antarctic Protection Act of 1990.[106]

Other concerns about the Antarctic environment also pricked congressional interest in 1990. Of particular concern was the problem of waste management practices at U.S. bases in Antarctica, as well as marine pollution, garbage disposal practices of U.S. ships visiting Antarctica, and the application of NEPA to federal activities in Antarctica. Legislation was introduced by Congressman Walter Jones in the House in an effort to compel the National Science Foundation to clean up its act.[107] In the Senate a companion bill was introduced by Senator Gore.[108] Although the environmental issues addressed by these bills were acknowledged as important, Congress also realized that there would be opportunities in the future to deal with them. It appeared that the ATCPs would soon be undertaking major efforts to adopt environmental protection measures for the treaty system. The more critical concern for the moment was the minerals issue; consequently, congressional action on both bills was deferred.

General Assessment

Congress should hardly be expected to assume a lead role in developing U.S. Antarctic policy. The congressional approach to foreign affairs in general, as well as to Antarctic affairs in particular, is strongly influenced by personal, local, or special interest considerations. Efforts to deal with the broad range of American interests in Antarctica take on far less significance. Two considerations appear especially decisive for explaining the lack of sustained influence by Congress on Antarctic policymaking.

The first factor is the episodic character of congressional interest in Antarctic affairs. Involvement by Congress in Antarctic policy formulation tends to be cyclical, ad hoc, short lived, and strongly influenced by the level of public or group interest in a particular Antarctic-related issue. As Antarctica gains prominence in the news or gets more vigorously pressed as a concern by various interest groups, the attention of Congress in Antarctica rises accordingly. Antarctic issues that are highly salient for powerful interest groups offer incentives for congressmen to seek participation in policy formulation and engage in oversight activities. Periodic interest in Antarctic affairs and pursuing actions to mollify public interest groups and attain short-term gains, nevertheless, undercut the ability of Congress to assume a more influential role in sustaining a coherent and comprehensive, long-term policy calculus for the Antarctic.

The second limitation concerns the decentralized character of the institution. The committee system, rendered more fragmented by the increasing proliferation of subcommittees, disperses responsibility and fosters jurisdictional overlap and duplication. Several disparate topics crosscut the jurisdiction of a number of congressional committees, and no committee in either the Senate or the House functions as a focus point for Antarctic affairs. Antarctica is viewed as just another amorphous foreign policy issue affecting U.S. interests abroad.

Nevertheless, the number of congressional committees and subcommittees that have been involved in Antarctic matters in recent years is quite impressive. Since 1980, at least seventeen public hearings have been convened by House committees, and seven others have been held by the Senate. These public hearings, which are punctuated with expert and official testimony, promote public awareness about Antarctica and can generate political debate, albeit often unfocused, over U.S. policies toward the Antarctic.[109]

Interest Groups/Nongovernmental Organizations

As alluded to earlier, a familiar theme in American politics is the influence of powerful interest groups on the formation of distinct government policies. From the pluralist perspective, national interest may not always be the overriding determinant in national policy; rather, decision making is affected by the tug-and-tangle of competing groups as they vie for influence in the domestic political process. Presumably, then, the outcome of the policy process can reflect policies and programs that benefit specific groups rather than the state as a whole.

Environmental interest groups have clearly become a force to be reckoned with in the domestic decision-making process, for they can apply significant political pressure.[110] Congressmen have been compelled to take positions on Antarctic environmental issues, and U.S. policymakers have increasingly sought to work with, rather than against, environmental interest groups in the formulation of U.S. Antarctic policy.[111]

Although these groups have clearly swayed government policymakers in certain directions on Antarctic policy, their activities have not produced major deviations from the pursuit of the full range of U.S. national interests in Antarctica. The limits of environmentalist influence can be seen not only in the number of occasions when the consultative parties have neglected to adopt the groups' policy preferences but also in that delegates from nongovernmental groups are rarely included in the informal afternoon discussions at consultative meetings and resource negotiations.[112]

Interest groups have played multiple roles in the U.S. Antarctic policymaking process.[113] They can sway the formulation of law and policy as government agencies prepare for law-making forums. Interest groups can also influence the implementation of law and policy as national laws are enacted or in the legislative review and enforcement of Antarctic-related international agreements. Interest groups stimulate public awareness, foster development of public opinion, and further educate the public on Antarctic issues. Public opinion is often a necessary foundation for being able to affect policy formation and implementation. Environmental activists with a large membership or constituency are usually the most effective in this latter strategy. They communicate to their members through newsletters, annual meetings, local chapter meetings, and the like to explain issues and bring them to the attention of the media. [114]

Preparing a constituency to act on Antarctic issues is not easy. Interest groups must work through academic and public affairs communities and through professional and trade association groups to organize meetings, give presentations at various conferences, seminars, and annual conventions, and to write articles in trade and professional periodicals. Interest groups also sponsor meetings, publish materials, and hold press briefings to put out their views on Antarctic matters.[115] In addition, there is the phenomenon of the dramatic event, for which Greenpeace International has become well known. The latter employs nonviolent action, such as holding organized public demonstrations at Antarctic meetings with members dressed in penguin costumes. In 1987, Greenpeace launched its highly publicized venture to establish a scientific research station in Antarctica.[116] Though the avowed purpose of this expedition was to demonstrate that Greenpeace could qualify for consultative status under the Antarctic Treaty, the domestic political purpose was clearly more immediate and attainable: To galvanize public and congressional attention on the plight of Antarctica and to promote the need for legislative action to prevent more extensive environmental abuse there.[117]

To influence U.S. Antarctic policy and law making more directly, interest groups in the United States have employed three main tactics. The lobbying process employs efforts to convince policymakers that a certain policy or objective should be adopted.[118] Lobbying involves single-minded pursuit of one's objectives through whatever means are suitable for a particular audience. Coordinating members to write letters, send telegrams, or make phone calls to public officials and law makers are commonly employed.

A second tactic used by Antarctic interest groups is to provide factual information and expert advice. Supplying factual data and expert analyses for government representatives is vital to garnering support for those initiatives.[119] During the 1980s, Antarctic-related interest groups in the United States contributed yeoman's work in furnishing information about

Antarctica to public groups, the media, other governments, and scholars, as well as to U.S. officials and congressional representatives.

Finally, interest groups can serve a catalytic function in the formulation of Antarctic policy. They can work to galvanize national agreement over Antarctic issues by sponsoring informal meetings among officials with opposing views so that they can exchange opinions more frankly than is possible in formal conference settings. The goal here is to find common ground acceptable to all sides. Interest group forums also permit government officials to interact with various outside experts whose views probably would not have been otherwise solicited.[120]

Interest groups in the United States have different access levels, both formally and informally, to the different branches of government. In the executive branch, interest groups may attend meetings of and present their views to the Antarctic Advisory Committee. The intent here is for policymakers to hear the group's views on U.S. Antarctic policy as it is being developed and implemented. Frequent contact with U.S. policymakers in different government agencies may allow interest group members further opportunities to comment on and offer friendly views about Antarctic policy issues. Industry and environmental groups, for example, are represented in the Antarctic section of the Public Advisory Group on Oceans. By the tenth consultative meeting in 1979, the Antarctic and Southern Ocean Coalition (ASOC) and individual organizations were sending people to lobby at consultative meetings and related sessions, distributing materials to journalists and the public in an effort to widen knowledge about and concern for the Antarctic environment, and lobbying individual governments.[121] The U.S. delegation to the London conference on seals in 1972 included environmentalists among the "advisors." Since then, environmentalists have also been included on U.S. delegations to consultative meetings. For example, the United States appointed members of nongovernmental groups to their delegations at the mineral and fuel resources talks during the 1980s.[122]

In the legislative branch, interest groups have turned to the committees responsible for funding, law making, and review of Antarctic policy as a main vehicle for influencing policy. Interest groups often are invited to testify before these committees, and they often consult with committee staff members on Antarctic issues as policy is being considered and formulated. In addition, interest groups can play a role in the convening and planning of hearings and can urge Congress to impose stronger enforcement of Antarctic-related agreements. The latter is particularly appropriate during passage of any implementing legislation necessary for making an instrument operative under U.S. law.

Finally, there is resort to the judiciary branch. In the United States, interest groups, especially environmental groups, have increasingly used the courts to halt perceived environmental threats, to delay commercial development, and to test whether national laws have been violated by various policies. Antarctica has not been the subject of great judical attention in the past, principally because other efforts through the executive and legislative branches have largely succeeded, making widespread recourse to courts unnecessary.[123]

Several environmental interest groups have worked to influence U.S. Antarctic policy in recent years. Perhaps foremost among these has been the Antarctica Project (and the Antarctic and Southern Ocean Coalition) and Greenpeace. Others include the Environmental Defense Fund, the Wilderness Society, and the former International Institute for Environment and Development. Groups in the scientific community have also been influential, among them the Polar Research Board and BIOMASS, particularly during the CCAMLR negotiations in the late 1970s.[124]

The most effective strategy used by environmental interest groups on U.S. policymakers is "the force of pure reason."[125] The ability to understand Antarctic issues and discuss them in a reasoned, knowledgeable manner has suasive ways with policymakers.[126] On the other hand, interest group attempts to intensify pressure by creating a highly charged political climate usually has limited impact, often negative.

The media is often presumed to exert a prominent role in shaping public opinion and hence foreign policy formulation. Interestingly, according to some U.S. Antarctic policymakers, the most influential media force pertaining to Antarctic policy in recent years has been *The National Geographic* magazine.[127] Other American policymakers give little credit to the daily press or media in general for positively influencing U.S. Antarctic policy. For example, press coverage of the minerals issue in the United States has been described as "atrocious" and driven by sensationalism rather than accurate reporting. As perceived by U.S. policymakers, such slanted misrepresentation of political or geophysical facts tends to limit the media's influence, not enhance it.[128]

Interest groups have more direct influence than the media on the formulation of U.S. policy. Though environmental interest groups certainly exert domestic political influence on the formulation of U.S. Antarctic policies, they have been less than successful in fully meeting their espoused environmental objectives. This failure may be attributable to the fact that the U.S. government maintains a well-articulated Antarctic policy that relates U.S. goals in the region to a wide range of interests. This approach clashes with that of environmental pressure groups, which assert a narrower perspective and focus mostly on a single interest. Environmental groups put a high value on environmental protection and exhibit a willingness to depreciate other interests for the sake of preserving the natural Antarctic environment.

Another significant force in Antarctic affairs is the polar scientific community.[129] Though less strident and less cohesive than organized interest groups, scientists may be among the most politically influential groups in the formulation of U.S. Antarctic policy. Membership on the Advisory Committee for Polar Programs enables the scientific community to influence Antarctic affairs. The NSF established this committee on October 1, 1977, to provide "advice, recommendations, and oversight concerning support for research and research-related activities in the polar regions area."[130] The committee, originally consisting of fifty-five polar scientists, functioned through subcommittees representing various scientific disciplines.[131] The committee was restructured in 1983. Today it consists of twelve polar scientists from each of the five scientific disciplines in which the Division of Polar Programs supports research. The committee's principal purpose is to provide oversight of management and program balance in the division's Arctic and Antarctic programs.[132]

Assessment

The domestic political influences with the greatest impact on the making of U.S. Antarctic policy are the executive branch, Congress, environmental interest groups, and the academic scientific community.[133] The executive branch clearly retains primary influence in the policymaking process. Since there is no Antarctic department per se, all governmental agencies with interests in the Antarctica have policymaking input through the APG interagency process. This process is important, for it balances, brokers, and mitigates bureaucratic interests and facilitates cooperative interaction to produce meaningful policy. The interagency process refines the

U.S. position substantively and makes choices within broader policy frameworks.[134]

The White House plays neither a pivotal nor important role in the routine formulation or framing of U.S. Antarctic policy. When, however, a highly salient issue is not resolved at the lower levels of government, as in the case of CRAMRA or the environmental protocol, then the National Security Council, Office of Management and Budget, and White House staff may get involved.

Congress does have certain authority that can impact U.S. Antarctic policy, namely the powers of oversight, legislation, appropriation, and conducting hearings. But the influence of Congress over U.S. Antarctic policy is only intermittent, usually when an Antarctic issue becomes publicly visible. Although certain congressmen may have great personal interest in Antarctic matters, little substantial interest in Antarctica resides throughout the institution.

Other political factors sometimes influence U.S. policymaking for Antarctic issues; for example, when partisanship between a presidency of one party and a Congress of the other, causes one branch to use Antarctic issues to score domestic political points with the American public. Linkages also occur between substantive Antarctic issues and external events, such as United Nations (UN) General Assembly meetings or the 1992 UN Conference on the Environment and Development.[135] Significantly, too, external political factors can influence the formulation of U.S. Antarctic policy. There is an overarching foreign policy component in the relationship with other consultative party governments that flows mainly from the Antarctic Treaty process, that is, from dealing with those other governments and from the workings of the Antarctic Treaty System mechanism itself.[136] Opinions and policy positions of other influencial governments in the Antarctica Treaty System can affect U.S. policy on Antarctic issues.

Conclusion

The preceding analysis leads to some general conclusions about the nature of U.S. Antarctic policy. Antarctic policy issues are generally formulated and defined through the U.S. government in a cyclical fashion. At least three stages can be discerned through which Antarctic policy initiatives pass in becoming policy outcomes. First, there is agenda set-

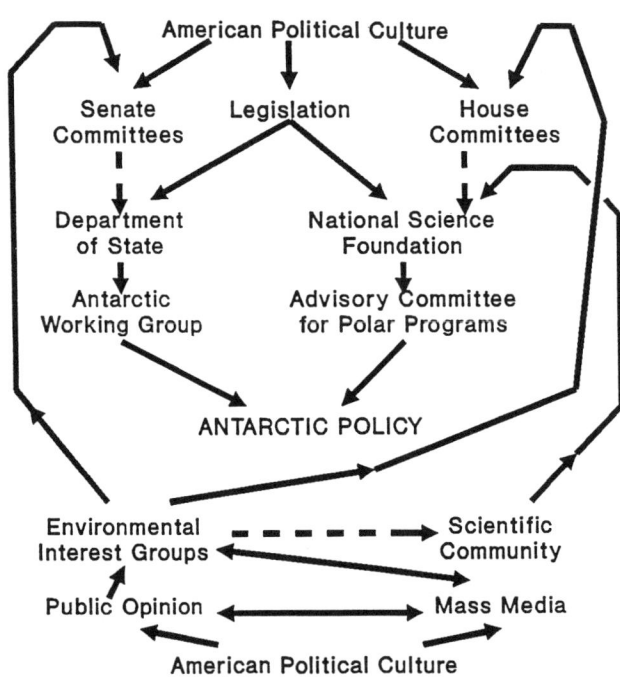

ting. Here an Antarctic item is placed for serious consideration before policymakers. For an item to become a public policy concern, it must first be an issue that policymakers are interested in addressing. Next, interested parties and representatives in Congress must be able to reach agreement on a policy, and, finally, some governmental agency—for example, the State Department—must have the incentives and resources necessary to carry out U.S. intent. Failure at any point during these stages can impair any policy from being effectively realized.

Clearly the ability to attract media attention can be an important preliminary step for governmental outsiders during the agenda-setting process. The strategy, tactics, and available resources of environmental groups foster use of the media for getting issues put on the public agenda. The broadcast media is driven by the desire to produce salable, interesting, attention-attracting news. Casting an Antarctic issue in a sensational, universal, or frightening manner can generate media attention and thus promote greater public awareness of the issue.

During the actual policymaking phase, various actions are taken by decision makers on the Antarctic policy item. In the United States, the focus of foreign policymaking falls mainly on the executive branch, especially the president, National Security Council, the Departments of State, Defense, and Commerce, and occasionally the Central Intelligence Agency. For Antarctic matters, the greatest weight for formulating policy generally falls on the Department of State and specifically on the Antarctic Policy Group.

Antarctic policy in the U.S. government is made adaptively, in a slowly evolving, accumulative fashion. Both practical and political reasons explain why this method so often prevails. Practically, it is impossible to consider all the alternatives to a decision and their conceivable ramifications. To be sure, the character of American pluralist society, the numerous parties involved in the policymaking process, and the inherent limitations on being able authoritatively to analyze and assess all the implications of foreign policy options strongly suggest that incremental policymaking may be inevitable. Politically, adaptation permits participants in the policy struggle to work from past agreements, with shared assumptions about possible policy outcomes. Likewise, policymaking inherently involves trade-offs, bargaining, and compromise. Adaptive decision making fosters outcomes that do not differ radically from past decisions on like issues. In this manner, the process of Antarctic policy formulation can go on with realistic expectations of what might be accomplished.[137]

Adaptation in U.S. Antarctic policymaking makes sense. This patent, bit-by-bit process simplifies governmental decision making. It facilitates action, permits assimilation, and fosters consideration of multiple views and facets of a policy so that it can be agreed upon by consensus.

The third stage in making U.S. Antarctic policy is implementation. It is important to realize that de facto policymaking on Antarctica may occur in any of the three branches of the U.S. government. Even inaction—the continuation of the status quo—is a form of decision making leading to policy formation. A critical point here is that resources must be available for implementing U.S. Antarctic policy. Resources most useful in this regard are the traditional ones of money, political support, information, and expertise that foster productive interagency action.

One must also consider the place of American society in the making of Antarctic policy. American society is pluralistic. In the United States, a democratic form of government theoretically engenders active involvement of the American people in determining what the government should do. Its institutional structures are designed to invite debate among the people, as well as among officials who govern and policymakers who formulate foreign policy. On Antarctic policy matters, these processes do, in the main, occur. Public policy is determined in large part through bargaining, compromise, and negotiation among various groups in society. This group activity emerges as a salient political variable influencing the public policy process.

An examination of U.S. Antarctic policy highlights the role of senior officials within the executive branch in the formulation and implementation of foreign policy. Congress and interest groups also figure prominently. One point, however, is fundamental: executive branch officials and institutions remain the key decision makers. The overall managing and planning strategies rest with the executive branch. Actual policy implementation is carried out through its agents, the secretary of state, the director of the National Science Foundation, and the secretary of defense. The critical importance of the executive branch lies, in part, in its ability to engage the full range of U.S. national interests, perceive trade-offs between group demands and established executive branch objectives, and steer the course of U.S. Antarctic policy in the desired direction.

Another conclusion is that U.S. Antarctic policies and the organizational framework for carrying them out are marked more by continuity than by change. In the main, the United States articulates an effective, long-standing, well-developed Antarctic policy with clearly delineated organizational arrangements for its execution. The evidence also shows that U.S. interests in Antarctica were present in American policy even before the 1959 Antarctic Treaty. Statements asserting U.S. Antarctic policy objectives have been reiterated on numerous occasions since then. This continuity helps explain the consensus among principal policymakers about U.S. national interests in Antarctica and the commitment to attain them. Responses of government officials and perusal of pertinent documentary materials strongly indicate that principal policymakers do concur on the main lines of U.S. Antarctic policy.

The claim of continuity in U.S. national interests toward Antarctica should not imply that these interests have remained static: Some significant reordering has occurred in the ranking of U.S. Antarctic goals. Foreign policies that fail to respond to changing global conditions or pertinent technological developments can undermine the pursuit of balanced and rational solutions to problems. National interests in a given geographical region or issue can change in priority without violating the continuity assumption. A shift in the ranking of U.S. Antarctic interests in the late 1980s and early 1990s—with economic interests becoming subordinated to environmental considerations—did occur and was prompted by two main factors. One involved the threat to the preservation of the Antarctic Treaty System that was posed by Australia's and France's refusal in 1989 to sign the minerals treaty. The second was improved knowledge about the sources and dangers of environmental degradation, which helped highlight the importance of keeping the Antarctic pristine and pollution free.

The United States remains committed in basic principle to the preservation of the Antarctic Treaty System. Other salient U.S. national interests tend to "grow out" of this basic policy: "To keep Antarctica free from international discord; to preserve freedom of scientific research; to ensure protection of the Antarctic environment and ecosystems; and to assure that Antarctic resources, if ever exploited, are developed under rigorous environmental safeguards, and that the U.S. have the opportunity to share in the benefits of resource activities in Antarctica, should those resource activities develop."[138] A final marked commonality in executive branch practice concerns the need for an active American presence in the Antarctic in order to exercise leadership in the ATCP decision-making process. The consensus on the nature of American national interests in the region and the consistent pursuit of those interests stand as the hallmark of U.S. Antarctic policymaking.

5 / Freedom of Scientific Research

PROMOTION and facilitation of scientific research has long been a leading U.S. national interest for the Antarctic region. During the 1960 Antarctic Treaty ratification hearings, the head for the U.S. delegation noted that a principal American objective in negotiating the agreement was to continue the freedom of scientific investigation and international cooperation that had prevailed during the International Geophysical Year (IGY) of 1957-58.[1] The Preamble to the Antarctic Treaty, which asserts among its major tenents the freedom of scientific research and international cooperation in scientific investigation, enshrines this American interest.

A recurrent theme throughout this study concerns the interrelationship among various U.S. national interests in the Antarctic. Antarctic science is no exception. James H. Zumberge's discerning remarks about Antarctic science are pertinent in this context. Testifying before Congress in 1984, Zumberge doubted that "science would be such a dominant force in the affairs of Antarctica without the political stability brought about by the Antarctic Treaty. International science in Antarctica is therefore facilitated by the Treaty."[2] That observation remains true today.

This chapter examines U.S. Antarctic scientific research to provide a clearer understanding of its significance as a secondary U.S. national interest. To this end, the chapter is organized into three sections. The first assesses key aspects of U.S. policy on scientific research in the Antarctic; identifies the major actors involved in the management of U.S. Antarctic science and their main responsibilities; evaluates the role that major environmental, nongovernmental organizations play in U.S. Antarctic research; and appraises the role of Congress, particularly concerning the funding of U.S. Antarctic science.

The chapter's second section considers the political and collaborative importance of Antarctic science to the United States in terms of two factors: the opportunity science affords for enhancing national prestige, including that of exerting an influential role in Antarctic decision making; and the significance of Antarctica as a global scientific laboratory. The final section examines the institutionalization of cooperation inaugurated by the Antarctic Treaty and explores the role it plays in promoting American scientific and other interests in the region. In this regard, attention is given to the benefits derived from the exchange of scientific information and personnel and to the stabilizing function served by international cooperation in the region. The chapter closes with some comments about the political and scientific importance of Antarctic science for the United States and the international community.

U.S. Antarctic Science and the Treaty System

Policy Objectives

The considerable, long-standing American interest in Antarctic science derives from the government's recognition of the multidimensional aspect of Antarctic research. The Antarctic region is a "natural laboratory" offering unequaled opportunities to scientists in many specialties for improving understanding of pressing regional and global issues. Long before the Antarctic Treaty was negotiated, the United States had acknowledged the importance of Antarctic science in its national objectives and had taken vigorous steps to put appropriate policies into effect.

As early as 1933, President Roosevelt had emphasized the importance of Antarctic scientific inquiry. Late that year, he wrote to Admiral Richard Byrd on the occasion of his second expedition to Antarctica and noted that "I am especially interested in the exhaustive study of weather on the Antarctic continent, . . . a weather maker for the greater part of the South American Continent. Your weather observations will undoubtedly be of great importance to . . . the scientific knowledge of world weather conditions."[3]

The International Geophysical Year that ran from July 1957 to December 1958 signaled a major turning point in Antarctic science. A cooperative enterprise involving sixty-six countries, the IGY included as a principal objective the comprehensive accumulation of knowledge about the Antarctic continent. During the IGY scientists learned more about the region in a few months than all the knowledge acquired since the circumnavigation of the continent by Captain James Cook in 1775.[4]

In March 1953, the National Academy of Sciences created the U.S. National Committee for the International Geophysical Year for the purpose of organizing and coordinating the participation of American scientists in this international endeavor.[5] The National Science Foundation played a major role in the planning of the U.S.-IGY program and obtained special funds from Congress for this purpose.[6] After four years of preparation and overcoming great logistics problems, U.S. scientists assumed their Antarctic posts.[7] Under the general direction of the National Academy of Sciences' National Committee for the IGY, the program was carried out by research institutions, government agencies, universities, and other organizations.[8] The U.S. participation was planned by leading American scientists and included projects in thirteen geophysical disciplines and areas of scientific activity, including cosmic rays, gravity, the ionosphere, meteorology, oceanography, solar activity, and seismology.[9] The IGY scientific discoveries were impressive. Also impressive was the extraordinary level of international cooperation that prevailed—cooperation that prompted energetic efforts to maintain and broaden such collaboration in post-IGY years.

Since the Antarctic Treaty's entry into force in 1961, Antarctic science has figured prominently as a U.S. national interest and has enjoyed direct government support. This interest has been articulated consistently since the definition of U.S. policy toward Antarctica. Indeed, a recurrent theme in U.S. Antarctic policy is the "major interest" in ensuring the maintenance of Antarctica as a laboratory for important scientific research.[10]

Scientific activities in Antarctica are not, however, devoid of political content. Government officials have consistently maintained that the successful pursuit of American interests in the Antarctic is dependent on an active and influential presence in the region. From the early 1960s, successive White House directives have reiterated that "an active and influential presence" requires conduct of scientific research, together with the maintenance of year-round stations and the availability of logistics support. Presidential Memorandum No. 6646 of February 5, 1982, on Antarctic Policy and Programs is a more recent affirmation of this policy.[11] In this memorandum, President Reagan directed that the Antarctic program be maintained at a level providing an active and influential presence in Antarctica, designed to support the range of U.S. Antarctic interests. Moreover, this presence should include "the conduct of scientific activities in major disciplines; year-round occupation of the South Pole and two coastal stations, and availability of related necessary logistics support."[12]

America's scientific activity in the region exceeds that of any other treaty member. Many observers recognize that the U.S. Antarctic program is the "premiere research program in the region."[13] Even during the late 1950s, American officials realized that the U.S. program was probably 50 percent larger than the Soviet program, which then ranked as the second largest program in the south polar region.[14] As recently as 1991, Peter Wilkniss, the director of NSF's Division of Polar Programs, said:

All the countries do not have the same amount of research in their bases going on. We certainly are the leader in this research from the ozone hole to understanding the whole ice sheet to the impact of looking out in the universe and to studying the deep water formation in the Antarctic. So I should say that next to us, nations like the United Kingdom, France, Soviet Union are a second tier of countries that invest in the research. Not quite as in the leadership role that we do.[15]

Bolstering U.S. scientific efforts is an effective transportation system. One former National Science Foundation official has observed that the U.S. air-operating capability in Antarctica is the best primarily because of its ski-equipped Hercules airplanes.[16] These airplanes can land on skis or wheels and give U.S. personnel sufficient flexibility to carry out field activities virtually anywhere on the continent.[17]

The Executive: Actors and Responsibilities

The U.S. Antarctic Program (USAP) embodies the government's strategy for scientific research and presence in Antarctica.[18] It is managed by the executive branch, currently under the NSF, and funded by Congress. The National Science Foundation (NSF) Act of 1950 created the NSF and designated it as the lead agency for development of U.S. basic and applied scientific research.[19] Its statutory mission made the NSF the obvious agency to manage the U.S. Antarctic Research Program. The foundation's close association with the American scientific community places it in a privileged position for planning and developing programs, considering the advice and research interests of government agencies as well as the views of the scientific community.[20]

The NSF has not always enjoyed full responsibility for managing U.S. Antarctic science, however. During the early years of USAP, the NSF shared responsibility with the Department of Defense. The foundation managed the U.S. Antarctic Research Program (for which the acronym USARP was used), whereas the Department of Defense, more specifically the Department of the Navy, was responsible for conducting operations in support of scientific or other programs in Antarctica. This situation changed in 1971, when responsibility for the entire program was transferred to the NSF. Since then, the foundation has managed and funded both the scientific and the logistics components of the U.S. program. At that time, reflecting the broader responsibility of the NSF, the program was rechristened the United States Antarctic Program (USAP).

The Division of Polar Programs (DPP) manages NSF-supported research in the Antarctic.[21] DPP's official responsibilities can be summarized as follows:

- Prepare the annual budget for review by executive branch agencies and appropriation by Congress.
- Develop scientific research goals for Antarctica, obtaining advice as needed from the scientific community.
- Review proposals for research projects, evaluate them in terms of relevance to program goals, scientific merit, and logistic feasibility, and grant funds for approved projects.
- Conduct detailed planning of logistics and provide logistics requirements and funds to the U.S. Naval Support Force Antarctica and to the U.S. Coast Guard.
- Manage U.S. stations, including engineering, construction, and maintenance.
- Contract with a commercial firm (currently ITT/Antarctic Services, Inc.) for operating the South Pole, Siple, and Palmer Stations, the R/V *Polar Duke*, and other services.
- Arrange cooperative scientific and logistics programs with other Antarctic Treaty members.
- Designate a senior U.S. representative in Antarctica with responsibility for on-site management of field research programs.
- Serve as the clearinghouse and source of information regarding

Antarctic records, files, documents, and maps maintained within agencies and nongovernmental organizations.[22]

The USAP is organized around long-range scientific program requirements that are developed, coordinated, and monitored by DPP. Long-range plans derive from the known research interests and capabilities of federal agencies, universities, and other private organizations and are devised while fully cognizant of U.S. national interests.[23] Plans receive numerous reviews, by the Polar Research Board, the Scientific Committee on Antarctic Research, and the Advisory Committee on Polar Programs, as well as by numerous individual scientists.[24] It is the Division of Polar Programs, however, that assumes the lead in formulating research initiatives, based on suggestions from working scientists, various governmental committees, and its own information, experience in logistics, and scientific research requirements.[25]

Following consultations with the Department of State and other government agencies and groups with expertise in Antarctic affairs, the division develops a comprehensive five-year program for scientific research. The plan is obviously more detailed and specific for the first year than for the fifth year.[26] Long-range projections are updated and revised annually to reflect past accomplishments and changes in scientific trends and to incorporate new items as appropriate.[27] This process is repeated every year so that, when the long-range plan is reviewed and updated, it is extended a year.

An important aspect of the five-year plan is its set of desirable and feasible goals that take into account the expected capability of necessary logistic supports. Since U.S. policy for Antarctica postulates a research program that can be responsive to future U.S. interests, flexibility remains a key consideration in DPP's planning for future research.

Each plan consists of three major elements that, though closely related, cover different facets. First, "Scientific Plans" involve the progressive updating of earlier long-range plans to ensure they reflect current research goals and objectives; second, "Other Activities" deals with miscellaneous U.S. interests in the Antarctic area or materials for nonscientific uses; and, third, "Development of Facilities and Techniques" concerns installations and methods to support U.S. activities "on the ice."

As the agency responsible for providing logistic support requested by the foundation, the Naval Support Force, Antarctica (NSFA), is intimately involved in the planning process. Following review of long-range plans, NSFA logistic experts develop a corresponding five-year plan for logistic support requirements. This plan includes an assessment of potential physical hazards and a determination of whether logistical requirements can be accomplished with available resources. Anticipated problems are discussed with DPP, and, if required, necessary modifications are made to scientific and logistics plans.[28]

Key for the U.S. presence in Antarctica has been the availability of icebreaker support to open supply channels to coastal stations and to perform other functions during the austral summer season (November through early April).[29] As the agency that provides essential icebreaker services and logistics assistance for science projects, the U.S. Coast Guard shares in the development of long-range plans.[30] The Department of State also reviews these plans in light of U.S. foreign policy objectives and suggests priorities.[31] Long-range plans are subject to approval by the Antarctic Policy Group, the interagency group headed by the Department of State, mainly to ensure that international considerations are taken into account before beginning any major scientific undertaking.[32]

Other agencies play important, though more limited, roles in Antarctic research. The National Oceanic and Atmospheric Administration (NOAA), an agency of the Department of Commerce, merits special mention. Included in NOAA's mission is responsibility for research in marine resources management. One direct responsibility includes fisheries, marine mammals (in conjunction with the Fish and Wildlife Service), and marine and estuarine pollution. In response to requirements of the Antarctic Marine Living Resources Convention Act of 1984 (Public Law 98-623), NOAA retains chief responsibility for research concerning the living marine resources of Antarctica.[33]

Another agency that plays a smaller but still critical role in United States Antarctic research is the United States Geological Survey (USGS), a component agency of the Department of Interior. As a scientific organization performing both basic and applied research to fulfill national needs, the USGS was particularly active in providing geologic information for development of policy positions on Antarctic mineral resources.[34]

Responsibility for topographic mapping in support of the U.S. Antarctic program also lies with the USGS. The United States maintains a program for aerial photography of sites presently or potentially suited for conducting scientific research.[35] Accurate maps and charts essential for planning, carrying out, and interpreting results of Antarctic scientific research, as well as for assisting logistics efforts.[36]

The NSF basically issues grants rather than operating research itself.[37] Under its modus operandi, the foundation receives proposals for projects from individual scientists in universities, private institutions, and government agencies. It traditionally funds research on the basis of outside peer review: Experts in a particular scientific field review the research proposals of other scientists to recommend grant support for those they consider promising.

A proposal is assigned to a program officer who oversees external reviews, evaluates reviewers' comments, and makes recommendations to award or decline it, taking into account other considerations, such as the relationship of the work to the field as a whole and other pending proposals, and the program's purpose and budget. . . . The proposal is sent to several people . . . identified by the program officer as knowledgeable on the topic. The reviewer receives standard instructions and forms and responds directly to the program officer. . . . All reviews are routinely sent to the proposer, but without attribution to individual reviewers.[38]

The foundation approves and funds projects deemed most worthy and capable of being supported logistically.[39] From those proposals found acceptable, the NSF develops a tentative, detailed research program for the current year, which falls within the general outlines of the long-range plan.[40] United States Antarctic science also takes place aboard research vessels, such as the *Eltanin*.[41]

The NSF reports on the preliminary results of research through the *Antarctic Journal of the United States* (AJUS). This journal is published four times a year and puts out an annual review issue that highlights the preliminary results, project-by-project, of the previous year's research in the Antarctic. Reports in the review issue reflect the diversity of U.S. Antarctic research, describe the fieldwork for that year, present preliminary analysis of data, and offer continuing studies of data previously acquired.[42]

Nongovernmental Organizations in U.S. Antarctic Science

THE POLAR RESEARCH BOARD.[43] At the national level, the most prominent nongovernmental organization involved in U.S. Antarctic science is the Polar Research Board. Established by the National Academy of Sciences[44] in 1958 as the Committee on Polar Research, the entity was

86 / EAGLE OVER THE ICE

renamed in April 1975 with the reorganization of the National Research Council. Its establishment was partly in response to a request from the director of NSF for assistance in the formulation of U.S. Antarctic scientific programs.[45] The purpose of the board is to stimulate and advance polar science in order to maintain and strengthen U.S. polar research capability.[46] Other major responsibilities derive from the board's role as the American national committee to attend meetings of the Scientific Committee on Antarctic Research (SCAR).

Since its founding, the board has served as a national advisory group on polar science and has assisted U.S. government agencies in developing and maintaining polar research programs that are responsive to scientific opportunities and national interests in the Arctic and Antarctic.[47] As a unit of the National Research Council,[48] "the Board is able to call upon the major talents of the scientific and engineering research communities of the United States and Canada to address both scientific and technical problems."[49] The board does not undertake research but rather issues comprehensive reports with recommendations for research activities in the polar regions. Its regional orientation and multidisciplinary character have resulted in a program of studies ranging from physical and life sciences to environmental and conservation matters.

The board meets semiannually to review its program, to develop U.S. positions on matters to come before SCAR, and to provide a forum for discussing programs and activities of governmental and nongovernmental entities in the Arctic and Antarctic. The group's diverse membership includes representatives from academia, industry, and government, with backgrounds in the biological, engineering, and physical sciences as well as expertise in polar environmental and societal issues.[50] The work of the board and its subgroups is supported by grants from the National Science Foundation, the National Oceanic and Atmospheric Administration, the Office of Naval Research, the Department of Energy, the Department of the Army, the United States Geological Survey, and the Andrew W. Mellon Foundation.[51]

NSF'S ADVISORY COMMITTEE FOR POLAR PROGRAMS.[52] In October 1977, the NSF established an Advisory Committee for Polar Programs to provide advice, recommendations, and oversight concerning support for research and research-related activities in the polar regions.[53]

More importantly, the committee serves as a source of independent advice to the staff of the Division of Polar Programs on both program management and program evaluation. It also determines special needs of the research community in polar programs and makes appropriate recommendations to the director of DPP.[54] The permanent chairperson of the committee is the director of the Division of Polar Programs. The group holds at least one annual meeting, which generally is open to the public. Various subcommittees represent different scientific disciplines. These subcommittees assist the DPP staff in the review and evaluation of selected research proposals as necessary. They also take part in evaluating the overall content and quality of DPP's research program.

SCIENTIFIC COMMITTEE ON ANTARCTIC RESEARCH. Most experts on Antarctic matters would agree that, at the international level, the most influential organization for scientific advice and coordination is the Scientific Committee for Antarctic Research, better known by its acronym SCAR. SCAR, a nongovernmental organization, was created in February 1958 as a committee of the International Council of Scientific Unions (ICSU) for the purpose of coordinating scientific activities in the south polar region.[55] SCAR accomplished its mission in such an exemplary way that it became a permanent apparatus of the ICSU.

As with the Antarctic Treaty itself, membership in SCAR is open to any state that establishes an ongoing scientific program in Antarctica.[56] Each member government of SCAR has a national committee for Antarctic research. For the United States, the Polar Research Board fills this function. SCAR's resolutions are transmitted by its delegates to their respective national committees and scientific unions for study, after which they are forwarded to appropriate governmental bodies for use in developing national Antarctic programs.[57] These resolutions are advisory and thus not binding on member governments.[58]

Although no formal link exists between SCAR and the consultative parties to the Antarctic Treaty, there is effective interchange. SCAR remains charged with furthering the coordination of scientific activity in Antarctica, with a view toward "framing a scientific program of circumpolar scope and significance."[59] More explicitly, SCAR coordinates research in Antarctica, sometimes initiates new research thrusts, responds to recommendations from the treaty parties, sponsors symposia and workshops, and maintains open lines of communication among its member states.[60] The work of SCAR is largely carried out by ten standing working groups, four subcommittees, and five groups of specialists.[61] Occasionally, these working groups promote or assist in collaborative studies with states where, through joint planning and execution, more effective use can be made of available resources.[62] SCAR also maintains groups of specialists formed to undertake specified tasks, often of an interdisciplinary nature. Though not permanent, these groups may have long life spans.[63] Perhaps SCAR's most valuable contribution is in providing a mechanism for exchange of information about national programs, plans, and achievements, and for organizing scientific symposia.[64]

COUNCIL OF MANAGERS OF NATIONAL ANTARCTIC PROGRAMS.[65] Each ATCP designates one or more individuals with responsibility for operational activities "on the ice." Until 1988 communication among national Antarctic programs was irregular and exchange of information was usually related to SCAR's Working Group on Logistics. The Council of Managers of National Antarctic Programs (COMNAP) was established in 1988 improve this situation.[66] COMNAP includes among its objectives the following: To provide a forum for exchanging information among the national operators and for seeking possible solutions to common problems; to provide input to SCAR responses to ATCPs' questions on operations and logistics; to review, with appropriate SCAR working groups and groups of specialists, projected programs requiring major international cooperation on operations/logistics; and to assist in the implementation of appropriate recommendations of the ATCPs.

COMNAP's subgroup, the Standing Committee on Antarctic Logistics and Operations (SCALOP), expedites the sharing of information and deals with technical implementation and whatever research support issues may arise.[67] COMNAP holds annual meetings. In each alternate year, it meets at the same time and place as does the SCAR. A major aim of COMNAP is to seek full advisory capacity to the ATCM on Antarctic logistic, operational, and environmental matters within its competence.[68] In 1995 the NSF funded about half of COMNAP's budget.

Congress: Oversight and Funding

The appropriations process provides Congress with substantial control over the operation of agency programs, since without funding these programs would flounder or cease to exist.[69] Congress adopts legislation establishing federal agencies and programs and sets their funding at certain levels. Then it enacts appropriation statutes that authorize agency spending. Congress thus addresses funding of USAP as part of the annual budget process for the federal government.[70]

Details of a particular agency's budget usually are worked out in the appropriations subcommittee that has jurisdiction over the agency. The House authorizing committee for the NSF is the Committee on Science (formerly the Committee on Science, Space and Technology).[71] In the Senate, two different committees share jurisdiction over NSF's funding: the Committee on Labor and Human Resources and the Committee on Commerce, Science and Transportation.[72]

A frequent source of uneasiness among some members of Congress has been the high cost of maintaining U.S. stations and support personnel. The south polar region is the most costly location on earth to conduct research because of its remoteness, hostile climatic conditions, and diversity and size of operational areas. Compared to the cost for logistics, funds necessary for direct scientific activity are relatively small.[73] During congressional hearings in 1979, Richard Atkinson, then director of the NSF, acknowledged that at first glance the balance between scientific and support activities seemed extremely uneven, with about 10 percent of the budget targeted for scientific activities and 90 percent for support activities.[74] Upon closer scrutiny, however, he noted that "those support activities in any other part of our budget would be viewed as part of the actual scientific program. A helicopter used at remote stations for biological research or glaciological research is as much a scientific tool as a telescope in the laboratory. So the distinction in our budget between support activities and scientific research is not the sharpest distinction one can make."[75] On the whole, appropriations committees have supported the NSF's budget requests for USAP.[76]

Less money and more cuts are what executive branch agencies can expect as a result of the Republican "Contract with America" and the Clinton administration's commitment to reduce the size of the executive branch bureaucracy. According to Representative Robert S. Walker, chairman of the recently re-named House Committee on Science, downsizing and budget cutting will not spare the "science" agencies.[77] Even so, it appears that the NSF will fare better than other agencies.[78] The Clinton administration is seeking a 3 percent increase in NSF's funding for fiscal year 1996.[79] It is not surprising that NSF Director Neal Lane characterized the $3.36 billion request as "good news in tight times."[80] The Antarctic research program and its operations and science supports would receive increases of 8.5 percent and 4.8 percent, respectively.[81] Funding for Antarctic logistical support would remain flat at $62.6 million.[82] In an era of severe budget constraints, this funding suggests continued congressional support for the means and ends of U.S. Antarctic science.

Political and Scientific Import of Antarctic Science for the United States

Antarctic Science: A Means to Further Political Goals

Scientific knowledge represents Antarctica's principal export, and it is likely to remain so in the future. One should not infer, however, that scientific activities in the region are therefore devoid of political content. Science serves a political role by expressing U.S. national interests in the Antarctic. The bases established on the continent during the 1957/58 IGY frequently were regarded as support for political/legal claims. Likewise, geopolitical motives are generally ascribed to the U.S. decision to build a station on the South Pole, the point where all Antarctic "sector" claims converge.

During congressional hearings on Antarctica in 1965, Admiral David M. Tyree, a former U.S. Antarctic projects officer, highlighted the political importance of Antarctic science for the United States. Tyree pointed out that the reason for the U.S. presence in Antarctica "is more basic than just science: it is to keep the United States in a position of leadership in the Antarctic both for our own security and in order that we may have a potent voice in the progressive development of this area of the world and to meet the future needs of mankind."[83] More recently, Richard Atkinson reiterated that "scientific research is the primary expression of the U.S. effort to maintain an active and influential presence in Antarctica."[84] This claim has lost none of its relevancy. Antarctic science continues to provide the United States with the opportunity to maintain a leadership position in polar research and exert strong influence over international Antarctic decision making.

Antarctic Science: A Source of Knowledge

In considering factors influencing U.S. scientific activities in the Antarctic, it would be misguided to stress only the political dimension. American scientific interests retain significance beyond that expressed in formal statements of policy.[85] The continent's geopolitical location and geophysical characteristics make it a unique laboratory for scientific research. Scientists flock to the Antarctic to pursue opportunities that exist nowhere else.

Three aspects of the Antarctic environment highlight its particular scientific importance to the United States. First, Antarctica exerts a major influence on the behavior of the world's oceans and atmosphere. The scientific study of this phenomenon increases understanding of the evolution of the world's climate. Second, Antarctica contains significant living and mineral resources, the scientific understanding of which is an important baseline for developing wise international management. Third, Antarctica provides a unique site for long-term baseline measurements of environmental pollution because of its remote location.[86]

If anything, the decades of research in Antarctica have accentuated the unique possibilities the region offers for the acquisition of scientific knowledge. This understanding is reflected in the focus and accomplishments of the U.S. Antarctic Program.[87]

The chief objectives of U.S. Antarctic research are to acquire better knowledge about the continent and adjacent marine areas, phenomena unique to Antarctica or best studied there, and the link between Antarctic phenomena and global processes.[88] Research regularly sponsored by the NSF consequently stretches across several disciplines, such as glaciology, meteorology, upper atmosphere physics, geology and geophysics, oceanography, and polar marine biology.[89] Fieldwork in each discipline often involves a combination of manned field studies and instrumented observations made from Antarctic ground stations, ships, aircraft, and satellites. Such studies have provided a significant return of scientific information in each discipline. Brief discussion of selected disciplines and examples of research supported by the NSF in these areas is therefore warranted.

UPPER ATMOSPHERE PHYSICS. The upper atmosphere in the polar regions has been called "earth's window to outer space," since many effects there are manifestations of deep-space phenomena.[90] Moreover, the physics of the upper atmosphere and near-earth space remain vital to national defense and communications.[91] The objective of most research concerning the upper atmosphere is to understand how energy emitted from the sun interacts with earth systems and how it produces phenomena such as magnetic storms, ionospheric disturbances, and the Aurora Austris.[92]

GLACIOLOGY.[93] Ice is central to the processes that determine climate, and it provides a unique record of climatic history over the ages. A major focus of research in glaciology has been to relate characteristics of the ice sheet to present and former global climates. Ice core drilling at Little America and Byrd Stations, for example, allowed for analysis of past climate and atmospheric constituents. This research enabled the NSF to discover past variations in atmospheric carbon dioxide and the appearance of pollutants such as lead in the atmosphere.[94] Working with Australian glaciologists, NOAA scientists have updated and refined the existing description of the Antarctic ice sheet. This work is important for paleoclimatic reconstruction as well as for providing a study of a possible carbon dioxide-induced warming trend. Moreover, NOAA's National Environmental Satellite Data and Information Service (NESDIS) collects and maintains a variety of Antarctic environmental data sets. For example, NESDIS prepares monthly sea surface temperature charts for various parts of the Southern Ocean. NOAA's World Data Center-A for Glaciology (Snow and Ice) also manages and distributes Antarctic data sets, including Antarctic Ice Concentrations and Southern Hemisphere Ice Limits.

METEOROLOGY.[95] Antarctica furnishes a major part of the global heat engine that determines world climate. The vast Antarctic ice sheet interacts with oceanic and atmospheric circulation to modulate global climate.[96] Accordingly, the behavior of the ocean/atmosphere system in Antarctica is expected to provide an early warning of climate change. The NSF's program in Antarctic meteorology centers on global climate, mesoscale systems, and atmosphere chemistry. The foundation's new Global Change Initiative calls for study of the earth as an interactive system of physical, chemical, biological, and geological processes, which should offer an integrated approach to understanding what is happening to global climate and what lies ahead.[97] The "ozone hole" over Antarctica has received considerable attention since 1985. NOAA, along with the NSF and NASA, remains at the forefront of international research on depletion of the ozone layer over Antarctica.

OCEANOGRAPHY.[98] The Southern Ocean makes up about 10 percent of the world's oceans and plays a major part in ocean mixing and global circulation. Its role in ventilating the world's oceans and supplying oceanic heat and moisture to the atmosphere makes the southernmost sea a key component of the global climate system.

USAP's research in oceanography began in 1962 with a systematic oceanographic survey using the USNS *Eltanin*, an ice-strengthened research ship. Through 1979 the *Eltanin* made fifty-five cruises, covering more than 500,000 miles in the Southern Ocean in connection with research in oceanography, geophysics, and biology. Current shipboard work extends and builds upon the *Eltanin* achievement using Coast Guard icebreakers and, beginning in 1984, a chartered, ice-strengthened research ship, the *Polar Duke*. This work has defined the physical and chemical oceanography of the Southern Ocean and has demonstrated the influence of Antarctic waters on the world ocean. Antarctic bottom water, for example, which is particularly cold and rich in nutrients, has been found to extend into the northern hemisphere, where it upswells to stimulate local biomass production. One of the long-term objectives of Antarctic research is to better understand the processes by which Antarctic bottom water is formed.

POLAR MARINE BIOLOGY.[99] Antarctica is also ideal for studying ecosystem structure and function, as well as for biological adaptations to extreme environments. Not only does such science retain value in its own right, it also remains essential for management and conservation of living resources in the polar regions and for understanding impacts of natural and human-induced environmental changes. The NSF's polar biology program aims to improve understanding of the relatively simple polar marine ecosystem to obtain a better understanding of ecosystems in general.

Other research concerns the ecology of specific marine organisms, especially krill. NOAA scientists have studied the relationship between krill abundance and ocean circulation features in the immediate vicinity of ice fronts.[100] NOAA has also made substantial contributions to the Biological Investigations of Marine Antarctic Systems and Stocks (BIOMASS), a program combining the efforts of several nations to improve their knowledge of Southern Ocean ecosystems.[101] Among its contributions to the First International Biological Experiment (FIBEX) in 1981–82, NOAA funded a National Marine Fisheries Service (NMFS) research project using acoustic techniques to survey krill populations. Achievements of FIBEX included acquiring a more reliable estimate of total krill abundance and new information on krill biology, ecology, behavior, and distribution.[102] Lessons learned in FIBEX were applied in planning the Second International BIOMASS Experiment—Phase I (SIBEX). SIBEX Phase I, designed to amplify work done in FIBEX on krill abundance and distribution, was completed in February 1984.[103] NOAA also participated in SIBEX-II, as well as in longer-term living resources research. One senior policy analyst at NOAA has even suggested that living resource research being conducted under the BIOMASS program may prove to be of considerable value in assessing the potential environmental impacts of Antarctic mineral activities.[104]

U.S. PERMANENT RESEARCH STATIONS.[105] Establishment of U.S. stations in Antarctica has not been haphazard, nor has it evolved without interagency coordination.[106] The United States has improved its network of useful stations by concentrating efforts in those Antarctic areas of greatest future interest.[107] Stations for scientific research in Antarctica are of two kinds: those that serve as both scientific observatories and staging bases to support field investigations (e.g., McMurdo, Amundsen/Scott, and Palmer); and stations that operate only as research sites. The United States presently maintains three active, year-round stations in Antarctica—McMurdo, the South Pole Station, and Palmer Station.

McMurdo Station, established in 1955 on Ross Island, is the world's southernmost post accessible by ship and the continent's largest scientific base and logistical facility. It is the logistics hub of the U.S. Antarctic Program, with a harbor, landing strips on sea ice and shelf ice, and a helicopter pad. Some 150 buildings, including a state-of-the art laboratory complex begun in 1988, can accommodate up to 1,200 people during the austral summer, especially from early November through the end of January. The winter population is about 180. Studies pursued at or near McMurdo include marine and terrestrial biology, biomedical work, geology and geophysics, glaciology and glacial geology, meteorology, and upper atmosphere physics.

Americans have occupied the geographic South Pole continuously since November 1956. The Amundsen-Scott South Pole Station was one of the earliest inland bases. The station was rebuilt in 1975 under a geodesic dome 53 feet high and 165 feet wide.[108] Inside, modular buildings support around 80 people in summer and about 20 in winter. The South Pole Station is supplied entirely by air from McMurdo 840 miles away. Although frequent flights come in during the summer season, the station remains isolated from mid-February to early November.

The South Pole Station's year-round scientific research efforts include glaciology, geophysics, upper atmospheric physics, meteorology, astronomy, and biomedical studies. The instruments and programs at the South Pole probe the remotest parts of the universe, monitor seismic activity in the interior of the sun, and track satellites in polar orbit.[109] NOAA maintains a two-person station at the South Pole, one of four baseline observation stations the agency operates in various parts of the world.[110] These stations, which comprise part of the global monitoring network of the World Meteorological Organization, measure atmospheric trace elements believed to exert potential impacts on climate. NOAA's South Pole research includes measurement of ozone, chlorofluorocarbons, and several components of solar radiation.[111]

Palmer Station, situated on a protected harbor on the southwestern coast of Anvers Island near the Antarctic Peninsula, is the only U.S. Antarctic station north of the Antarctic circle.[112] The station consists of two large and three small buildings, a helicopter pad, and a dock. Construction was completed in 1968, replacing a prefabricated wood structure built in 1965. Around forty people occupy Palmer in the summer. The winter population numbers about ten. Palmer Station is superbly located for biological studies of birds, seals, and other components of the marine ecosystem. It has a large and extensively equipped laboratory with seawater aquaria. Meteorology, upper atmosphere physics, glaciology, and geology have also been pursued at and around Palmer. Many Palmer-based research projects are undertaken in conjunction with the U.S. research vessel *Polar Duke.*

TEMPORARY U.S. RESEARCH SITES.[113] Temporary facilities are erected to support research performed away from the stations. Small tent camps serve as bases for field studies in areas where research can be accomplished on foot or by snowmobile. Huts are erected for larger summer projects expected to continue over one or several seasons at the same location. Resupply and transport are accomplished by airplane, helicopter, tracked vehicle, or hovercraft from McMurdo Station. Larger camps, comprised of prefabricated canvas and wood structures, support large-scale research projects that involve forty to sixty scientific personnel from November through January. A large camp might be set up and maintained using ski-equipped Hercules airplanes from McMurdo Station.

Institutionalization of Cooperation

The Antarctic Treaty institutionalizes the spirit of international cooperation in scientific research created during the IGY. International scientific cooperation is also institutionalized in the related agreements and instruments augmenting the treaty. This cooperation finds expression in several ways: exchanges of personnel among stations, joint planning and execution of large-scale science projects, and shared use of ships and aircrafts. Significantly, the United States has pursued collaborative projects with most Antarctic Treaty states. In the words of a Department of State official, the 1959 Antarctic Treaty has established the basis for

international cooperation in scientific research in the area, guaranteeing the freedom of research in return for the commitment among the parties to the treaty to share the results of the research, and to share in advance among the countries undertaking the activity plans for projected activities to allow for coordination and in fact for mutual cross-fertilization of research activities.[114]

These developments clearly serve U.S. national scientific interests in the polar south.

Cooperative ventures in scientific research, environmental protection measures, and exchange of scientific and technical data are commonplace in Antarctica.[115] While preparing a response to the UN secretary general's request in 1984 for views and information on Antarctica, the Department of State found that, throughout the history of the U.S. Antarctic Program, an estimated nine hundred to one thousand foreign scientists from some thirty countries had worked with the program.[116] The BIOMASS research program also demonstrates the scope of interstate collaboration in the Antarctic.[117] One of the largest cooperative efforts was the decade-long International Antarctic Glaciological Project.[118] Begun in 1971, Australia, France, the former Soviet Union, the United Kingdom, and the United States were involved in this joint effort to measure more precisely and determine more accurately the dynamics of the ice sheet overlying East Antarctica.

Polar Experiment (POLEX)-South, an international collaborative research program inaugurated in 1975, could reach beyond the scope of the glaciological project. Designed to examine the dynamics of climate in southern latitudes, POLEX-South specifically aims to integrate and expand existing national research programs on the atmosphere, oceans, and ice in Antarctica. The Dry Valley Drilling Project (DVDP), another cooperative activity, was conceived as a joint project involving Japan, New Zealand, and the United States. Completed in 1976, the DVDP embraced a five-year effort involving geophysical exploration, geological reconnaissance, and bedrock drilling in the dry valleys to the west of the McMurdo Station and in McMurdo Sound.

Cooperation as a Stabilizing Factor

Perhaps nowhere but in Antarctica have governments holding disparate interests and conflicting political philosophies demonstrated that a regime of mutual cooperation can work successfully. During the Cold War era, while government officials from East and West states often found it difficult to negotiate, the United States and the Soviet Union, as well as Argentina, Chile, and Great Britain, collaborated on Antarctic matters without signs of discord. Indeed, the low level of political tension promoted by the Antarctic regime, coupled with the capacity of participant governments to bypass intense international conflicts, remains nothing short of remarkable. Tucker Scully put it well when he said that

> one of the most important characteristics of the Antarctic Treaty system is that it has applied a unique conflict resolution/conflict avoidance approach to issues in Antarctica. The participants have basic differences over the legal and/or political status of Antarctica. Yet, they cooperate peacefully there. If one looks at the countries that participate in the treaty system and the state of their relations, there are participants that have no diplomatic relations working effectively together within the Antarctic Treaty System. There are nations whose relations are not all that they could be, including nations who in fact have been in armed conflict elsewhere and yet are still working together within the Antarctic Treaty system.[119]

The success of the Antarctic Treaty remains a vivid demonstration of the capacity of several states to engage in cooperative action. The demands of global science cannot be accommodated today without collaboration on a worldwide scale. Such a collaborative atmosphere remains critically important for U.S. Antarctic interests. It minimizes opportunities for conflict in the region; it emphasizes the need to identify issues and propose solutions to those issues; and it fosters the development of flexible mechanisms for negotiating and implementing necessary controls on human activity in Antarctica.

Conclusion

Scientific research in Antarctica constitutes a sizable U.S. program in terms of the quality and variety of the research and of the funds allocated to it. For the United States, the pursuit of science remains an important objective in itself. Antarctica, the world's last great wilderness, plays a vital role in sustaining life on this planet. The region is salient for global scientific research, which itself may be vital to earth's survival, especially for research dealing with the greenhouse effect and depletion of the ozone layer.[120] Scientists increasingly appreciate Antarctica's place in regulating the world's climate, the ocean currents, and the sea level.

The Antarctic Treaty not only recognizes the unique role of scientific endeavor, but it has encouraged, regulated, and facilitated scientific cooperation and research and the exchange of scientific information. Three decades ago, former Assistant Secretary of State for International Organization Affairs Joseph J. Sisco eloquently stated the value of the Antarctic Treaty for advancing U.S. interests in Antarctic science.

For the United States, the Treaty means that we may utilize our capability to the fullest. We are not confined to any specific area. We may seek the knowledge we desire in any part of the continent. Scientific knowledge, or the beneficial uses of Antarctica indicated by such knowledge, recognizes no political boundaries.[121]

Antarctic science is also critical to the U.S. policy goals of maintaining an influential presence in the region and of exercising a leadership position in Antarctic policymaking. The spirit of reasonable compromise and practical scientific cooperation that has prevailed in Antarctica remains an important vehicle for promoting broad U.S. national interests in the region. U.S. Antarctic research must profoundly apply to the contemporary situation, thus providing the United States with the opportunity to maintain a leadership position in science and technology on global environmental problems.

Equally important is to ensure that the results of U.S. Antarctic research are widely publicized and shared, thus benefiting the entire global community. Even governments outside the treaty framework can thus benefit from Antarctic research. In this sense, it can be said that U.S. Antarctic science is a "global good." The international community has benefited in other ways from the American pursuit of its interests in Antarctic scientific research; take, for example, the research on what is popularly known as the ozone hole. The discovery of the springtime depletion in stratospheric ozone in Antarctic has become one of the most crucial scientific and environmental discoveries of this century. The Protocol on Substances that Deplete the Ozone Layer, adopted in 1987 in Montreal—which mandated significant reductions by parties in the use of chlorofluorocarbons and halons—directly stemmed from this scientific discovery. The protocol thus demonstrated that a state's pursuit of its national interests has complemented, and even furthered, global interests.

Freedom of scientific research and the continuation of scientific cooperation in the Antarctic are prominent national interests held by the United States. The high accord given to scientific research was reaffirmed at the Antarctic Treaty Special Consultative Meeting in 1991 in the "Protocol on Environmental Protection to the Antarctic Treaty." Article 3 of the protocol states:

Activities shall be planned and conducted in the Antarctic Treaty area so as to accord priority to scientific research and to preserve the value of Antarctica as an area for the conduct of such research, including research essential to understanding the global environment.[122]

Recent developments in Antarctic affairs intimate that scientific research will remain the predominant human activity in the frozen south. If so, science will also be a principal force driving U.S. foreign policy there, and American interests will be better served if framed within the context of international peace and cooperation rather than rivalry over natural resources or national territorial imperatives.

6 / Environmental Interests

FEW GOVERNMENTS have devoted as much attention to environmental policy as has the United States. The United States stands as an international pace-setter in environmental protection, demonstrated through a diverse, dynamic set of environmental policies and controls. One need only examine executive branch directives and congressional legislation over the past twenty years to realize how pioneering and innovative U.S. environmental policies have been. Laws such as the National Environmental Policy Act of 1969 (NEPA),[1] the Marine Mammal Protection Act of 1976, and the Endangered Species Act of 1976,[2] among others, comprise a comprehensive set of policies for sound use and protection of the environment. The same government activism seen in efforts to safeguard the national environment are present in the international arena, particularly with regard to the Antarctic.

U.S. leadership in Antarctic environmental affairs stems from the American determination—unmatched by other Antarctic Treaty Consultative Parties (ATCPs)—to devote considerable attention and resources to protecting and maintaining the Antarctic environment. United States government officials have repeatedly underscored environmental protection as a fundamental U.S. interest in the Antarctic.[3]

This chapter surveys U.S. efforts to protect the Antarctic environment with emphasis on what has become, or is evolving towards, policy. It consists of five sections. The first examines conservation measures adopted by the ATCPs, giving particular attention to the role played by the United States in their promotion. The next section assesses U.S. law and policy for protecting the Antarctic environment. Section three deals with the policy, organization, and legislative context of U.S. Antarctic environmental activities in the executive and the legislative branches. The focus on the executive branch centers on overall policy, especially on formulating and implementing goals and objectives and the substance of policy directives and actions to promote environmental protection in the Antarctic. The congressional role in Antarctic environmental affairs has taken two courses. One is in serving a facilitating function, helping to promote executive branch goals and objectives for the region. The other course is as a constraining or inhibiting body, often acting as a brake on goal realization.

Section four surveys the role of prominent public interest groups involved in Antarctic affairs, including their access to Antarctic policymaking and their policy priorities and prominent manifestations. A brief fifth section highlights the convergence of Antarctic science and environmental policy. The main conclusion drawn here is that executive branch officials and institutions have been the most influential in shaping U.S. Antarctic environmental policies and actions. A second conclusion suggests that these policies are marked by consistency and continuity in their emphasis on conservation and environmental obligations in the Antarctic.

Environmental Protection and the Antarctic Treaty System

The Antarctic Treaty contains no specific reference to preservation of the environment. The only allusion to the environment is that of the responsibility of the parties, mentioned in Article IX, for the "preservation and conservation of living resources in Antarctica." Nonetheless, interest in preserving and protecting the Antarctic environment has remained especially important to the United States, as well as to other ATCPs. Pursuant to Articles IX and XII of the Antarctic Treaty, the ATCPs have dealt with Antarctic resources and regulated their possible exploitation through the development of several recommendations, agreed measures, and three special conventions. One noted commentator on Antarctic affairs has pointed out that all twelve original parties to the Antarctic Treaty easily agreed that measures should be taken to protect the Antarctic environment.[4] The undisturbed polar environment was regarded as a major asset to research, particularly regarding climate change and the dispersion of atmospheric pollution or radioactive fallout from atomic tests. Also, protection of the environment was seen as important in its own right. Governments in the southern hemisphere particularly feared that disturbing the Antarctic environment could lead to serious climate changes in their part of the world.[5]

The architects of the treaty were obviously aware early on of conservation priorities, for by 1966 they were seeking to augment it with special safeguards.[6] Concern even extended to the effects of tourism on scientific research, conservation, and the operation of Antarctic bases.[7]

Recommendations Adopted at ATCP Meetings

Supplementing the treaty and conservation conventions are recommendations adopted at ATCP meetings, which enter into force upon the unanimous approval (i.e., through consensus) of the participating governments. As one commentator has noted, the ATCPs developed some environmental measures long before such considerations became an international issue and before the 1972 United Nations Conference on the Human Environment.[8] Impressively, by 1979 the United States had adopted all 118 recommendations approved by the first nine ATCP meetings.[9] By 1995, at least 204 recommendations addressing meteorology, siting of bases, telecommunications, waste disposal, postal services, designation of especially protected areas, tourism, facilitation of scientific research, nongovernmental expeditions, logistics, rescue operations, safety, and exchange of information had been adopted by the United States, as well as by the other ATCPs.[10]

Continuing the trend, at the fifteenth consultative meeting in 1989 the ATCPs devoted special attention to issues such as waste management at Antarctic stations; marine pollution prevention, control, and response; environmental monitoring; and new types of protected sites.[11] Nearly 80 percent of the agenda for the 1989 consultative party meeting was dedicated to environmental issues.[12] Also in 1989, SCAR issued "Waste Disposal in the Antarctic," a report that recommended various improvements and changes in national practice.[13] Protection of the Antarctic environment became a priority at the sixteenth consultative meeting (1991). Participants at that meeting agreed "to continue and, when appropriate, expand programs to detect and monitor global environmental change and to establish monitoring programs to verify predicted effects and detect possible unforeseen effects."[14] The ATCPs subsequently used those environmental guidelines as the basis for their authority to draw up agreements governing living resources and minerals.[15]

The dynamic role played by the United States in promoting conservation measures at ATCP meetings is recognized outside the confines of the federal bureaucracy. As Lee A. Kimball, an expert on Antarctic affairs, recently declared, "[T]he United States has been a primary mover behind most of the environmental protection initiatives adopted at international treaty meetings in the last few years."[16] No question exists that the United States has manifestly shaped environmental measures adopted by the ATCPs.

Agreed Measures

The first concrete expression of concern for conservation measures among the ATCPs came in 1964 with the Agreed Measures for the Conservation of Antarctic Flora and Fauna.[17] Interestingly enough, the Agreed Measures are appended to a recommendation of the third consultative meeting (1964), but they remain distinguished as a self-contained text.[18]

It was the United States that initially recommended the adoption of the Agreed Measures.[19] Their purpose was manifold: to protect native birds, mammals, and plant life on the continent, safeguard against the introduction of nonindigenous species, prevent water pollution near the coast and ice shelves, and preserve the unique character of natural ecological systems. Besides listing specially protected species, the Agreed Measures allow the killing or capturing of native species only under a special permit. The protective measures established then were augmented by the introduction of the concepts of Specially Protected Areas (SPAs) and Sites of Special Scientific Interest (SSSIs).

SPAs are the most highly protected areas in Antarctica. Antarctic Treaty members designate SPAs "to preserve their natural ecological systems."[20] To enter the areas, individuals need a permit issued by a treaty member nation; however, SPA permits are issued only for scientific purposes that cannot be served elsewhere and that will not harm the natural ecological system. At least twenty-three sites were designated SPAs by 1996.[21]

SSSIs can be established to protect research where there exists a clear risk of interference or where there are sites of exceptional scientific interest.[22] SSSIs do not require a permit for entry. The sites are provided, however, with management plans that control the uses to which they may be put. Although SSSIs have been consolidated into the annex structure provided by the 1991 Madrid environmental protocol, by 1996 at least thirty-six SSSIs had been approved by the ATCPs.[23]

These networks of SPAs and SSSIs attest to the emphasis that ATCPs have placed on balancing the needs of science with long-term conservation. Indeed, the Agreed Measures rank among the most comprehensive and successful international instruments for wildlife conservation on land ever negotiated.[24]

Article VII of the Agreed Measures requires the United States to minimize harmful interference with the normal living conditions of any native mammal or bird by American citizens. Activities resulting in harmful interference include allowing dogs to run free; flying helicopters or other aircraft in ways that would unnecessarily disturb bird or seal concentrations, or landing close to such concentrations; use of explosives close to bird and seal concentrations; or any disturbance of bird or seal colonies during the breeding period.[25] Article XI requires the U.S. to ensure, if feasible, that the crew members of foreign flag vessels used in U.S. expeditions to Antarctica comply with these requirements.

In 1967, the Antarctic Policy Group (APG) issued policy guidance for the management of U.S. tourism, including making tourists subject to the provisions of the Agreed Measures.[26] The following year, the APG promulgated a policy requiring U.S. citizens also to comply with the provisions of the Agreed Measures concerning SPAs and SSSIs.[27] The adoption of controls over specially identified species and areas means that permits for the collection of such species or of organisms in designated areas will be issued only for compelling scientific reasons, and then only when collections do not jeopardize the existing natural ecological system or survival of the species.[28] The political effort involved in achieving compliance with the Agreed Measures illustrates well the seriousness with which the ATCPs view their responsibility to protect the Antarctic environment.[29]

Seals Convention

The Convention for the Conservation of Antarctic Seals (CCAS), a supplemental agreement to the Antarctic Treaty, was adopted at the seventh ATCP meeting, held in 1972,[30] and entered into force in 1978.[31] It applies to the seas south of 60° south latitude but requires reporting catches made in the pack ice even north of this latitude. This regulation marked a change in the Antarctic Treaty System, because it was now applying rules to the high seas.[32] As M. J. Peterson notes, the Seals Convention also marked the first use of a procedural innovation that would soon become characteristic: the "regulation of a specific activity by a distinct treaty to which states not party to the Antarctic Treaty could accede even if they had not or did not accede to the latter."[33] The CCAS seeks to limit the commercial exploitation of six species of seals in order to maintain optimal levels of the animal population. An annex to the CCAS specifies measures for conservation and for the scientific, rational, and humane use of seal resources. It also specifies maximum allowable annual catches.

On December 28, 1976, President Gerald R. Ford signed the instrument of U.S. ratification of the convention, the fifth country to do so.[34] In 1977, the United States approached the other ATCPs urging their ratification.[35] As a result, Belgium and the former Soviet Union both ratified. Canada and Brazil acceded to the convention in 1991.

The CCAS has been successful in preventing a recurrence of the extermination of some species of Antarctic seals.[36] The convention did not establish an independent secretariat but assigned the enforcement function to SCAR. The convention provides, however, for meetings of the parties every five years after commercial sealing begins, or at the request of any party if SCAR has indicated that sealing operations are causing substantial harm to any species.

Commentators have noted that the CCAS is exceptional in its explicit recognition of ecological values and its liberal protective measures. Moreover, the Seals Convention represents a "virtually unprecedented instance of intergovernmental regulation" instituted prior to commercial use beginning on a large scale and before a particular species becomes threatened."[37] The CCAS was promulgated at a time when there was no significant contemporary commercial harvesting of seals.[38] Due largely to the foresight of the ATCPs, scientific research now indicates that Antarctic seals have not been adversely affected by human activities in the Southern Ocean.[39]

Marine Living Resources Convention

Another appendage to the Antarctic Treaty System is the Convention for the Conservation of Antarctic Marine Living Resources (CCAMLR). The

United States played an instrumental role in the CCAMLR. The Antarctic Policy Group took the lead in seeking an international convention that would provide for the effective conservation of Antarctic marine living resources.[40]

At the ninth ATCP meeting in 1977, the ATCPs unanimously adopted a recommendation regarding Antarctic marine life. This recommendation, which reflected U.S. policy objectives, embraced an ecosystemic approach and urged that a comprehensive regime for the conservation of Antarctic marine living resources be concluded by 1978.[41] The Department of State's press release issued prior to the tenth ATCP meeting attests to the American commitment to an ecosystem approach. In it, the department stressed its desire that fishing be properly regulated "to ensure the health not only of harvested population but also of dependent and related species, including whales, and of the ecosystem as a whole."[42] The opening remarks of Lucy Wilson Benson, the head of the U.S. delegation to the tenth ATCP meeting, reiterated this commitment.[43]

The CCAMLR features several innovative proposals advanced by the APG. The APG advocated that the conservation regime be embodied in a separate international convention and supported the participation of nontreaty parties and international organizations having direct interest in Antarctic resources.[44] The most notable of the APG's proposals, however, pertained to adopting an ecosystemic approach for a conservation regime.[45]

As elaborated by U.S. executive branch officials, an ecosystem approach would involve comprehensive ocean management. It requires that the harvesting of a certain species, such as krill, be related to the consequences for other dependent species in the ecosystem.[46] The ecosystem approach should promote the "rational use" of all Antarctic marine life, inclusive of creatures on the continental shelf and bird life.[47] It was this approach that became the conservation core of CCAMLR.

Some knowledgeable observers have rightly pointed out that the CCAMLR goal of ecosystem management imposes rather demanding requirements for information. As in traditional species-by-species management, both direct fishing and the activities of predators must be taken into account.[48] Also, and especially since the convention commits the parties to restoring depleted stocks, managers must account for indirect impacts from fishing other species or stocks.[49]

The CCAMLR was signed in Canberra on May 20, 1980, and came into force on April 1, 1982.[50] The United States became a party in February

1982, as the seventh ratifying government.[51] The CCAMLR applies not only to the region south of 60° south latitude but also to the entire ecosystem south of the Antarctic Convergence. The preamble to the CCAMLR reflects the ecosystem approach, for it emphasizes "the importance of safeguarding the environment and protecting the integrity of the ecosystem of the seas surrounding Antarctica."[52] Harvesting must be done according to stated conservation principles; no harvested population may be allowed to decrease below those levels that ensure its stable continuation; ecological relationships between harvested, dependent, and related populations must be maintained and depleted populations restored; and changes that cannot be reversed within a decade must be prevented.[53]

The direct link between CCAMLR and the Antarctic Treaty is Article IV of the former, which reaffirms the decision to set aside the thorny issue of territorial claims. The CCAMLR also incorporates by reference other basic provisions of the treaty. For example, Article V of the Marine Living Resources Convention specifically recognizes the "special obligations and responsibilities of the Antarctic Treaty Consultative Parties for the protection and preservation of the environment of the Antarctic Treaty area."

To implement the basic objectives of the convention, the ATCPs established a permanent Commission at Hobart, Tasmania, a Scientific Committee (in which all Commission members are entitled to participate), and an Office of Executive Secretary. Members of the Commission are drawn from the original signatories to the Antarctic Treaty (Article VIII(2)(a)), from acceding members if they are "engaged in research or harvesting activities" (Article VIII(2)(b)), or from representatives of regional economic integration organizations. Critical to the operation of the Commission is its method of decision making. On matters of substance, consensus is required (Article XII(1)); otherwise, a simple majority may suffice (Article XII(2)).

The Commission's powers encompass a wide range of resource management functions, such as facilitating research; publishing and maintaining a record of all conservation measures in force; monitoring the activities of states with respect to the convention; implementing a system of observation and inspection; and, most important, the formulation, adoption, and revision of conservation measures. Among its regulatory powers are the right to specify the quantity of species caught, designation of protected species, and specification of size, age, or sex of harvestable species. The Scientific Committee acts as a consultative body to the Commission. The duties of the executive secretary remain unspecified but essentially consist of performing functions assigned by the Commission.

The convention is not without critics. Some are particularly skeptical that the mechanism can fully achieve its goals. F. M. Auburn, for example, finds that, in theory, the ecosystem may be safeguarded but practical application will probably not bear out the promise of conservation.[54] This prediction is partly because the Scientific Committee is composed of government representatives and has no powers of decision making.[55]

The Minerals Convention

The several successful components of the Antarctic Treaty System foreshadowed the need for establishing an agreement to regulate mineral exploration and exploitation in Antarctica. The third resource agreement, the Convention on the Regulation of Antarctic Mineral Resource Activities (CRAMRA), evolved out of six years of intense negotiations during the 1980s. The convention was adopted by a special consultative meeting in Wellington in June 1988,[56] and the United States signed it on December 2, 1988.[57] Any hope for consensus among the ATCPs for its entry into force, however, vanished in the spring of 1989. On May 22 Australia announced that it would not sign CRAMRA but instead would pursue the urgent negotiation of a comprehensive environmental protection convention.[58] On August 18, 1989, the Australian and French prime ministers announced in a joint statement that neither government would support the minerals accord.[59] Since agreement on the convention had to be unanimous among the ATCPs for its entry into force, their actions effectively scuttled it.[60]

CRAMRA was intended to fill a significant gap in the Antarctic legal system. Scientists and mineral experts have recognized that exploitation in Antarctica could only occur in the distant future. At the same time, U.S. foreign policymakers thought that a disorderly competition to extract minerals would do serious damage to the Antarctic environment, to scientific activity there, and to the spirit of collaboration that has been the hallmark of the Antarctic Treaty System. Also, as Philip Quigg has observed, a majority of the ATCPs had become more aggressive in pursuit of national interests more narrowly defined.[61] For some particularly energy-dependent nations, "Antarctica appeared to be the last hope of slaking their thirst for oil."[62] South Africa had felt intensely vulnerable to an oil cutoff much longer than other nations. Japan's energy predicament was without parallel; and an Australian official remarked that "every drop of oil that may be found in its claimed territorial waters belongs to Australia."[63]

Environmental protection was touted as a prominent rationale for developing CRAMRA,[64] and on balance, government agencies on the APG cited the convention as a visible means of extending effective protection to the Antarctic environment. For the United States, a key objective of CRAMRA was to determine whether or not resource activities were acceptable in light of environmental concerns; another objective was to ensure that such activities could be undertaken in strict conformity with rules specified by the convention.[65] No less important was the desire to ensure U.S. access to potential strategic minerals in Antarctica, should mining go forward under the CRAMRA regime. As some commentators have pointed out, the United States has consistently adhered to the principle of "free non-discriminatory access."[66] As with scientific research, ATCPs should have the right to go anywhere, and "spoils will go first and fastest to the nation(s) with the most advanced technology."[67] The alternative, according to U.S. government officials, would result in mineral resource activities occurring pursuant to national standards, which might not necessarily reflect the importance of Antarctica to the global environment.

Regardless of U.S. policy objectives or national rationales, U.S. support of CRAMRA was overwhelmed by the swell of international opposition among the ATCPs during 1989 and 1990. Although the United States did not abandon its concern for equality of opportunity to Antarctic mineral resources, its concern for accommodation among the consultative parties became critical. The overwhelming concern among U.S. foreign policymakers was that an uncooperative stance and the fueling of antagonism might wreck the treaty, thus endangering the whole range of American national interests in the polar south.

Environmental Protocol

The refusal by Australia and France to sign CRAMRA set in motion negotiations for a new legal instrument for ensuring the protection of the Antarctic environment. Following negotiating sessions in Viña del Mar, Chile (December 1990), and Madrid, Spain (March and June

1991), on October 4, 1991, representatives of the twenty-six consultative parties adopted by consensus the Protocol on Environmental Protection to the Antarctic Treaty.[68] This pivotal agreement, which builds on the Antarctic Treaty and related agreements, extends and improves the effectiveness of the treaty system to preserve the region's environment.[69] The protocol designates Antarctica as a natural reserve, devoted to peace and science, and prohibits all mineral resource activities, except for those related to science. It establishes environmental principles to guide planning and management of activities and sets up rules for environmental assessment and waste management. Fifty years after the protocol enters into force, any ATCP may request a conference to review its operation.

The United States signed the protocol in October 1991, and the Senate gave its advice and consent in early 1992. The remaining stumbling block is in obtaining appropriate implementing legislation, which would complete the ratification process for the United States. As manager of the U.S. Antarctic Program, the NSF has been working with other agencies and environmental groups to fashion legislation for implementing the protocol.[70]

Antarctic Environmental Protection: U.S. Policy and Law

Overall Policy

As with other American interests, protection of Antarctica's environment has been articulated consistently as a U.S. national policy since the late 1950s.[71] For example, President Johnson stated in 1965 that the United States supported "the preservation of unique plant and animal life" in the Antarctic.[72] In 1970, President Nixon stressed the need to protect the environment and develop suitable measures "to ensure the equitable and wise use of living and non-living resources."[73] In 1978 Patsy Mink, chairwoman of the APG, asserted that an essential element of U.S. Antarctic policy remained the "protection of the environment and preservation of the ecosystem from undue harm."[74] In his environmental message to Congress on May 23, 1979, President Carter stressed the need for the preservation of Antarctica's environment.[75] This commitment has thus often been reaffirmed.[76]

Addressing the anticipated benefits of a minerals agreement, in 1987, John Negroponte, then chairman of the APG, indicated that an agreement would ensure that the full range of U.S. Antarctic interests, including protection of the environment, were met.[77] More recently, in response to a letter from some members of Congress, the administration declared in 1990 that "the United States is committed to ensuring that human activities in Antarctica do not result in harmful impacts upon its environment."[78] In official policy rhetoric and declarations, then, the U.S. government has often asserted its commitment to protection of Antarctica's environment.

Administrative Arrangements

Several executive branch agencies participate in the management of U.S. Antarctic environmental affairs: the National Science Foundation (NSF), the Department of State, the Department of Interior's Geological Survey (USGS), the Department of Commerce's National Oceanic and Atmospheric Administration (NOAA), the Environmental Protection Agency (EPA), the Marine Mammal Commission (MMC), the Council on Environmental Quality (CEQ),[79] and the Department of Transportation's Coast Guard. The NSF, however, holds the leadership position in coordinating U.S. operations in Antarctica, including environmental considerations. The NSF's responsibilities include ensuring compliance with the environmental protection measures contained in the Antarctic Treaty and supplementary instruments, and developing measures to protect the environment, such as environmental monitoring and assessments.

The Antarctic Conservation Act of 1978 specifically directs the NSF to monitor the activities of American citizens in Antarctica; to ensure protection of native flora and fauna and preservation of the ecosystem; and to develop and issue regulations to control pollution in the Antarctic. When the legislation was passed, the NSF implemented the regulations for conservation of fauna and flora but decided to defer development of pollution control regulations. It announced that it would issue a separate regulation at a later time governing the discharge and disposal of pollutants in Antarctica.[80] Four years later, the NSF issued an internal directive on pollution control in lieu of formal regulations.[81] The foundation explained that formal regulations to control pollution were not needed at the time "because major activities in Antarctica other than scientific research are unlikely for the next several years," and "essentially all U.S. citizens who would be subject to pollution control regulations are within the U.S. Antarctic program."[82] The directive, entitled "United States

Antarctic Program Directive No. 84-1," became effective on October 1, 1983.

Some environmental groups have sharply criticized the NSF's weak compliance with the act, particularly its pollution control provisions. The Environmental Defense Fund (EDF), for example, has viewed the foundation's delay in issuing regulations as a deliberate attempt to circumvent compliance with the act. Moreover, EDF rightly criticizes the directive for not identifying which substances, if any, the NSF has determined to be pollutants under the act.[83] EDF and others have opposed assigning to the NSF responsibility for environmental matters on grounds that the agency's mission does not qualify it for the task.[84] Others disagree. A former NSF official has argued that the foundation is uniquely qualified to discharge these responsibilities:

> The expertise necessary is readily available to the NSF either from their own staff, from the pool of scientists and engineers with polar experience, or from other sources to assist in the development of the various regulations, plans, inventories, audits, inspections, monitoring programs, and assessments called for. . . . No other Federal agency has the ability to tap this available Antarctic and polar expertise as quickly and with a proven record of previous interaction.[85]

The Department of State is also assigned a principal role in U.S. efforts to protect Antarctica's environment.[86] Through its Bureau for Oceans, and International Environmental and Scientific Affairs, the department formulates U.S. positions on relevant international agreements and participates in a variety of environmentally related organizations. The department thus exerts considerable influence on U.S. policy positions concerning international pollution issues such as ozone depletion, protection of the oceans, and Antarctica. Also, the Antarctic Marine Living Resources Act of 1984 makes the secretary of state responsible for establishing a system of observation and inspection within the CCAMLR area (i.e., the area south of the Antarctic Convergence).

NOAA has become increasingly involved in Antarctic environmental matters. When NOAA was created in 1970,[87] it was made responsible for exercising "leadership in developing a national oceanic and atmospheric program of research and development."[88] In 1984 it assumed an expanded regulatory and management role for specific marine activities. The Antarctic Marine Living Resources Act of 1984[89] directs NOAA to design and conduct a directed research program in consultation with the NSF, the Department of State, and other appropriate agencies.[90] Its work was intended to supplement other research in Antarctica by U.S. scientists supported by the National Science Foundation.[91] The act also calls for the Coast Guard to provide the necessary icebreakers to support NOAA's research.

Other agencies play a more limited role. EPA's role in national environmental matters is understandably significant.[92] Concerning Antarctica, however, EPA's contribution consists mainly in advising the NSF and other APG agencies on technical and procedural aspects of environmental protection. The Council on Environmental Quality (CEQ) administers the process for preparing and reviewing Environmental Impact Statements (EISs) within the executive branch.[93] As the "expert" in the area of environmental assessments, CEQ has provided valuable advice to the NSF. The Marine Mammal Commission, in turn, is concerned with conservation and protection of the fifteen species of mammals found in Antarctic waters, as well as other aspects of the ecosystem that impact them.

Added Efforts at Protecting the Antarctic Environment—
Environmental Impact Statements

The National Environmental Protection Act (NEPA) of 1969[94] prescribes detailed Environmental Impact Statements (EIS)[95] for all major federal actions. Whereas NEPA's EIS requirement is clearly applicable to domestic activities of executive branch agencies, a long-standing debate has focused on whether NEPA applies to their actions abroad as well. That is, does U.S. environmental legislation pertain to the activities of U.S. nationals outside U.S. territory? Almost every agency has taken the position that NEPA is not exportable, including the Department of State (going back to 1970) and the Department of Defense.[96] Although the NSF has expressed its commitment to apply to Antarctica no less stringent protection than afforded by U.S. domestic environmental statutes,[97] it has concurrently resisted congressional pressures to apply NEPA to American activities on the continent. NSF's opposition stems from the fact that similar requirements exist under Executive Order 12114 and the Antarctic Treaty recommendations.[98] The Justice Department provided a legal perspective on the issue. In a memorandum to the president's counsel, the Office of the Attorney General advised that, upon issuance of a promised executive order addressed to international environmental problems, the department "will take the legal position that NEPA does not require the preparation of an environmental impact statement where agency actions abroad do not have a significant impact on the United States environment or the global commons."[99]

In response to the controversy, President Carter issued Executive Order 12114, "Environmental Effects Abroad of Major Federal Action," on January 4, 1979. The order established procedures governing the preparation of EISs for federal actions abroad.[100] It also attempted to balance extraterritorial environmental considerations with foreign policy concerns by allowing agencies discretion in determining whether or not to prepare an EIS. The express intent of E.O. 12114 was to further the purposes of NEPA "with respect to the environment outside the U.S., its territories and possessions."[101] Federal actions abroad that significantly affect "resources of global importance" require an environmental assessment from the responsible agency.[102] The order also introduces two assessment documents less stringent than EISs — concise reviews of the environmental issues involved and bilateral or multilateral environmental studies. For actions affecting the environment of other nations, the order permits agencies to choose the less demanding assessment documents.[103]

In compliance with E.O. 12114, the NSF has issued procedures for evaluating the environmental effects of USAP actions in Antarctica. These procedures, which were described in a proposed rule published in the *Federal Register* on July 10, 1990, involve a review of proposed actions to identify those with the potential to significantly affect the Antarctic environment.[104] This review is documented with an Environmental Action Memorandum (EAM). If a proposed action is excluded or exempted under the order, no further documentation is required. If the proposed action falls within the ambit of the order, a further review is required to determine potential environmental effects on a global commons, in any area outside national jurisdiction, or on foreign soil.

The preliminary analysis, usually referred to as an Environmental Impact Assessment (EIA), considers "whether the action: (1) adversely affects public health and safety through the environment; (2) has highly uncertain environmental effects; (3) involves unique or unknown environmental risks; or (4) together with other actions the effects of any one of which are individually insignificant, will have cumulatively significant environmental effects."[105] An EIS will be prepared if it is determined that potentially significant impacts might occur.

On June 20, 1980, the NSF issued a Programmatic Environmental Impact Statement (PEIS) on the U.S. Antarctic Program.[106] The PEIS, made available in draft form for public comment in August 1979, described the program and evaluated programmatic activities for a five-year period starting that year. It summarized program objectives, assessed possible environmental impacts, and described alternatives to the present program. The PEIS was reprinted and redistributed in 1984 with no changes. Since its publication, many changes in USAP activities have occurred. As a result, NSF officials produced a Supplemental Environmental Impact Statement (SEIS). The SEIS seeks to update baseline descriptions of the program and of the Antarctic environment presented in the PEIS and to assess the environmental significance of proposed program actions and alternatives.[107] A draft SEIS was made available for public and agency comment in January 1991;[108] the final version was issued in October 1991.

The final SEIS—on the environmental impacts of scientific research—incorporated the comments received during the public review period.[109] It also revised the NSF's environmental protection policy for operating its facilities and conducting research in the Antarctic; it specified full implementation of the Safety, Environment and Health initiative begun by USAP in fiscal year 1990; and it streamlined support operations to maximize efficiency and reduce personnel.[110] As the SEIS indicates, "the proposed action would use U.S. environmental laws and regulations as guidance for best management practices."[111] "To reduce program costs and environmental impacts, a long-term effort aims to reduce the number of military and civilian support contractor staff not directly needed for support of science at USAP stations. This would reduce materials and energy consumed, as well as wastes generated at McMurdo Station."[112]

Safety, Environment and Health Initiative (SEHI)

The need for enhanced environmental, health, and safety controls was pointed out in a 1988 study by the National Science Board,[113] and in a subsequent report entitled *Safety in Antarctica*.[114] Both reports identified the need to change the practices of the USAP. The latter report included seventy-two recommendations, some of which suggested administrative actions such as the appointment of a safety, environment, health (SEH) officer within the DPP, while others involved significant additional funding for actions such as major cleanups at USAP stations.[115] In response to these recommendations, the NSF proposed a $180 million, five-year SEH Initiative.[116] The initiative was initially funded in FY 1990, designating

$8.3 million to conduct studies of environmental problems that had been intractable and to implement initial actions on environmental and safety matters for which immediate needs were apparent.

Environmental Protection Agenda

The increased U.S. presence at Antarctic stations prompted the NSF in April 1987 to initiate an agenda designed to correct past environmental problems and anticipate future ones,[117] while simultaneously ensuring the integrity of scientific research conducted in Antarctica.[118] The agenda also directed the Office of General Counsel (OGC) to review environmental protection laws applicable to Antarctica to assess the NSF's compliance with these laws. The OGC report, "A National Science Foundation Strategy for Compliance with Environmental Law in Antarctica," was issued on December 29, 1989, and recommended actions to bring U.S. activities into compliance with environmental law. In response, DPP prepared an implementation plan and schedule for completing the recommended actions.[119]

The actions outlined in the plan included "removing from the continent all hazardous and toxic wastes from U.S. stations, installing equipment to prevent and clean up fuel spills, treating waste at McMurdo and Palmer stations, developing 'cradle-to-grave' waste management for all U.S. stations, and cleaning up contaminated areas near stations and field camps."[120] Funding for these actions would come from the dollars allocated to NSF under its major, multi-year Safety Environment and Health Initiative.[121]

USAP Environmental Awareness Program

Since 1980, NSF's Division of Polar Programs has taken several steps to promote environmental awareness among field personnel, official program visitors, and tourists to mitigate adverse environmental impacts.[122] For example, the division has updated an environmental reference guide on the Antarctic Conservation Act of 1978 and has developed an audiovisual briefing that is required viewing for all USAP personnel and visitors before they leave for Antarctica.[123] During the 1990–91 austral summer season, the division also began to offer "environmental briefings" to personnel arriving at USAP stations as part of its general orientation program.

The DPP's 1988 "Environmental Protection Agenda" proposed that biennial environmental inspections be conducted at USAP facilities by designated environmental inspectors. The explicit purpose of these inspections would be to appraise and report on local environmental problems at each USAP station and to counsel station management personnel on requisite environmental safeguards.[124] The DPP also proposed establishment of an awards program for accomplishments in Antarctic conservation by individuals of USAP,[125] as well as negative sanctions for violators.[126]

The Office of Polar Programs (formerly the Division of Polar Programs)[127] also supports a variety of studies that promote a better understanding of environmental impacts at USAP stations. Among these are studies on the benthic communities and oceanography in Winters Quarters Bay and on bird and seal populations in and around McMurdo and Palmer Stations.[128] In the late 1980s, NSF also established long-term ecological research sites in the vicinity of Palmer Station to study the effects of natural interannual variability of sea ice on marine phytoplankton, krill, krill-feeding fish, birds, and mammals.[129] More recent research on environmental impacts at USAP facilities has focused on the effects of the 1989 oil spill offshore Palmer Station by the Argentine vessel *Bahia Paraiso*.[130] USAP was not only heavily involved in cleanup operations during and after the accident, but it also sponsored studies to monitor impacts from the spill.[131]

The Role of Congress in Antarctic Environmental Affairs

Congress has been far more active in Antarctic environmental policy than in making decisions about scientific and security interests. Congressional action on environmental policy has been instrumental in implementing executive branch objectives through funding and legislation. On occasions, however, congressional action has constrained the ability of the executive branch to accomplish stated goals and objectives. The Convention on Mineral Resource Activities is a case in point. It might therefore be useful to examine the congressional role in Antarctic environmental affairs in terms of both its facilitating and inhibiting functions.

Congress as Facilitator

Ratification and enforcement of international legal agreements often involves enactment of corresponding domestic law. As a facilitator, Congress has been essential for instituting appropriate mechanisms to promote executive branch objectives, such as revising existing law or enacting legislation that provides a statutory basis for executive branch policies and goals.

For example, the National Science Foundation had applied the Agreed Measures as interim guidelines since their adoption in 1964 at the third Antarctic Treaty Consultative Meeting. Executive branch officials, however, recognized the need to enact specific legislation for enforcing the guidelines and controlling the activities of U.S. citizens in Antarctica. As a result, on May 23, 1977,[139] the Department of State presented legislation to both houses of Congress as S.R. 1691 and H.R. 7749.[133] The department urged prompt enactment of this proposed legislation "in view of the interest of the United States in supporting efforts for the protection of the Antarctic environment."[134]

To implement the Agreed Measures and other Antarctic Treaty recommendations for the United States, Congress enacted the Antarctic Conservation Act of 1978.[135] The act provides for the conservation and protection of the fauna and flora of Antarctica and of the ecosystem upon which it depends, consistent with the Antarctic Treaty, the Agreed Measures, and Recommendation VII-3 of the eighth Antarctic Treaty Consultative Meeting.[136] Section 6 of the act designates the NSF as the lead agency for pollution control, monitoring, and enforcement in the Antarctic. Section 5 directs the NSF to establish a regulatory system, including the issuance of permits, to control the taking of plants and animals native to Antarctica and the introduction of non-native species into the region.[137]

Environmental protection in the Antarctic has also been substantially aided by the enactment of the Antarctic Marine Living Resources Convention Act of 1984 (AMLRCA).[138] The AMLRCA applies to all species of marine animals, including birds, found south of the Antarctic Convergence and makes it unlawful for any U.S. national to engage in harvesting those species or to conduct associated activities in violation of the convention.[139] The act authorizes the secretary of state, with the concurrence of the director of NSF, to appoint the U.S. representative to the Commission for the Conservation of Antarctic Marine Living Resources and its Scientific Committee, to determine when a conservation measure is unacceptable to the United States, and to agree on behalf of the United States to the establishment of a system of observation and inspection.[140]

Congress as Constrainer

Congressional action has, on occasion, constrained the ability of the executive branch to accomplish stated goals and objectives. With respect to Antarctica, this inhibiting function of Congress is exemplified by attempts to alter the direction of U.S. Antarctic environmental policy. Institutional weaknesses affecting congressional policymaking have tended to amplify the inhibiting role. The 1988 minerals agreement, in particular, became a polemic issue, as the executive branch worked to promote the convention and Congress and environmentalists sought to scuttle it.

The debate over the minerals agreement highlights certain institutional weaknesses of Congress that, when reflected in legislative proposals, can limit executive branch freedom of action. First, congressional actions tend to be influenced by personal or special interest considerations. The minerals agreement was a highly salient issue for influential environmental groups, thus offering incentives for congressmen to participate in policy formulation. Another institutional weakness is that congressional policymaking often seems to be single-issue oriented. In the context of Antarctic policy, this single mindedness has resulted in piecemeal consideration of related interests rather than an attempt to deal with the broad range of U.S. Antarctic interests.

The committee system, which at least since the early part of this century has been the de facto locus of policymaking, largely accounts for congressional fragmentation.[141] The several legislative initiatives prompted by the CRAMRA agreement illustrates the fragmented nature of congressional activities. These initiatives not only revealed a lack of specific knowledge about the whole range of U.S. Antarctic policy, including the U.S. international obligations as a leading treaty consultative party, but they also demonstrated Congress's tendency to view interests in isolation.[142]

Most, if not all, the congressional initiatives related to CRAMRA suggest that Congress perceived a wide disparity between its goals and those of the executive branch on Antarctic environmental and resource issues. Despite administration officials' claims to the contrary, Congress reacted to the executive's support of CRAMRA as a poorly disguised desire to open up Antarctica to mineral operations. This perception prompted members of Congress to press for legislative proposals that reflected its environmental goals and objectives. Following announcement by France and Australia on August 18, 1989, that they would no longer support CRAMRA, Senator Albert Gore and Representative Wayne Owens introduced Joint Resolution 206. This resolution called on the administration to discard CRAMRA and initiate immediate negotiations aimed at a new agreement among the ATCPs.[143] As passed, Joint Resolution 206 urged executive branch officials to ensure that the new agreement recognized Antarctica as a "global ecological commons" and included a permanent prohibition on mineral activities.[144] A second resolution, S.R. 186, was introduced on September 26, 1989, by Senator Jesse Helms. It acknowledged the significance of the region's ecosystem to the global environment and stressed the need to maintain the Antarctic as a natural laboratory and wilderness area. The Helms resolution did not propose specific actions. It simply asserted that the protection of Antarctica must be a basic consideration in decisions relating to all activities conducted in Antarctica.

Senate bill 2575, introduced on May 3, 1990, by Senator John Kerry, became a companion piece to H.R. 3977, the "Antarctic Protection and Conservation Act of 1990," which had been introduced by Representative Silvio Conte.[145] The bill banned prospecting, exploration, and development of Antarctic mineral resources by U.S. citizens. It also urged the executive branch to begin negotiations on a new agreement to preserve the Antarctic environment, prohibit mineral resource activities by all nations, and designate the continent an "International Reserve—Land of Science." On June 26, 1990, the House held hearings on H.R. 3977 and H.R. 4210, the "Antarctic Environmental Protection, Clean-Up and Liability Act of 1990."[146] H.R. 4210 also aimed at providing increased protection for the Antarctic environment. The "Antarctic Protection Act of 1990," sponsored by Representative Conte and Senator Kerry, was signed into law by President Bush on November 16, 1990, as Public Law 101–594.[147] The act imposes an indefinite ban on Antarctic mineral resource activities by U.S. nationals. The new legislation "makes it unlawful for any person to engage in, finance, or otherwise knowingly provide assistance to any Antarctic mineral resources activity."[148]

The constraining role of Congress is also evident in its efforts to make domestic environmental laws, specifically NEPA, directly applicable to federal activities in Antarctica. In its original form, H.R. 3977 extended

NEPA to U.S. activities in Antarctica. The NEPA provision was dropped from the bill because of strong executive branch opposition. The executive branch rationale for opposing this provision was that the extension of NEPA to USAP would have allowed private litigants to seek injunctions and halt important research projects, notwithstanding NSF's compliance with environmental impact statements.[149]

H.R. 4514, the "Antarctica World Park and Protection Act of 1990," also demonstrated the problems associated with congressional decision making. The proposed bill directed the secretary of the interior to prepare a plan for "the establishment of an Antarctica World Park dedicated to scientific research and public enjoyment compatible with the preservation of such values."[150] It also required that "until completion of the plan, no citizen, national, officer, or agency of the United States shall carry out any activities (including any discussion or negotiations on behalf of the United States with other nations or international organizations) related to or affecting Antarctica except in a manner consistent with the purpose of this Act and the policy set forth in section 1(b)(6)."[151] The bill further prohibited "any prospecting or other mineral exploration or development activities in Antarctica"[152] and extended jurisdiction of U.S. environmental laws to the Antarctic.[153]

H.R. 4514 was uniformly opposed by the executive branch agencies with involvement in Antarctic affairs. During congressional hearings on the legislation, a representative from the Department of Interior stated that the department did not "support a unilateral resource inventory and a plan by the National Park Service."[154] Clearly, the bill was viewed as lacking specific knowledge of U.S. Antarctic policy or of U.S. international obligations. The Department of State thus contended that the bill could create "unilateral action on the part of the United States that cannot guarantee reciprocity from other nations with an interest in Antarctica." The department also strongly objected to provisions in H.R. 4514 that sought "to prohibit activities by U.S. citizens that do not take account of the international process necessary to resolve issues relating to Antarctic resources."[155] The Department of State reiterated its concern that the unilateral approach of the legislation "would run counter to the pattern of international cooperation under the Antarctica Treaty system" and signal a reduction in the U.S. "commitment to the Antarctic Treaty itself."[156]

NSF objections to H.R. 4514 also reflected concern over policy proposals conceived with little or no appreciation for existing conditions and obligations. One chief objection voiced by NSF to the House bill was that the "unilateral U.S. imposition of a management plan over the entire continent" could prompt serious questions about the exercise of U.S. jurisdiction over Antarctica. "Questions concerning these sensitive issues are expressly reserved for discussion by the Contracting Parties under Article IX of the Antarctic Treaty," and enactment of H.R. 4514 "appears contrary to the letter and spirit of this Treaty provision to which the United States is a signatory."[157] The NSF also opposed the bill because, by assigning major responsibilities to the Department of Interior, it directly conflicted with the program role of the foundation and the USAP.[158]

Implementing Legislation for the Environmental Protocol

Recent congressional efforts to legislate U.S. environmental policies in the Antarctic relate to the passage of implementing legislation to effect ratification of the 1991 Protocol on Environmental Protection.[159] Conflict between the executive branch and the environmental community over stipulations in implementing legislation for the Antarctic environmental protocol has made the ratification process more protracted. Three versions of implementing legislation for the protocol were left over from the 103d Congress, and these likely will be re-introduced into the 104th Congress. The first of these, H.R. 1066, "The Antarctic Environmental Protocol Act," was previously introduced by Representative Garry Studds and is in the main supported by the U.S. environmental community. The second, S.R. 1427, "The Antarctic Scientific Research, Tourism, and Marine Resources Act of 1993," was introduced by Senator John Kerry and also enjoys support from U.S. environmental organizations. The third is H.R. 3532, "The Antarctic Environmental Protection Act of 1993," which is the Clinton administration's bill introduced by Representative Rick Boucher.

Regardless of the version of implementing legislation adopted, the U.S. Congress must address certain considerations if environmental protection is to be effectively applied to U.S. nationals in the Antarctic. First, the legislation must set out clearly and unequivocally what activities are wholly prohibited in Antarctica and what activities may be performed with a permit. The conduct of any activity inconsistent with the protocol must be made unlawful. This action is essential for upholding the environmental principles in Article 3 of the protocol, which became binding upon U.S. citizens with its ratification. Essential also is that the implementing legislation stipulate that U.S. citizens are banned from conducting any mineral resource activity in Antarctica, a prohibition that would implement the protocol's fifty-year moratorium on mineral resource activities.

Second, implementing legislation should restrict any person from transporting passengers to Antarctica unless bound by agreement to comply with the marine pollution provisions of Annex IV to the protocol and the U.S. Act to Prevent Pollution from Ships.[160] This provision should apply to all tourist companies in the United States that sail to the Antarctic, ensuring that American companies comply both with U.S. obligations under the environmental protection protocol and the International Convention for the Prevention of Pollution from Ships, 1973/78 (MARPOL).[161]

Another significant provision should allow for certain activities, otherwise banned, to be conducted provided a permit is obtained from the appropriate agency, presumably the Department of Commerce (through NOAA). Such activities might include the taking of specially protected species such as native mammals, birds, and plants, as well as entering specially protected areas, discharging untreated sewage, or operating an incinerator. Presumably such activities would be permissible in an emergency to prevent the loss of human life or to protect the safety of a ship or aircraft. Implementing legislation should also require permits for expedition activities by Americans in Antarctica, as well as for construction and operation of new facilities under the U.S. Antarctic Program.

A vital component of U.S. implementing legislation must concern environmental impact assessment (EIA) of activities in Antarctica. Article 8 and Annex I of the protocol obligate parties to conduct prior environmental assessments of proposed activities, including scientific research activities, tourism, and all other activities for which advance notification is required, inclusive of logistical support activities. For the United States, such environmental impact assessments should be made consistent with the EIA procedures set out in the National Environmental Policy Act of 1969 (NEPA).[162]

For the United States, federal government agencies and their employees remain the principal U.S. nationals operating in the Antarctic. Among U.S. nationals, these persons are most likely to have significant impact upon the Antarctic environment. Their activities in Antarctica should therefore be regulated by U.S. law, as provided for in NEPA. Implementing legislation should provide for that compliance and bring

U.S. standards for evaluation and preparation of EIAs for Antarctica more closely into line with standards provided for in the protocol.[163] Important also is to provide a remedy for environmental groups against perceived persistent violations of the protocol, namely through judicial review by U.S. federal courts.

This latter point underscores the need for implementing legislation to assign primary responsibility to a U.S. federal agency for carrying out compliance with the protocol. Traditionally, chief responsibility for oversight and enforcement of U.S. legislation in Antarctica has been given to the National Science Foundation. In the past, however, the NSF has been less than effective in implementing environmental review procedures in the Antarctic.[164] If U.S. implementing legislation is to ensure that the NSF will perform review procedures as mandated by the protocol, those activities should not be shielded from judicial review. Implementing legislation that provides for a mechanism like NEPA should ensure more effective undertaking of review procedures in Antarctica and therefore contribute to better safeguarding of the Antarctic environment. In its legislation implementing the protocol, the United States should also require a review that is subject to federal judicial enforcement.

Provision must finally be made for assessing civil penalties for violations of the act and for setting criminal offenses for knowing violations. Enforcement in the Antarctic of the legislation's provisions, including those against marine pollution in Annex IV to the protocol, should fall to the U.S. Coast Guard.

In the five years since the protocol was signed, significant compromises have been made between the executive branch and environmental groups toward obtaining acceptable implementing legislation. Perhaps most significant, environmentalists have conceded that the National Science Foundation should continue to serve as the lead agency in charge of the U.S. Antarctic Program, with appropriate legal responsibility over the activities of the scientists it is supporting. Though there is still concern over the implications of self-supervision of NSF's activities in the Antarctic, recent efforts by the foundation to improve waste disposal and recycling operations, as well as to upgrade fuel handling and storage facilities, call for praise. In addition, NSF's Waste Regulations issued on June 29, 1993, bring the agency into compliance with the Antarctic Conservation Act of 1978. For environmental groups, the principal qualification offsetting NSF's continued self-regulation and self-enforcement, however, turns on the adoption of legally enforceable provisions to check against possible environmental abuses. To this end, NOAA, EPA, the Coast Guard, and the Marine Mammal Commission should be aptly involved in developing regulations and oversight of American activities in Antarctica.[165]

The U.S. environmental community wants several points remedied in the administration's proposed legislation before it is acceptable as implementing legislation for the protocol. First, environmentalists are concerned over the lack in H.R. 3532 of any mandatory obligation for NSF to modify, suspend, or revoke a permit if conditions change or the permit is violated. Such a mandate, conservationists contend, should be inserted to ensure the implementation of Article 3 of the protocol, which obligates the United States to modify, suspend, or cancel an activity that threatens or impacts the Antarctic environment or associated ecosystems.[166]

A second concern focuses on the need to give the Environmental Protection Agency a greater regulatory role, especially in acting on permits for waste disposal and reviewing impact assessments. Self-regulation by NSF could be offset by the EPA having some oversight of the process, including the opportunity to review decisions on the need for EIAs and the level of assessments.[167] Third, no means are provided in the administration's legislation to ensure compliance with the protocol by a scientist or scientific program. Environmentalists therefore advocate inclusion of a provision permitting citizens' suits if the NSF refuses to bring action against an alleged violator.[168]

A fourth concern relates to the environmental impact assessments section of the administration's bill and the belief that it could foster noncompliance with the NEPA. The wording in H.R. 3532 would exempt "Antarctic joint activities" from the EIA process. Importantly, "Antarctic joint activities" are defined as "any federal activity in Antarctica which is . . . conducted jointly or in cooperation with one or more foreign governments." Such a caveat could obviously create a gaping loophole for activities associated with scientific research.[169] Fifth, environmentalists would like the implementing legislation to modify provisions for on-site inspections; that is, a requirement should be inserted that would compel the NSF to respond to and correct violations reported by an inspection team.[170]

Finally, concern has surfaced over the exemption in Annex IV dealing with prevention of marine pollution. Annex IV articulates rules to control the discharge of oil, noxious liquids, garbage and plastics, and sewage from ships in the Antarctic. Article 11 of that instrument, however, contains a sovereign immunity clause, exempting "any warship, naval auxiliary or other ship owned and operated by a State and used, for the time being, only on government and non-commercial service." Environmentalists want the American implementing legislation to require that all vessels operating in the Antarctic over which the United States has jurisdiction comply fully with Annex IV, notwithstanding the exemption.[171]

Until these disparities are mollified, obtaining acceptable congressional legislation for implementing the protocol appears elusive. Regrettably, though not surprisingly, in an era when the U.S. Congress is more sensitive to and preoccupied with domestic social and economic matters, reconciling discrepancies in important legislation pertaining to U.S. foreign policy in Antarctica has received less than adequate attention and priority.

The Role of Non-governmental Organizations (NGOs)

Growing public concern in the 1970s over the protection of the environment was also evident in the activities of nongovernmental organizations (NGOs)[172] concerned with Antarctic affairs.[173] No other U.S. Antarctic interest has received as much attention from NGOs as that of conservation. Organizations such as Greenpeace, the Environmental Defense Fund (EDF), the Cousteau Society, the Antarctica Project, the Antarctic and Southern Ocean Coalition (ASOC),[174] and to a lesser extent, the World Resources Institute, have been particularly active in focusing public and congressional attention on the Antarctic environment. For this reason, this section looks at the avenues available to NGOs for influencing U.S. Antarctic environmental policies and programs, their policy priorities, and the chief manifestations of their efforts concerning Antarctica.[175]

NGO Access to Antarctic Policy-Making

A principle implicit in American politics is that organized interests affected by public policy should have an important role in shaping that policy.[176] In seeking to influence Antarctic policymaking in accordance with their objectives, NGOs interact with executive branch officials and members of Congress by: providing comments on rules and regulations proposed by the executive branch; serving as members on U.S. delegations to ATCP meetings; monitoring environmental practices in the Antarctic; serving as members on the Antarctic Section of the Department of State Advisory Committee on Oceans; working with Congress or lobbying; and testifying during congressional hearings. The courts generally have also been a promising arena for environmental advocacy.[177] The mid-1960s changes in litigation law gave environmentalists new leverage in governmental decision making.[178] Through changes in the rules of standing, "executive branch agencies were gradually converted from experts above the fray to defendants forced to explain in detail to federal judges the rationale behind their decisions."[179] Finally, the influence wielded by NGOs is enhanced by their educational efforts, press conferences, and seminars, all of which provide them effective avenues for communicating their views and gaining allies.

The American political process provides NGOs with avenues for influencing U.S. Antarctic policymaking in ways perhaps unequaled to what is available to their counterparts in Western Europe.[180] A comment made by Markus G. Schmidt in another context applies equally well to the Antarctic. He notes that "access to valuable information puts American NGOs in a favorable position in comparison to their counterparts in Western Europe, where public-interest groups often experienced difficulties in finding out what actual government policy . . . was."[181] Assistant Secretary of State Patsy Mink highlighted this point in 1978 when she stated:

Consultations with conservation groups in particular have been beneficial in the policy formulation process, and their adviser role on U.S. delegations dealing with Antarctic matters has been solicited and accepted. This is a departure from the practice of exclusion of public members prevalent as late as 1976 and still exercised by almost all other consultative parties.[182]

Thus, contrary to claims of exclusion made by some environmentalists, executive branch officials have provided NGOs with opportunities to convey their concerns on the Antarctic. For example, the ATCP conference held in February 1972 to negotiate a convention for the conservation of Antarctic seals included as advisers representatives from several conservation groups in the United States and abroad.[183] Policymaking on CCAMLR provides another case in point. The process involved not only coordination among interested executive agencies but also incorporation of views of the interested public and the Congress. For instance, the Department of State held an initial public meeting on December 20, 1977, at which representatives of NGOs presented their views on a proposed conservation regime for Antarctic marine living resources. Another meeting was held in February 1978. To provide greater access to public views, the department amended the charter of its Oceans Affairs Advisory Committee to include Antarctic matters and set up an Antarctic section to advise the APG on regional matters, including Antarctic resource and environmental issues. The section was to consist of fifteen to twenty members drawn from various public sectors.[184]

Although environmentalists have numerous ways and means for influencing U.S. Antarctic policy, their credibility may be undermined by the lack of reasonable objectivity in their criticism about the efficacy of the ATS or intentions of the ATCPs.[185] Generalizations about NGOs must be treated with caution; still, organizations such as Greenpeace, the EDF, and the Antarctica Project have been particularly aggressive in their determination to revise radically the terms by which the Antarctic is managed. In this regard, the observations of two Norwegian commentators seem appropriate:

> Quite apart from pointing to obvious faults on which all can agree, the environmentalists are also prone to charge the ATS with deficiencies so serious as to undermine the ATS's entire existence. To this group it is not reform but a transformation of the manner in which Antarctica today is managed which alone will suffice. This calls not only for a revision of the aims and purposes of the ATS but also for a transfer from the present to a new lot of the managerial responsibility for the continent.[186]

The stakes of such sweeping transformation should give pause for serious reflection.

NGO Policy Priorities and Manifestations

The activism of Antarctic NGOs became increasingly intense in 1983. This change stemmed from the opening of the Antarctic minerals negotiations, when Greenpeace launched in earnest its international campaign to preserve Antarctica.[187] Joining forces with ASOC, Greenpeace adopted the slogan "World Park Antarctica" to symbolize the goals of its campaign.[188] The world park notion is not new; it dates back to the Second World Conference on National Parks held in 1972.[189] Since then the concept has become closely associated with Greenpeace and ASOC's

efforts to have the continent designated as a world park.[190] According to Greenpeace, Antarctica as a world park would be an area
- where wilderness values are paramount;
- where there is full protection of flora and fauna (terrestrial and marine) and the environment;
- of peace, free of nuclear and conventional weapons, and all military activities;
- of scientific research that encourages scientific cooperation between scientists of all nations.[191]

One difficulty with the world park concept is the considerable uncertainty remaining about what form of protection that notion should entail. More importantly, to many observers Antarctica presently exists as a de facto world park, where science is the primary activity and export. The environmentalists' goal of transforming the continent into a "world nature park" has also brought them into conflict with basic research scientists, who might occasionally need to modify or destroy portions of the natural environment to perform their experiments. For instance, biologists sometimes must kill seals or penguins for their studies; geophysicists may need to carry out blasting operations; and supply bases need to be constructed on the continent. Interference with the ecosystem might even be required for conservation purposes but would still be rejected by radical environmental organizations.

The world park option was presented as an alternative to CRAMRA, partly because Greenpeace, ASOC, the Cousteau Society, and the Environmental Defense Fund perceived U.S. executive branch attitudes as being "pro-mining."[192] Not surprisingly, they portrayed executive branch support for the minerals agreement as such in the media and before the Congress. The repeated claim by U.S. government officials that CRAMRA "does not open Antarctica to mining, or the development of oil resources, nor does it presume that mining or oil extraction definitely will take place," was dismissed as political rhetoric.[193] Also challenged was the U.S. government claim that the key purpose of the agreement was to establish "an agreed international process for determining whether mineral resource activities should ever take place in Antarctica, if interest in minerals development were to emerge in the future, and for regulating any activities determined acceptable."[194] NGO environmental groups in the United States saw the government as pro-minerals development, not as pro-conservationist.

Throughout the debate on CRAMRA, organizations such as ASOC, the EDF, and the Cousteau Society persistently maintained that "the Minerals Convention must be recognized for what it is: an agreement to regulate minerals activities. It is not a conservation regime. The primary function of the Convention is to establish minerals operations as a legitimate activity in Antarctica."[195] Part of the explanation for this perspective comes from the polarization of the debate about CRAMRA, a point highlighted in Lee Kimball's remark that "at times the issue of minerals development in Antarctica has been characterized that if you are for ratification of CRAMRA, you are for minerals development in Antarctica, and if you are against ratification of CRAMRA you are against mineral development in Antarctica."[196]

Another possible explanation for the Greenpeace decision to critically assess the Antarctic environmental situation can be traced to the austral summer of 1985–1986. At that time Greenpeace launched its first Antarctic expedition "to identify first-hand actual environmental problems in evidence on the continent."[197] Between 1986 and 1992 Greenpeace undertook six annual expeditions, establishing the only nongovernmental base on the continent. Greenpeace's oversight activities consisted of surveys of environmental conditions at national research stations, including documenting waste disposal practices and the general level of compliance of base operations with the treaty system environmental requirements. Members also conducted interviews with station personnel regarding environmental procedures and policy, and instituted a program of soil, water, sediment, and biological sampling. Overall, Greenpeace concluded that poor compliance with treaty recommendations on environmental matters was the rule rather than the exception.[198] Although its base was closed in 1992 for fiscal reasons, Greenpeace leaders still maintain that recent heightened environmental sensitivity by many governments operating in Antarctica stems in part from the World Park Base experiment.

The Environmental Defense Fund (EDF), another leading environmental organization, can be distinguished from other environmental groups in that it stresses legal and advocacy actions, backed by an understanding of the scientific and legal bases of issues.[199] The EDF, for example, has maintained that NEPA applies to the Antarctic and has testified to this effect on various occasions in Congress. In 1991 EDF took its case to court, charging the NSF with procedural failures under NEPA.[200] In 1993 the court handed down its decision—that NEPA did indeed apply to U.S. nationals in Antarctica, even though they were literally outside the borders of the United States.[201] EDF was also instrumental in helping draft legislation in 1990 that prohibited Americans from conducting mineral exploration and exploitation in Antarctica. EDF's influence on the Antarctic Treaty System has been enhanced by participation of its staff members on U.S. delegations to international meetings. Thus far, an EDF attorney has participated in meetings on the Antarctic Seals Convention, the Antarctic Marine Living Resources Convention, and in the recently concluded special meeting of the ATCPs on Comprehensive Environmental Protection.

Conclusion

The record, overall, tends to support the claim that the United States has exercised a leading role in efforts to protect the Antarctic environment. Executive branch officials responsible for the USAP have shown a longstanding interest in promoting environmental practices in Antarctica. Apparently solutions for previously recognized environmental problems are being addressed; other potential problems are being prevented as a result of more environmentally sound practices. It is true that scientific activities in Antarctica are enhanced by a more pristine environment and that the incentive for ecological protection is real for U.S. scientists there. It is also true that pressures by environmental NGOs have contributed to heightened awareness of environmental problems and to the need to take preventive and restorative measures.

Antarctic policy cannot be reduced to a single interest. It involves a cluster of interests, each of which has its own structure and dynamics. For that reason, the actors affecting U.S. Antarctic environmental policy (the executive branch, the Congress, environmental groups, and other ATCPs) all play distinct roles in policymaking and exercise varying degrees of influence. A salient characteristic of Antarctic policymaking—and this comment applies to environmental policy as well—remains the dominance of the executive branch. Of all the actors involved, executive branch officials are the ones who have provided long-term continuity and coherence in pursuit of the full range of American goals and objectives in the Antarctic. Although other actors have wielded considerable influence, their actions have not, in most cases, been decisive. One example highlights this point. The driving rationale behind U.S. executive branch officials' abandonment of CRAMRA was the potential political damage to (and perhaps even the collapse of) the Antarctic Treaty System. The refusal of Australia and France to sign the minerals agreement placed severe strains on the vitality of the entire treaty system, which supports the full range of U.S. Antarctic interests. In abandoning CRAMRA, the United States reestablished consensus, restored needed stability to the ATCP group, and preserved its position of leadership and influence throughout the Antarctic Treaty System. Since access to minerals or hydrocarbons on the continent was neither in the near future nor even conceivable, no economic costs were at stake. The pro-environmental decision was thus all the more palatable to U.S. policymakers.

Global interests also benefit from American support of environmental considerations in Antarctica, especially in regards to the region's value for scientific research. For example, the near pristine nature of the Antarctic provides a baseline against which to measure air pollution in more populated regions. The sensitivity of the Antarctic may thus serve as an early detection system for changes in the atmosphere of the earth. This sensitivity allows physical, biological, and chemical investigations to be carried out early enough so that remedial measures might be taken if necessary.

U.S. environmental goals and objectives in the Antarctic share several characteristics with U.S. scientific and economic interests. Environmental interests are long standing and have been pursued resolutely by U.S. foreign policymakers. Likewise, successful realization of U.S. environmental interests continues to be dependent on the preservation of the Antarctic Treaty System and the use of Antarctica for peaceful purposes only. Obtaining a comprehensive environmental regime for the regulation of Antarctic activities would have been far more difficult without these two overriding interests being firmly fixed. In fulfilling the objectives of these two policy interests, the United States also contributed to the eventual strengthening of the Antarctic Treaty System as an instrument of environmental security for the polar south.

7 / Geostrategic Interests

PRIOR TO THE 1970s, the Antarctic attracted attention from only scientists, scholars, and policymakers with specialized interest in the remote south polar region. During that decade, however, the image of Antarctica changed from that of a frozen wasteland to that of a region containing superabundant supplies of nonrenewable resources, particularly minerals and fuels.[1] Claims about the vast wealth of Antarctica's nonrenewable resources have been highly speculative and greatly exaggerated. Geological information about the continent remains incomplete, which renders inconclusive any reliable estimates on the magnitude of mineral resources.[2] Although it is difficult to estimate the availability of mineral resources in and around Antarctica, such deposits very likely do exist, particularly hydrocarbons.[3]

Two fundamental aspects of contemporary U.S. geostrategic interests in the Antarctic stand out. First, the desire to obtain gains from potential exploitation of Antarctic resources has been tempered by, or more accurately made subordinate to, a commitment to protect the Antarctic environment. Second, geostrategic interests in the Antarctic have evolved into a complex blend of environmental, resource, and security concerns. Consequently, this chapter examines United States Antarctic geostrategic interests from the vantage points of these particular concerns.

We first examine the extent of Antarctica's living and nonliving resources, the threat posed by overexploitation, and the prospects for developing Antarctic minerals, particularly hydrocarbons.[4] The analysis then surveys the evolution and content of U.S. policy toward Antarctic resources. The environmental facet of U.S. geostrategic interests is assessed, especially to determine whether the dominant concern of American policymakers in pressing for a minerals agreement was actually more environmental and less economic. We evaluate the hazards presented by exploration and exploitation of hydrocarbons on the fragile Antarctic environment, and address the security dimension of U.S. geostrategic interests. During the analysis, two seemingly disparate but related concerns are highlighted: first, the desire for uninterrupted energy supplies at reliable prices, and second, the prospects for a stable international political environment. The last section looks at the economic aspects of hydrocarbon development and the problems associated with the commercial prospects of exploitation, including the harsh Antarctic setting and the state of available technology.

Antarctic Resources: An Assessment

Living Resources

Whereas development of nonliving resources remains only a future possibility, exploitation of Antarctic living resources has occurred since the first explorers sailed along the edge of the frozen continent. Historically, the United States has evidenced little interest in exploiting Antarctic marine living resources. It has demonstrated instead a firm commitment to devising regulatory mechanisms for their protection and preservation. The seas surrounding Antarctica are richly productive. Relatively few species of marine life are found, but their numbers abound. The biologically productive Antarctic waters contain rich populations of whales, seals, fish, and crustaceans such as krill, lobsters, and crabs.[5] Fifty species of Antarctic birds, including penguins and albatross, feed on the marine organisms.[6] These birds nest and breed on the coast, but they rely on the sea for sustenance, consuming nearly as much krill as do whales.[7]

An important distinction between the Antarctic marine ecosystem and that of marine environments elsewhere is the dominance of krill in the former.[8] Krill swarms constitute the principal link throughout the natural economy of the Southern Ocean.[9] Krill is the primary food source for baleen whales, some 15 to 30 million crab-eating seals, over 50 million penguins, and numerous flying birds. Adelie penguins, which comprise the largest biomass inhabiting Antarctica, feed primarily on krill. Fish and squid also eat krill, which links them to the few higher branches on the rather simple Antarctic marine food chain. Coastal areas serve as sites for breeding, resting, and recovery for seals and birds, which derive their food from the sea.[10]

Mineral Resource Assessment

Although Antarctic mineral wealth remains a matter of conjecture and inference, evidence suggests that its potential could be impressive. Speculation about Antarctica's mineral potential revolves around the Gondwanaland hypothesis.[11] There is evidence that South America, Antarctica, India, Africa, and Australia were linked as one landmass as early as 200 million years ago, when the present continents and islands began to drift apart. The assumed relationship between Antarctica and the other continents of the southern hemisphere has prompted geologists to believe that mineral deposits comparable to those currently being exploited elsewhere may exist in Antarctica.[12]

Of the nonliving resources that have attracted attention, hydrocarbons head the list. Speculation suggests that the Antarctic contains substantial hydrocarbon reserves, with fields comparable in size to those offshore Mexico, the Alaskan slopes, and the North Sea;[13] but it is only speculation. In terms of minerals known to exist in Antarctica, only coal and iron have been found in accumulations sufficiently large to be termed deposits. Occurrences of a wide range of hard minerals have been recorded, including copper, molybdenum, chromium, nickel, cobalt, tin, uranium, titanium, manganese, lead, and zinc.

During 1972 and 1973, the NSF-sponsored research ship *Glomar Challenger* attracted international attention when it discovered traces of the hydrocarbon gases methane, ethane, and ethylene in three of four holes drilled in the Ross Sea.[14] Methane is common in deep-sea cores, but ethylene often occurs with petroleum.[15] Hypothetical projections were advanced about the potential size of offshore Antarctic oil and gas fields. A U.S. Geological Survey report in 1974 estimated deposits of 45 billion barrels of oil and 115 trillion cubic feet of natural gas,[16] though it considered only a third of this quantity recoverable.[17] Spectacular claims were made for a "Middle East" in the Antarctic, including an assertion that the oil reserves reported by the U.S. Geological Survey almost matched "the proven reserves of the entire United States."[18] In 1979 a representative of Gulf Oil stated that the oil potential of the two most highly prospective areas in the Ross and Weddell Seas approximated 50 billion barrels, but it could far exceed that estimate.[19] Since then, however, geological assessments suggest more conservative estimates. In 1983 the U.S. Geological Survey summarized the present situation accurately by observing that "no known petroleum or mineral resources occur in Antarctica."[20] A 1990 Congressional Research Service study concluded that, "while at present there are no known mineral resources in Antarctica that are commercially developable, it would seem reasonable to presume that they exist."[21] As of 1996, no known petroleum or mineral deposits have been discovered in, on, or around Antarctica.

Although, short of drilling, the presence of hydrocarbons cannot be stated with certainty, geological evidence is compelling that major deposits probably do exist on the continent or its shelf.[22] Seismic surveys have been conducted throughout the seasonally ice-free portions of the Antarctic continental margin, but drilling has been done in only a few areas.[23] John B. Anderson has used these data, together with extrapolations drawn from marine geophysical surveys, the geology of neighboring Gondwana continents, and information gained from geological studies of coastal regions, to assess the hydrocarbon potential of five different sectors of the Antarctic continental margin.[24] The overall conclusion he reaches is startling: "Large oil and gas fields, and possibly even some giant fields, probably exist on the Antarctic continental margin."[25] He points out, however, that logistical obstacles and the geophysical hazards that must be overcome to exploit these resources are immense.[26]

Evolution and Content of U.S. Antarctic Resources Policy

The national policies formulated when the Antarctic Treaty was being negotiated in 1958–59 have basically shaped U.S. Antarctic scientific and environmental interests since. To a large degree, this condition also applies to American economic interests. The prospect that exploitable quantities of nonrenewable resources might be discovered in Antarctica was considered as early as 1939 in a U.S. Department of State study.[27] Again, in 1959 a report issued by the National Security Council noted that the strategic and economic potentials of Antarctica were uncertain, and the present-day advances in science and technology had to be borne in mind.[28] Elaborating further on U.S. interests, the report emphasized the importance of developing uniform, nonpreferential rules applicable to all states for any future development of Antarctic resources.[29] These statements clearly recognized the U.S. interest to be in retaining an advantageous position for developing the area's economic resources, should new conditions emerge.

During the 1957–58 IGY, little was known about the mineral resource potential of Antarctica. This uncertainty made it easier during the Antarctic Treaty negotiations to circumvent an issue that surely would have been difficult, or perhaps even impossible, to agree upon. American policymakers were not ready to risk agreement on a treaty in pursuit of some unknown Antarctic riches. Testifying before Congress in 1975, Dixy Lee Ray, assistant secretary of state and chairwoman of the Antarctic Policy Group (APG), observed that during preparatory talks in 1959, it became clear that resource questions were so contentious that the issue should not be addressed at the treaty conference.[30] For that reason, the treaty eventually negotiated and signed on December 1, 1959, was silent on the question of resources. No provision in the treaty text expressly regulates mineral policies. Antarctic resources, however, were discussed during the U.S. Senate debate over advice and consent of the treaty.[31]

As with other Antarctic interests, consistency and continuity have marked U.S. policy on Antarctic resources. As an original consultative party and very active participant in the management of Antarctic affairs, the United States has persistently felt that it should retain the option to share in any possible future exploitation of Antarctic mineral resources, particularly hydrocarbons. Rather than begin any mineral resource activities, however, U.S. policymakers have sought to "hold open the door to the possible development of any resources that might be discovered."[32]

With respect to the economic potential of the region, an executive document drafted in 1962 advised that the United States stay ready to benefit from possible future uses of Antarctica, be they political, scientific, economic, or otherwise.[33] The document also emphasized the need to prevent any use of Antarctica that might be detrimental to these interests. Later, a 1970 review of U.S. Antarctic policy produced a presidential statement on resources. In concluding this review on October 13, 1970, President Nixon reaffirmed the resource-related objectives of U.S. policy, including that of developing appropriate measures to ensure the equitable and wide use of living and nonliving resources.[34] Another formal statement followed in 1975. At that time, Dixy Lee Ray, chair of the APG, affirmed that "[n]ondiscriminatory guaranteed access by the United States and others for exploitation purposes to any part of the Antarctic

Treaty area, except especially protected areas, is critical to avoid prejudicing our underlying juridical position."[35]

American policymakers have also stressed the critical role of environmental safeguards in any strategy devised for the development of Antarctic resources. Implicit in this view is official acknowledgement of the critical need to protect the fragile Antarctic ecosystem. For this reason, U.S. policymakers have frequently insisted that, if conditions ever permitted exploitation of Antarctic mineral resources, this activity must be conducted only under appropriate environmental safeguards.[36] Such views were also reflected in a Department of State press release issued prior to the tenth Antarctic Treaty Consultative Party meeting in 1979.[37] The press release, which urged other ATCPs to begin work on a minerals regime, emphasized that any agreement should account for environmental consequences and include strict ecological safeguards. No less significant for U.S. national interests, it added that the United States should have the opportunity to share in nondiscriminatory access to benefits if mineral activities were to prove acceptable.[38]

American resource interests in Antarctica also were highlighted in a 1983 report submitted by the Office of Management and Budget to the Committees on Appropriations of the U.S. Senate and House of Representatives. The report, "The U.S. Antarctic Program," included among priority interests that of "providing an opportunity for U.S. private industry to exploit Antarctic resources if and when it becomes feasible and appropriate."[39] In testimony before Congress in 1984, R. Tucker Scully, senior official of the Department of State, broadened U.S. economic interests in Antarctica to include discovering the resource potential of the region. He also reaffirmed American interest in ensuring that resource activities, should they ever take place, not be detrimental to the environment and in providing the United States with the opportunity to participate in development activities on reasonable terms and conditions.[40]

In 1986 David A. Colson, then deputy assistant secretary of state for oceans and international environmental and scientific affairs, reiterated that U.S. economic interests in the Antarctic remained virtually unchanged. This position is clear from his assertion that the United States was committed to ensuring that any mineral resource activities be environmentally safe; however, nondiscriminatory access should be provided for the country to all areas of Antarctica in which mineral resource activities may be determined acceptable.[41] In 1990 Curtis Bohlen, chairman of the APG during the Bush administration, explained the interrelationship between environmental and economic interests in U.S. Antarctic resources policy: "It is not the United States policy to hold open the opportunity to exploit minerals in Antarctica. It is more a question of if there is ever to be an agreed exploitation of minerals in Antarctica, then we want to be sure that U.S. companies have equal access with other countries on a non-discriminatory basis. I think that there is a subtle difference there."[42] This subtlety in position may well have nurtured misperceptions about American objectives among environmental groups.

The Environmental Dimension

The development of natural resources in the Antarctic carries with it certain environmental costs. These costs might be evident in the areas of prospective mineral activities in and around the continent, in the harvesting of living marine resources in circumpolar seas, and in the likelihood of environmental hazards from offshore petroleum development infecting the Southern Ocean.[43] Each of these circumstances generates important implications for U.S. Antarctic policy in general and for economic policy in particular.

The Minerals Convention

Political, environmental, and economic factors motivated U.S. policymakers to negotiate a regime for Antarctic minerals. From the political perspective, the U.S. objective was to preserve the principles and purposes of the Antarctic Treaty.[44] Entry into force of a regulatory convention, that is, CRAMRA, would remove mineral resources as a contentious issue and would guarantee that questions concerning mineral resources would be treated according to Antarctic Treaty principles. Environmentally, ensuring protection of the Antarctic ecosystem continued to be a salient feature of American policy. Concerning the economic factors, a minerals regime promised to keep the door open for American business to participate in any future mineral resource activities, should parties to the regime determine that such activities could take place without detriment to the Antarctic environment.[45]

The Antarctic Treaty makes no provision for the regulation of mineral resource development. The need to fill this vacuum stimulated American policymakers to support the establishment of rules of conduct, including adequate environmental safeguards, concerning development of Antarctic mineral resources. A minerals agreement would remove any possibility of unregulated mineral activities in Antarctica.[46] Similarly, negotiating a regulatory minerals regime while the commercial development of mineral resources seemed remote allowed the ATCPs to take proper account of environmental and political concerns. For these reasons, American policymakers ostensibly came to regard a minerals convention as "the most balanced and environmentally sound framework that could be achieved to deal with possible mineral resource activities."[47]

Minerals negotiations were launched at the fourth special Antarctic Treaty Consultative meeting (ATCM) in 1982, with the United States assuming a leadership role.[48] As promoted by U.S. policymakers, the regime would not presume whether or not mineral resource activities should take place. It should, however, provide the necessary machinery for applying strict environmental criteria if and when such activities were determined to be possible.[49] The resultant convention met these tests. The Convention on the Regulation of Antarctic Mineral Resource Activities (CRAMRA) was adopted in June 1988 and opened for signature in November 1988. Nonetheless, the refusal of Australia and France during 1989 to sign the convention precluded CRAMRA from entering into force. Even if Australia and France had signed the agreement, opposition in the U.S. Senate dimmed the immediate possibility that the U.S. Congress would adopt legislation for implementing CRAMRA.

Congressional opposition to CRAMRA can be explained partly by the failure of Bush administration officials to convince congressional leaders that environmental concerns motivated its desire for an agreement. These concerns were also not apparent to environmental groups such as Greenpeace, the Antarctic and Southern Ocean Coalition, the Environmental Defense Fund, and the Cousteau Society. That misconceptions existed about the administration's intentions concerning Antarctic minerals became evident from media and congressional hearings statements. The prevalent view among congressional leaders and environmentalists suggested that the administration favored exploitation of Antarctic minerals, regardless of the implications for the region's environment.

R. Tucker Scully, an active participant in the CRAMRA negotiations,

Geostrategic Interests / 141

denounced this portrayal of the administration's position when he forcefully stated:

> The Convention does not presume that mineral activities will take place but establishes an internationally agreed system for determining the acceptability of such activities should they be proposed in the future. The U.S. policy regarding Antarctica is intended to preclude human activities that may harm the environment. And we have sought . . . agreement that would protect the Antarctica environment and associated ecosystems. CRAMRA actually prohibits exploration and development of minerals unless there is a consensus that such activities can be accomplished in an environmentally sound manner.[50]

CRAMRA is now legally moribund. Still, examination of selected aspects of the convention within the context of U.S. policy objectives remains instructive.

CRAMRA's approach to the management of mineral resources resembled that of the Convention for Marine Living Resources, but it contained more elaborate binding principles. Environmental principles in CRAMRA are not mere exhortations; they are binding rules. A curt comment summarizes the essence of the agreement: "Any mineral activity is prohibited unless it is permitted."[51]

CRAMRA's stringent rules for mineral activities included ensuring the availability of the appropriate technology and procedures for safe operations, as well as monitoring the environment to detect adverse impacts. They also called for the capacity to respond effectively to accidents. The restrictions imposed by CRAMRA's Article 4 reinforce the claim that the minerals treaty was essentially a framework to protect the Antarctic environment, rather than one to exploit the continent's mineral resources. Article 4 would forbid any mineral activity that could possibly affect the quality of the air and water or produce changes in the atmospheric, marine, and land environments. CRAMRA also called for a system of inspections and the development of environmental impact assessments to determine potential impacts on the environment.

The regime established special institutions to carry out various aspects of the convention supported by the United States. A Commission, consisting of all ATCPs, was the plenary decision-making institution. The Commission had the authority to review decisions of special "regulatory committees" to ensure that they were consistent with CRAMRA's environmental standards.[52] The Commission was also to designate areas where mineral resource activities would be prohibited, adopt measures to protect the environment, and decide whether to grant requests for identifying particular areas for possible mineral exploration and development.[53] The Commission would be aided by a Scientific, Technical, and Environmental Advisory Committee,[54] and also by a Secretariat.[55]

The other institution relevant for environmental protection was the regulatory committee. CRAMRA required establishment of regulatory committees for each geographic area identified by the Commission for possible minerals exploration.[56] The functions of the committees would have included drafting detailed terms and conditions governing minerals activities, considering applications for permits by operators, approving management schemes, and monitoring minerals activities.[57] To promote compliance with established standards, CRAMRA made the activities of profit-seeking, nongovernment entities (e.g., companies, joint ventures, and consortia) the responsibility of each state party.[58]

Living Marine Resources

Indiscriminate harvesting of krill stands as a chief threat to the Antarctic ecosystem.[59] Serious fishing operations for krill began in 1972–73, conducted mainly by the former Soviet Union and Japan.[60] The former Soviet Union has traditionally been the major Antarctic fishing nation and accounts for 84 percent of the total harvest of krill and finfish.[61] Japan ranks second with an annual catch of around 14.5 percent.[62] Chile, France, the Republic of Korea, and Poland also account for notable shares of the annual catch.[63] Japan's catch of krill in the treaty area has fluctuated over the years from between 80,000 and 112,000 tons annually.[64] According to a Japanese report to CCAMLR, during the November 1986 to April 1987 season, Japanese fishermen made their largest catch ever, taking 78,360 tons of krill in the treaty area and another 29 tons outside the area. This amount represented 20 percent of the world's total catch of 376,527 tons.[65] The total catch of krill in the 1988–1989 season was 395,000 metric tons, compared to 370,663 metric tons during the 1987–1988 season and 376,456 metric tons in 1986–1987.[66]

One inherent weakness of the Antarctic ecosystem is its great dependence on a single organism, krill, making overharvesting of the crustacean an important concern. For that reason, CCAMLR's conservation efforts have focused considerably on krill, even though the size of the Southern Ocean's krill population is as yet undetermined and estimates vary considerably. Some specialists believe that the krill stock in the

Antarctic could supply 10–20 million metric tons a year on a sustainable basis; others estimate the sustainable yield at 200 million metric tons a year.[67] The fact that the catch of all marine living resources worldwide in recent years has been around 70–100 million metric tons makes krill of significant economic interest for some countries.[68]

At the ninth meeting in 1990 of the Commission and the Scientific Committee established by the CCAMLR, the Commission sought to adopt a precautionary approach to the development of the krill fishery, pending the availability of more precise information on its potential yield.[69] The krill fishing nations did not support this effort, arguing that current scientific evidence did not warrant imposition of regulatory measures.[70] At the same time, the Commission endorsed the principle that, in the absence of essential data, very conservative catch limits should be set. At that meeting, the Scientific Committee suggested several useful conceptual bases for a management policy for the krill fishery, particularly in light of the need to avoid adverse effects on krill predators in keeping with the ecosystem standard.[71] In 1991, the precautionary approach for krill harvesting was adopted by the Commission, and a ceiling for krill catches was fixed at 1.5 million metric tons a year.

Seals have long been hunted in Antarctic waters. Fur seals are still recovering from overexploitation, and their harvesting, along with five other species of seals, is regulated by the Convention on the Conservation of Antarctic Seals. The number of whales, another important Antarctic resource, had so diminished by 1932 that the International Whaling Commission was established to regulate whaling through voluntary annual quotas.[72] Although these quotas have been disregarded by Norway and Japan, most nations have abandoned Antarctic whaling.[73] The fact that Antarctic marine resources are being harvested within the limitations of two associated instruments—the Seals Convention and the Marine Living Resources Convention—tempers any further concerns that these resources will be overexploited.

Environmental Hazards: Blowouts and Oil Spills

Serious degradation to marine biota may result from offshore petroleum development.[74] Offshore oil exploration and production and tanker traffic account for two very serious sources of oil leakage in the marine environment. Iceberg collisions also present significant hazards to petroleum development on the Antarctic continental shelf.[75] Icebergs can damage a drilling rig at the sea surface or damage supporting structures and other seafloor structures when scraping along the sea bottom. A major spill or accident that might cause minimal damage to the ecosystem elsewhere could have a major impact on the biological balance in and around Antarctica.[76] The dependence of so many predators on krill leaves the entire ecosystem vulnerable to variations in the level of the krill population. Hence, decreases in krill caused, for example, by the effects of accidentally spilled crude oil would produce harmful effects throughout the region's ecosystem.[77]

Given the operational difficulties likely in the Antarctic, major damage to the environment could result before cleanup could be effected from a major blowout.[78] Should a blowout occur during the Antarctic winter, there would be little possibility of capping it or drilling relief wells for six months or more, during which oil would gush continuously.[79] Even if only temporary, such damage could possibly impact thousands of square miles of the offshore coast.[80]

Tanker operations and accidents account for most of the oil entering the ocean.[81] Tanker spills are harmful, though not in direct proportion to their size. One of the largest spills occurred in 1978 when the *Amoco Cadiz* ran aground and released nearly 60 million gallons of crude oil along the coast of France.[82] Its effect on marine mammals was minimal, however, compared to the spill from the grounding in March 1989 of the tanker *Exxon Valdez* off Prince William Sound in Alaska.[83] The *Valdez* spill spread roughly 10 million gallons of oil through Prince William Sound and into the Gulf of Alaska.[84] This spill, and later efforts at cleanup, generated skepticism about the ability of oil companies to afford proper protection to the environment. The grounding on January 28, 1989, of the Argentine supply vessel *Bahia Paraiso* two kilometers offshore Palmer Station dramatically highlighted the disastrous environmental effects of oil pollution on Antarctic marine life.[85] Roughly 150,000 gallons of diesel fuel spilled into the ocean.[86] Scientists at the station reported krill being washed up on local shorelines and estimated that a 100 percent mortality rate of South Polar Skua chicks would occur in the nearby nesting areas.[87] The NSF reported that surveys of local Adelie penguin rookeries revealed that 80 percent of the birds had been exposed to the spill.[88]

Oil spilled in polar regions presents an even greater environmental threat than in milder areas because frigid temperatures retard the breakdown of oil and prolong recovery of affected species.[89] Depending on sea and ice conditions, spilled petroleum may spread over or under the ice, on the water, or throughout several media at once.[90] Spilled oil in polar regions tends to accumulate at the ice edge, where marine life congregate.[91]

Oil spills can cause the fur of Antarctic seals to lose its insulating qualities.[92] All marine mammals spend considerable time at the surface, swimming, breathing, feeding, or resting, thereby enhancing the possibility of contact with a surface slick, water-in-oil emulsion, or tar balls.[93] Some baleen whales forage at the surface, a behavior called skim-feeding. When feeding in an area of a slick or tar balls, whales may foul their feeding apparatus. The feeding strategies of marine mammals could lead to ingestion of oil-contaminated food, because most of the prey organisms can accumulate petroleum hydrocarbons in their tissues.[94] These detrimental environmental effects obviously produce serious repercussions—not only on the health of indigenous living resources but also on the sustainability of the local ecosystem.

The Security (Strategic) Dimension

Economic opportunities for the United States in Antarctica are grounded in the region's nonmilitarized status;[95] that is, the prohibition against conducting military activities, establishing bases, or implanting weapons systems in the Antarctic (the peaceful-uses-only stipulation of Article I of the Antarctic Treaty) open the door for securing the region's economic potential. The Antarctic security dimension has evolved over the last half century from one of serious concern over the possible militarization of the continent to its status today as the largest demilitarized zone on earth. This fact has afforded the United States—and all states, for that matter—considerably more opportunities for greater economic activities throughout the region.

The following discussion about the strategic dimension of U.S. policy in the Antarctic addresses security concerns. Unlike the Arctic's pronounced status among U.S. security interests, the south polar region is viewed as much less critical, especially since the Antarctic Treaty's entry into force.[96] Geographical location partly accounts for this attitude. U.S. strategic interests in the Antarctic are less urgent because of the continent's remote location from American shores and because the region's hostile climate renders it less desirable for geostrategic purposes.[97]

146 / *EAGLE OVER THE ICE*

Pre-Treaty Era: U.S. Antarctic Security Interests

The strategic dimension of American interests was not irrelevant to U.S. Antarctic policy in the decades preceding the Antarctic Treaty. On several occasions before World War II, serious international disputes arose over Antarctic issues, and twice these disputes erupted into the use of force. In fact, alarm over the possibility of a major disturbance elevated the military neutralization of Antarctica to a preeminent place on the agenda of American foreign policymakers during the 1959 treaty negotiations.

Three concerns assumed particular salience in the American resolve to find an international solution to the problem of international conflict in the Antarctic. First was the need to diminish the threat posed by military uses of the continent and its surrounding oceans; second, the United States sought to temper the dispute among allies regarding overlapping claims to Antarctic territory; and third, demilitarization would reduce anxiety that the Cold War might spread to Antarctica.[98]

Military Uses of the Continent

In 1940, Secretary of State Cordell Hull proposed extending the Monroe Doctrine to the South Pole.[99] This move was prompted primarily by the presence of Nazi Germany in Antarctic waters during the late 1930s. Germany's activities in the area had alarmed U.S. foreign policymakers and allied governments, who wished to prevent any hostile power from gaining a strategic foothold in Antarctica.

On December 17, 1938, Nazi Germany dispatched an expedition "to secure for Germany her share in the approaching division of the Antarctic among world powers."[100] The expedition's seaplanes mapped Queen Maud Land, dropped spearlike shafts with emblazoned swastikas on the ice sheet, and mounted a shore party to raise the flag of the Third Reich.[101] The Germans named their new territory "NeuSchwabenland."[102] The land would be developed, declared the expedition's formal report, "unhindered by the rights of sovereignty of other nations."[103] To the consternation of the United States and its allies, the German press spoke glowingly about imminent formal claims. With the invasion of Poland, however, the Third Reich abandoned territorial aspirations in Antarctica and shifted its attention from photogrammetric flights over the continent to submarine warfare around it.

German raiders operated quite successfully in the South Atlantic during World War II. They "seemed to appear from nowhere, preying upon allied ships in the Indian Ocean and other southern waters and then vanishing when warships went in pursuit."[104] German raiders sank several hundred thousand tons of shipping, including the Australian cruiser *Sydney* and the *City of Bayville*, the first U.S. vessel to be sunk in the war.[105] They also mined the ports of Sydney, Melbourne, Adelaide, and Hobart and captured a substantial part of the Norwegian whaling fleet off the disputed coast of Queen Maud Land.[106] Norway's desire to protect its whaling industry and to forestall a German claim to Antarctic territory strongly influenced that government's decision in 1939 to proclaim title over Queen Maud Land.[107] Germany ignored this claim. It would acquire NeuSchwabenland indirectly through invasion of Norway in 1940.

In the decade preceding the Antarctic Treaty, U.S. foreign policymakers and scholars publicly voiced concerns over the military potential of the Antarctic. Their common worries were chiefly five: namely, that the Antarctic might serve as (1) a base for surface or submarine raiders preying on commerce in the Indian Ocean or the areas around Cape Horn and the Cape of Good Hope; (2) a base for launching attacks on southern hemisphere states; (3) a base for long-range bombers or missiles carrying nuclear weapons; (4) a site for weather and satellite tracking stations; and, (5) a possible storage depot or refuge in the event of global nuclear war.[108]

For years maritime states have realized the strategic significance of the Drake Passage—it is the only direct open sea passage between the Atlantic and Pacific—and have regarded the southern seas as important for transoceanic shipping.[109] The United States has regarded unimpeded access through the Drake Passage as an important freedom of navigation.[110] Prepared by the Policy Planning Staff of the Department of State on June 9, 1948, a paper entitled "Antarctica" asserted that the Drake Passage "could be of strategic concern to the United States."[111] The paper, designated PPS-31, indicated that in the "event of the closing of the Panama Canal, this passage would become an important sea route and hostile naval or air units based on either side of it could interfere with the passage of U.S. naval or commercial shipping."[112]

Responding to perceptions about the geostrategic value of the Drake Passage, several Americans argued forcefully for American and allied control of the passage. During congressional hearings in 1954, Captain Finn Ronne contended that the United States should take an active stand in the Antarctic to counter the threat of communist aggression spreading to the Antarctic and to prevent the Soviets from controlling the Drake Passage.[113] In 1957, Walter Sullivan opined in the popular journal *Foreign Affairs* that Antarctica's "vastness provides a sanctuary from which aircraft could dominate the waters, that apart from the vulnerable Panama and Suez Canals, provide the only ready links between the Atlantic and Pacific and the Atlantic and Indian Oceans."[114] That view had considerable impact on U.S. policy planners during the early Cold War years.

The debate over ratification of the Antarctic Treaty also records American concern over maintaining unimpeded access to the Southern Ocean. Senator Clair Engle stated that if the Panama or the Suez Canals were closed in a future war, "Western shipping would be detoured into the southern oceans around Cape Horn, the Cape of Good Hope, and Australia. Whoever controlled Antarctica would have a great asset in the control of these vital southern routes."[115] Whether Western apprehensions over denial of access in the southern seas were grounded in reality remains largely conjectural, for there was only scant detailed examination of the issue.[116] For some commentators, however, the geostrategic value of the Drake Passage retains its significance even today. Its importance becomes obvious with the realization that, other than the vulnerable Panama Canal, it is the only ocean link between the Atlantic and the Pacific. To the United States, with its ocean-to-ocean land mass and worldwide military commitments, both of which require a reliable, continuing capability to move naval units rapidly from ocean to ocean, the Drake Passage remains especially meaningful. Reliance by the United States solely on the Panama Canal to ensure this capability is generally recognized as impossible. Not only is the canal extremely vulnerable to sabotage and nuclear missile attack, but its locks can no longer accommodate most of the navy's major ships.[117]

Antarctica's value for space, satellite, and communications activities also figured prominently as an American concern in the pre-treaty era. Writing in 1959, Phillip C. Jessup and Howard J. Taubenfeld noted that "the strategic importance of Antarctica might be found in bomb-testing or missile-launching sites or the unique advantages of the South Pole for contact with satellites on polar orbits."[118] These issues continued to surface prior to the ratification of the treaty. In 1960, for example, Senator Engle maintained that the advances in military technology had increased the strategic importance of Antarctica in the defense of the free world. Antarctica was viewed as not only "advantageously situated for the observation and conduct of space activities," it could also be used as a long-range missile base.[119] L.V. Berkner, president of Associated Universities, also referred to Antarctica's strategic advantages. Antarctica held promise as an important link in transportation between East and West, just like the Arctic. "It commands the Southern Hemisphere for missile control. Its South Pole is a unique location for control and interrogation of north-south space vehicles on every rotation."[120] When the possibility of military buildup in Antarctica was raised during the Antarctic Treaty ratification hearings, Ambassador Herman Phleger's response reflected the traditional views of that time. There was nothing to prevent any of the states involved in Antarctica from considering military installations, he opined; and, he noted, this view was shared by southern hemisphere countries: "It would be quite within the capability of some of these parties to go down there and install launching pads."[121]

The acute logistical problems of overcoming this wide and stormy expanse of Antarctic seas significantly devalues the strategic importance of the Drake Passage. Also, save for a minor interest in tracking satellites, strategic military uses appear unrealistic given their highly visible nature and the fact that they can be accomplished far more effectively elsewhere, such as from nuclear-powered submarines. As Jozef Goldblat pointed out, advances in military technology have greatly decreased the strategic potential of Antarctica.[122] Similarly, Peter J. Beck observed that the various threat scenarios connected with the south polar region display an ignorance of strategic considerations such as those related to both the range and operational effectiveness of contemporary submarines and missiles.[123]

Disputes Among Allies

The fact that Chilean, Argentine, and British claims to the continent overlap has historically complicated the sovereignty situation. Prior to the Antarctic Treaty, conflicts over claims to Antarctic territory produced notable tensions in a region valued by the United States. The three antagonists—U.S. allies under various security treaty arrangements— also had extensive political, economic, and other significant relationships with the United States. As the most active participant in scientific research in Antarctica, the prospect of discord among America's allies presented a direct threat to U.S. scientific and other interests in the region. American policymakers were acutely aware that tension might escalate into open conflict in the absence of a mutual agreement. The Department of State's commitment to reach a resolution of the problem became abundantly clear in a 1948 paper, which emphasized that U.S. national interest would be well served if the dispute could be settled in a manner acceptable to the countries involved.[124]

The years following World War II saw exaggerated, ominous international posturing[125] and a marked deterioration in British-Argentine relations.[126] By late 1948, fear of a clash was so intense that Chile, Argentina, and Britain agreed to forego sending warships south of 60° south latitude except for normal relief expeditions.[127] Incidents continued to occur, however. Britain was not prepared to abandon its Antarctic claims to other equally bellicose antagonists. As a result, rivalry over claims led to several minor military episodes, prompting real concern in the United States.[128]

In 1955, Britain submitted a unilateral application to the International Court of Justice for a resolution of the Antarctic claims.[129] The following year the court dropped the application from its agenda because Argentina and Chile refused to let the Court decide, asserting that sovereignty was so obvious that no third power would be qualified to settle the question.[130] Recurrent clashes among the three countries over Antarctic claims posed risks of escalating armed conflict. Agreeing to remove Antarctica from political contention would serve all states' broader interests in the region. The mechanism that essentially achieved these goals was the Antarctic Treaty. The treaty represented in 1959, and still represents today, formal assurance that Antarctica will remain peaceful and will not, in the words of the treaty's Preamble, become "the scene or object of international discord."

Superpower Rivalry

The prospect of superpower rivalry spreading to the continent magnified the strategic security concern for the United States.[131] By the late 1940s, increased Soviet activity in the Antarctic had aroused U.S. anxieties over the geostrategic designs motivating Soviet involvement there.[132] The Truman administration also feared that this rivalry could precipitate an arms race in the Antarctic, ultimately leading to implanting or testing nuclear weapons there.

Denying access to the continent and circumpolar islands to adversaries, especially to the Soviet Union, was important to American policymakers. One particular concern was that the Soviet Union might set up base camps in the unclaimed portion of the continent where the United States had sought to establish a firm legal basis for asserting a future claim.[133] A major U.S. policy objective thus became to deny the Soviet Union the opportunity to participate in an Antarctic settlement or administration.[134] A related objective aimed to discourage any Soviet attempt to become a territorial claimant on the continent.[135] For U.S. policymakers, national security considerations dictated that control of the Antarctic be exercised by friendly powers and be denied to antagonistic powers. A 1950s NSC memorandum specified the basic U.S. objectives as: (1) orderly progress toward a solution of the territorial problem of Antarctica, which would assure control by the United States and friendly parties; (2) freedom of exploration and scientific research for the United States and friendly parties; and, (3) access by the United States and friendly parties to natural resources discovered in Antarctica.[136]

Following the International Geophysical Year, the Soviet Union showed few signs of departing the continent, though no military presence had apparently been introduced.[137] The new active role by the Soviet Union made interested Western governments increasingly aware of the need for international accommodation to ease Cold War tensions over the continent. The United States viewed nonmilitarization as an effective way to remove the Soviet military threat to the southern hemisphere. The demilitarization provisions of the Antarctic Treaty removed the continent from the arena of superpower strategic rivalry, saving all concerned the expense of maintaining defenses against real or supposed aggressive designs.

The Antarctic's strategic irrelevance, however, should not be taken for granted.[138] American interests would be seriously jeopardized were Antarctica to lose its demilitarized status under some new legal regime. Under such conditions, Antarctica could become an area of rivalry among states striving to establish a military foothold in the region. The 1982 conflict between Britain and Argentina transformed the South Atlantic into an area of acute international concern and raised anxieties over the potential spillover of hostilities into Antarctica. If anything, the Falklands/Malvinas War revealed that, at least for Argentina and Great Britain, islands in the South Atlantic are valued considerably in military, economic, diplomatic, and human capital. It is reassuring, however, that even during the Falklands War Argentina and Great Britain continued to cooperate together in consultative party negotiations.[139]

The fact that sovereignty claims to portions of Antarctica persist also suggests that nonmilitarization might be compromised were claimant states to attempt to sustain or defend their titles there. Should the Antarctic Treaty collapse, one might reasonably predict that the claimant states would hasten to shore up and protect their claims, and even resort to military means if necessary. Such a scenario no doubt would jeopardize continued scientific and economic opportunities for the United States in the Antarctic.

The Post-Treaty Era: Legal/Political Dimensions of U.S. Security Interests

International law is essential to foreign policy in that it provides mechanisms, norms, rules, and procedures for interaction by members of the international community. Law operates not simply as a constraint on state conduct; it also functions to direct and thereby facilitate interaction among sovereign states.[140]

In few areas of the world has the rule of law, as embodied in international agreements, been as auspicious for providing political stability as in the Antarctic. United States policymakers consider the Antarctic Treaty's broad prohibition on military activities the sine qua non for successful pursuit of all U.S. national security interests in the region.[141] This general security goal has become an overriding consideration for the United States and for the other treaty parties, a factor that has prompted them to cooperate on potentially destabilizing economic matters such as access to natural resources.[142] Augmenting the treaty's importance for U.S. national security interests is the instrumental role it has played in the progressive development of international law,[143] including subsequent auxiliary agreements adopted by the consultative parties.[144] Two treaty provisions that contribute to more orderly Antarctic political and economic situations are the dispute settlement procedures and the compliance inspection mechanism.

Settlement of Disputes

Article XI of the Antarctic Treaty requires the contracting parties to resolve any dispute over the interpretation or application of the treaty among themselves by means of negotiation, inquiry, mediation, conciliation, arbitration, judicial settlement, or other peaceful means. Article XI also exhorts states to refer unresolved disputes to the International

Court of Justice.[145] Article 25 of the Convention on the Conservation of Antarctic Marine Living Resources contains similar provisions. Some scholars find these mechanisms weak and ineffective, primarily because they do not require compulsory settlement of disputes.[146] Although valid, this criticism ignores the fact that states frequently will not agree to third-party settlements. Notably, however, member states have not found it necessary resort to this provision. Throughout the life of the Antarctic Treaty, when problems have arisen between ATCP members, they have been resolved through diplomatic negotiations without the need of other dispute settlement techniques.

Compliance Inspections

All nations need credible assurance that other states will honor their international obligations. Monitoring is one device that enhances confidence among parties that consensual rules are being followed. For this purpose, the Antarctic Treaty guarantees in its Article VII unlimited, unannounced, on-site inspections. The treaty authorizes all contracting parties to conduct inspections—even utilizing aerial observation—in all areas of Antarctica, including stations, installations, and equipment. Inspections serve to verify compliance with the nonmilitarization provisions, to monitor the ban on nuclear explosions, and to ensure that the prohibitions on disposal of radioactive waste are respected.[147] The ultimate goal, of course, is to deter and uncover violations.

Linking the nonmilitarization of Antarctica with inspection opportunities was a U.S. proposal.[148] The United States prevailed in its insistence on an unlimited right of unilateral inspection.[149] Also vital from the U.S. perspective was that the Soviet Union agreed to these inspection rights.[150] On a number of occasions, American officials have stressed the importance of inspections. As Ambassador Herman Phleger remarked more than three decades ago, the inspection provision "will not only serve to protect the parties against any violation of the treaty, but should also prove a valuable source of practical experience in the detailed process of international inspections."[151] More recent statements from U.S. officials attest to the continued importance of Article VII, especially from an arms control perspective. "The Treaty establishes for each contracting party the right to designate observers who have free access to all parts of Antarctica, including all stations, to carry out inspection activities."[152]

Inspection provisions also serve as checks on compliance with regulations, such as those designed to protect the Antarctic environment. Environmental inspections by consultative parties, however, have only recently been initiated and are still relatively uncommon. The 1972 Convention on the Conservation of Antarctic Seals incorporates certain provisions on verification, but they are of a general character.[153] To bolster enforcement, the Convention on the Conservation of Antarctic Marine Living Resources set up a general system of inspection to be implemented by the Commission, its decision-making body. The CCAMLR provides for "observation and inspection" of vessels engaged in scientific research or harvesting of marine living resources in the area to which the convention applies. It was only in 1989, however, that an inspection system was approved.[154] That system of inspection came into operation during the 1989–90 season; the U.S. performed the first inspection when it boarded the Japanese krill fishing vessel *Aso Maru* in March 1990.[155]

Similarly, the 1988 Convention on the Regulation of Antarctic Mineral Resource Activities (CRAMRA) included an inspection system to enforce compliance with its provisions. The inspection system for the minerals regime would have involved not only observers under Article VII of the Antarctic Treaty but also observers designated by the CRAMRA Commission or relevant regulatory committee.[156]

Inspections have proven useful in ways unforeseen by the treaty's drafters. In addition to benefits inherent in the inspection process, one commentator has observed that an inspection team "with adequate technical and linguistic expertise serves a representational function, provides a special opportunity for the exchange of information, and, by hauling mail and people from station to station, helps replace suspicion with amity."[157] Thus far, in carrying out more than one hundred inspections, the process has served the treaty system well.

The Economic Dimension

United States foreign policymakers regard Antarctic minerals, particularly oil, as resources of last resort. They subscribe to the view that, for the foreseeable future, neither oil nor natural gas deposits of commercial value exist in the Antarctic region. Technical, economic, environmental, and political factors combine with market and supply forces to question even the long-term prospects of exploitation. These factors, as well as the

1991 environmental protocol's fifty-year ban on the development of Antarctic mineral resources, might suggest that U.S. economic interests in the frozen south have diminished as geostrategic policy considerations.

One should be wary, however, of dismissing as irrelevant U.S. economic interests in the region. Although political and legal circumstances presently disallow mineral development and Antarctica's present economic utility is minimal, future economic possibilities—even if only in the very distant future—should not be discounted. Periodical changes in the price, demand, or availability of Antarctic hydrocarbon resources could coalesce to markedly transform the economics of Antarctic energy resources. American oil production has been falling slowly since 1986, when the price for crude oil collapsed.[158] A third of the U.S. oil supply was imported in the mid-1980s.[159] The mainland United States has been heavily exploited, though, and new fields are hard to find.[160] As a result, in 1994, with its oil production at a forty-year low, the United States for the first time imported more than 50 percent of its crude oil.[161] Some members of Congress view the continued and rising U.S. dependence on foreign oil as a threat to American security.[162]

It is conceivable that Antarctic hydrocarbons one day may be critical to U.S. energy needs, and it may be technologically possible to exploit them without posing substantial risks to the environment. Recent improvements in technology have produced some expectation that the problems of drilling for oil in hostile environments may some day be overcome.[163] An oil industry consultant recently commented that "they're doing things out there due to new technology that five years ago nobody thought were economically possible."[164]

To some, protection of the Antarctic environment and development of mineral resources can coexist without any serious harm to the former. For example, during a congressional hearing in 1990, a Department of State official took issue with the "premature" assumption that mining and protection of the unique Antarctic environment are incompatible.[165] Though acknowledging that protection of the fragile Antarctic environment is paramount, he pointedly noted that mineral activity has occurred in other fragile environments with successful implementation of safeguards.

Oil extraction, production, transportation, refining, and distribution require advanced technology and large amounts of capital. The United States may be the only country in the world that possesses the technological and financial capability to develop and exploit Antarctica's hydrocarbons. American companies have been active for some time in exploring and developing the oil potential of the Arctic, and they have acquired significant expertise for operating in ice-infested waters and drilling at deep ocean depths. To some, however, economics rather than technology looms as the main factor limiting hydrocarbon development in polar regions.[166] The costs of operating in the Arctic imply that petroleum development in the Antarctic will not be economically competitive for a considerable time, particularly at today's crude oil prices.[167] According to analysts in the Office of Technology Assessment, petroleum exploration and production in the Arctic may cost ten to one hundred times more than comparable operations in the Gulf of Mexico.[168] Field development costs also run much higher in the Arctic. A 50 million barrel field in the Gulf of Mexico costs $100 to $200 million to develop.[169] The development of a commercial Arctic field may range from $6 billion in the Beaufort Sea to more than $10 billion in the Navarin Basin.[170] For revenues to cover development costs in the Antarctic a discovery would have to approximate giant (i.e., 500 million barrels) or supergiant (in excess of one billion barrels) proportions, or reserves comparable to the Prudhoe Bay Field on the north slope of Alaska.[171] The prospects for such at present are unknown.

The profit motive, however, may not be the only incentive to develop Antarctic resources. As some commentators have suggested, the possibility exists that in the future nations that are resource poor, like Japan,[172] may subsidize operations that would not produce sufficient return for private companies.[173] Likewise, consultative parties with claims to Antarctica might initiate activities in their sector territories to bolster substantiation of their claims after the fifty-year ban.[174] Worth noting here is that, though France refused to sign the Wellington minerals convention in 1989, that government has since substantially increased its support for geophysical research in the Antarctic continental margin.[175] Such increased research must suggest certain economic implications.

Hydrocarbons Exploration and Exploitation in a Harsh Climate

The harsh polar environment compounds the problem of conducting geophysical surveys and carrying out hydrocarbon development in the Antarctic.[176] The massive ice sheet, the climate, and attendant logistical problems effectively preclude year-round activities. The thick mantle of glacial ice overlays 90 percent of the potential mineral deposits.[177] All these factors severely impact the operational efficiency of men and machines.

Three major climatic zones are found in Antarctica: the frigid interior, the stormy coastal zone, and the somewhat milder but still difficult Antarctic Peninsula. Work outside during the austral summer in the interior of the continent is difficult, inefficient, and dangerous. Travel for short distances on foot or by vehicle for longer trips to outlying work areas is feasible under strict safety rules, except during major storms. Winter work outside in the same areas is practically impossible. Extensive year-round drilling or mining operations anywhere in Antarctica would require artificially maintained environments.[178]

The vast continental margin offshore Antarctica has long been considered a potential location for large oil fields. Even so, several factors render the Antarctic continental shelf among the most unsuitable regions in the world for hydrocarbon exploitation.[179] The shelf has an average water depth ranging between 400 and 800 meters and is covered by thick pack ice during most of the year. Huge icebergs with drafts of several hundred meters drift freely on the shelf, and the seas that surround the continent are among the most turbulent in the world.[180] In this environment, the cost of developing and exploiting oil could easily exceed the value of recovered commodities.[181] If oil were discovered in commercially recoverable quantities, the problems of storage and export would still have to be resolved. Antarctica's location remains remote— around 600 miles—from all other southern hemisphere land masses. This isolation is magnified by the seasonal belt of pack ice that can extend as far out as 950 miles from Antarctica. The continent's isolation and its great distance from supply bases and markets present major logistic, technical, and economic problems that cannot be easily overcome. For example, a major difficulty in Antarctic exploration and mining operations is the physical transport of people, fuel, supplies and equipment to the continent and of waste and mining products out.[182] Transporting goods inland to a site of operations would compound these problems.[183]

Available Technology

Experience in the Arctic suggests some optimistic prospects for hydrocarbon development in the polar south. Mineral and fuel extraction in the Arctic has expanded rapidly over the past fifty years.[184] The technological accomplishments resulting from operations in Prudhoe Bay—the largest single oil field found in North America—have provided considerable know-how concerning the discovery and development of hydrocarbon deposits in polar climates.[185] Any petroleum and mineral resources in Antarctica will undoubtedly be discovered by applying knowledge, techniques, and experience gained in parts of the world where development is taking place under conditions of hostile climate and terrain.[186] Arctic expeditions and research programs could thus serve as the bases for planning future exploration, mining, and petroleum production in Antarctica.[187]

Improvements in mining technologies pioneered in the Canadian Arctic have made drilling for oil and extraction of minerals feasible where previously thought impossible.[188] Drilling for oil in Arctic waters, for example, "has always been difficult due to scouring of icebergs on the sea bottom, which can damage the wellheads."[189] But experience in Canada—especially in the Beaufort Sea, Sverdrup Basin, Baffin Bay, and Labrador—has shown that these and other difficulties can be overcome.[190] Indeed, some analysts with the Office of Technology Assessment find that economics rather than technology limits Arctic petroleum development.[191]

Transportation of Antarctic energy resources would require numerous support services. For surface marine transportation, minimum requirements would include those necessary for operations in the Arctic: icebreaker escort, weather and sea ice reconnaissance and forecasting, navigation and communication aids, search and rescue capabilities, oil-spill surveillance and cleanup capacity, and emergency towing and repair facilities.[192] One important role served by icebreakers is to escort commercial shipping through ice. In this regard, William Westermeyer observed in 1984 that the U.S. polar class icebreakers *Polar Star* and *Polar Sea* are capable of operating continuously through six feet of ice at a speed of three knots.[193]

Even though "the techniques might be similar to those developed elsewhere, petroleum development in Antarctica will require special research efforts and technology that are not commonly used in other parts of the world."[194] For example, experience in the Arctic is with multi-year sea ice (i.e., frozen salt water). The properties of that ice differ substantially, however, from the properties of ice formed by freezing freshwater. In Antarctica, there are both fresh water masses that move seaward from the mainland ice cap and a peripheral zone of sea ice that forms, expands, and contracts annually.[195] There are other important differences as well:

> Weather and iceberg conditions in the South are less predictable than those in the North, and once extracted, Arctic oil is much nearer to large consumer markets in the North than would be the case with Antarctic oil. In addition, the kinds of artificial offshore islands employed in the Arctic would be more difficult, if not impossible, to use in the Antarctic, where the continental shelf lies under much greater water depths. . . . Moreover, the July 1988 explosion . . . and resultant deaths on the North Sea oil rig *Piper Alpha* . . . gives rise to increasing concern that safety standards for advanced production technologies are not yet adequate.[196]

In the final analysis, the economics of petroleum production in the Antarctic are not cost-effective, now or in the foreseeable future, either in terms of extraction/drilling technology, transportation, or logistics. For coming decades, therefore, the prospects for hydrocarbon exploitation is not likely to figure in the geostrategic considerations affecting U.S. policy in the Antarctic.

Conclusions

The United States has maintained a long-standing geostrategic interest in the possible development of Antarctica's mineral resources, particularly hydrocarbons. As with other U.S. Antarctic interests, the official position of U.S. foreign policymakers on Antarctic resources has exhibited significant consistency and continuity.

A recurrent theme throughout this study has been the interrelationship among various U.S. national interests in the Antarctic. Geostrategic interests also reflect this interaction. They can be viewed in terms of three factors: (1) economic, especially in terms of profit orientation; (2) security, particularly energy security; and, (3) environmental, that is, preservation of the region's environment. American foreign policymakers' urgency to negotiate an agreement regulating the commercial development of Antarctic minerals was not economically motivated. The reasons were primarily political and environmental. Nevertheless, many members of Congress and environmentalists viewed CRAMRA as encouraging mineral exploitation in the Antarctic region and placed considerable pressure on policymakers to abandon the convention. Opposition from congressional and interest groups was not, however, the determinant factor in the U.S. decision to abandon CRAMRA. The decisive factor was that the consensus necessary to bring the convention into force internationally had been lost. Rather than jeopardize the stability of the entire Antarctic Treaty System, U.S. foreign policymakers decided to abandon CRAMRA and move toward negotiating a comprehensive Antarctic environmental protection protocol. The continued vitality of the Antarctic Treaty System, which accommodates the full range of U.S. Antarctic interests, deserves no less.

As with other U.S. Antarctic interests, American geostrategic interests have exhibited significant consistency, with development of Antarctic minerals being subordinated to environmental concerns. The Antarctic Treaty has proved a resilient legal instrument that has promoted well American security interests in the region. Orderly legal processes have provided American foreign policymakers with a politically secure environment in which to pursue the multiplicity of U.S. national interests in the south polar region, including nonmilitarization, nonnuclearization, and dispute settlement.

8 / Ideological Interests and Ecopolitics

UNITED States foreign policy in the Antarctic has been directly affected by the pressures of international politics on the Antarctic Treaty System (ATS). For nearly four decades, the south polar region has been governed by this legal arrangement. Since the early 1980s, however, certain ideological and political challenges to the legitimacy of the Antarctic Treaty System have arisen. Chief among these have been the "common heritage of mankind" notion (CHM). There has also been, however, a rapid increase in the number of ATCP members in the ATS, as well as an appeal to internationalize Antarctica under an environmental ethic for the polar south. These challenges tested the political vibrancy of the treaty regime itself, as well as the resiliency and resolve of ATCP member governments, in particular, the United States. American national interests are thus being politically pricked and directly affected. This chapter examines these challenges to U.S. foreign policy and explores how each has affected U.S. national interests in the frozen south.

The Common Heritage of Mankind

During the 1980s, an international debate arose over extending the legal-philosophical concept of a common heritage of mankind to Antarctica. Indeed, implicit in the CHM notion was an ideological challenge to the propriety of the Antarctic Treaty, a challenge that gained momentum in the UN General Assembly during the mid-1980s. American policymakers began to perceive the CHM idea as a Third World means of "redistributive justice," that is, of restructuring the international economic system to benefit developing countries, at the expense of the United States. Concern was heightened over the fact that the CHM concept had attained tangible legal substance through conclusion of the Agreement on Celestial Bodies (the "Moon Treaty") in 1979[1] and the UN Convention on the Law of the Sea in 1982.[2] These two agreements fixed the common heritage of mankind as a principle of international treaty law, as well as a potential legal precedent for Antarctica. Were the CHM notion to be applied persistently and compellingly by the majority of states to Antarctica, the ability of the United States and other ATCPs to justify their lawful presence in the polar south could be seriously eroded. The concept of CHM thus came to represent a perceived threat to U.S. interests in the region.

Evolution of the CHM Concept

Many of the new states that emerged from decolonialization during the 1960s questioned traditional international law,[3] including the legal regime for the oceans.[4] These new states sought to use their collective political power through the Group of 77 in the United Nations to correct perceived inequities in the world economic system by establishing international control over resources beyond the limits of national jurisdiction.[5] The so-called "New International Economic Order" (NIEO) movement emerged with an ideology that advocated general reordering of the world economic system on terms more beneficial to the developmental aspirations of newly independent Third World countries. At the core of the NIEO was the concept of the "common heritage of mankind."[6]

The common heritage notion proffers a novel approach to economic development of natural resources beyond the limits of national jurisdiction. The concept emerged as the major impetus galvanizing negotiations for a Third United Nations Conference on the Law of the Sea. Arvid Pardo, then the representative from Malta to the United Nations, is often credited with originating the proposal to apply common heritage to the seabed and its resources in an address delivered to the UN General Assembly in November 1967. Less well appreciated is that, the year before Ambassador Pardo's speech at the United Nations, President Lyndon Johnson articulated a similar notion. As President Johnson averred in a July 1966 statement,

Under no circumstances, we believe, must we ever allow the prospects of rich harvest and mineral wealth to create a new form of colonial competition among the maritime nations. We must be careful to avoid a race to grab and hold the lands under the high seas. We must ensure that the deep seas and the ocean bottom are, and remain, the legacy of all human beings.[7]

Ambassador Pardo echoed these words in his famous 1967 address before the UN General Assembly.[8] Under Pardo's proposal, the seabed and its resources beyond the limits of national jurisdiction would not be subject to national appropriation but would be declared the "common heritage of mankind."[9] The financial benefits derived from exploitation of these resources would be "used primarily to promote the development of poor countries."[10] Similarly, on May 23, 1970, President Nixon proposed that all states adopt a treaty under which all national claims over the seabed beyond the exclusive jurisdiction of states would be renounced and resources be regarded as the common heritage of mankind.[11]

On December 17, 1970, without a single vote in opposition, the UN General Assembly adopted Resolution 2749 (XXV), which endorsed the common heritage concept and proposed an international regime for the seabed to ensure equitable sharing in the benefits derived from resource exploitation.[12] The CHM concept was first incorporated into the Informal Single Negotiating Text of the Third Conference on the Law of the Sea,[13] and came to supply a essential ingredient in the eventual 1982 Law of the Sea Convention.[14]

Ambassador Pardo's use of the term "common heritage of mankind" suggested several elements. Among the most critical ideas were that (1) the seabed beyond national jurisdiction should be excluded from national control; (2) the United Nations should establish an international agency to regulate the exploitation of mineral-rich manganese nodules found on the seabed; and, (3) any profits derived from exploitation should be used to benefit all mankind, in particular, developing states.[15] The universalist, redistributive ideas in this conception mean that chief responsibility for regulating exploitation of and access to resources is assigned to international organizations.[16] As such, the CHM notion maintains that no state is legally entitled to unilateral exploitation of resources until an appropriate international regime is established to control and manage these activities.

No general agreement on the legal parameters of common heritage yet exists. Clearly, though, Ambassador Pardo's proposal generated legal implications for global commons beyond the limits of recognized national jurisdiction.[17] His initial suggestion, coupled with UN General Assembly Resolution 2749, prompted subsequent initiatives that eventually produced a new legal regime for the oceans.[18] Pardo's address also encouraged attempts to extend application of the CHM concept to other areas, both on earth and in space[19]

Proponents of a New International Economic Order (NIEO) maintain that the common heritage of mankind concept has entered into customary international law with the General Assembly's adoption of Resolution 2749.[20] This view is not widely shared.[21] The portion of the Law of the Sea Convention containing the CHM provision neither expresses nor codifies customary international law.[22] The CHM notion remains more an emergent principle rather than an internationally accepted customary rule. As such, it does not yet generate specific legal requirements and detailed consequences for all states. Still, the CHM concept retains certain identifiable goals. As Gillian Triggs has observed,

First, it rejects the notion of domination of property of natural resources either by a particular state as sovereign, or by legal persons as owners of property. Thus, it is analogous to the notion of *res communis*. Second, it seeks to safeguard a common heritage for future use, and emphasizes conservation of resources, in the sense of sustainable use and exploitation of those resources, and environmental protection. Third, the principle aims for an equitable allocation of the resources and benefits, with particular attention to the needs of developing States. Finally, it contemplates a legal regime to formulate precise rules to give effect to these general objectives.[23]

International law scholars have grappled with defining the "common heritage of mankind," but no universally acceptable definition has been produced. The concept was born in ambiguity and, as set out in legal documents, still "carries no clear judicial connotation, but belongs to the realm of politics, philosophy, morality, and not law."[24] A number of difficulties inherent in using the concept undermines its influence. One such difficulty arises when the CHM concept is "combined with the international case law which constitutes part of the global constitutive process of authoritative decision-making."[25] The roots of the definitional conundrum can be traced to Ambassador Pardo's inability in 1967 to define precisely and authoritatively the new concept he was introducing into international law.[26]

Lack of precision invites various interpretations and misinterpretations. As a consequence, although both developing and industrialized states agree that resources outside the limits of national jurisdiction may be subject to common heritage implications, each group views the CHM notion and those implications differently.[27] This confusion became evident during the protracted negotiations on the law of the sea.

Most developing countries consider the CHM concept as a legally binding prohibition against unilateral exploitation of resources in areas designated a common heritage.[28] These states overwhelmingly support regulating access to and exploitation of resources exclusively through a collectively owned and supragovernmentally controlled corporate authority. They also favor requiring transfer of technology to that supragovernmental corporation.

By contrast, the United States and other Western industrialized governments interpret common heritage to mean the right of all countries to exploit resources outside national jurisdiction, "subject to the corresponding duty not to interfere with the correlative duties of others to engage in such exploitation."[29] Steven Burton put the matter even more explicitly:

To the industrialized countries, common heritage means that the area may not be appropriated by any state, but should be open for exploitation by any state and its nationals under reasonable conditions established by treaty, including some taxation for the benefit primarily of developing countries. To the Group of 77, common heritage suggests that the area may be neither appropriated nor exploited by any state, and should be exploited only by an international enterprise, established by treaty and governed by the majority of all nations.[30]

In testimony before Congress in 1979, Ambassador Elliot L. Richardson, special representative of the president to UNCLOS III, pointed out that U.S. policy accepts the concept of the common heritage of mankind in principle, but that the U.S. government felt the concept generated no legal consequences.[31] Elaborating on this interpretation, Richardson posited:

We regard it [the common heritage principle] as essentially an abstract concept conveying the idea that the world community has, in a sense, an undifferentiated share of these resources. But we do not accept the proposition that for any nation or its citizens to exploit this common resource, it is necessary to await the creation of some new international machinery.

We say it is our position that the common heritage is both shared and available to the world community under the terms of preexisting international law, and that any constraint upon access, therefore, can only come about as the result of some new international regime to which we become a party.[32]

Given this position, the U.S. Deep Seabed Hard Minerals Resources Act of 1980[33] evidences support for the common heritage notion, "with the exception that this principle would be legally defined under the terms of a comprehensive international law of the Sea Treaty."[34] The implication here is plain: "The concept of a common heritage has legal substance only within the precise terms of the treaty which employs it and then, obviously, in relation to those states which are party to it."[35] "To employ the phrase a 'common heritage' in the absence of such a treaty, or readily identifiable customary rule, is to invoke a moral ideal or political aspiration which has no independent legal content."[36]

Notwithstanding the validity of these objections, inclusion of the common heritage concept in the 1982 Law of the Sea Convention reserves a notable place for it in international law.[37] What the notion lacks in conceptual clarity and legal soundness is compensated for in ideological vim and political vigor. The CHM notion enjoys substantial rhetorical support among developing governments, which have used it as a quasi-legal justification for bridging economic disparities between the poor countries of the southern tier and the more industrialized, rich states of the north.

Antarctica as "Common Heritage"

Third World demands for greater equity in sharing resources and benefits in the Antarctic arose during the 1980s. Antarctica became a region where application of the common heritage concept was hotly debated, and the United Nations became the forum for that exchange. During the mid-1980s, efforts intensified within the UN General Assembly to criticize the lawfulness and fairness of the Antarctic Treaty System.

Since 1984, the question of Antarctica has been formally placed on the agenda of the General Assembly. In meetings of the First Committee, which debated the merits and substance of resolutions pertaining to Antarctica, as well as in the subsequent plenary sessions, calls were made repeatedly during the 1980s to abandon, or at least modify, the Antarctic Treaty System so that the legal status of Antarctica might be transformed into the common heritage of mankind. Led by Malaysia, critics of the ATS persistently advocated that the "international community" should take "collective actions under the auspices of the United Nations for an open, accountable and equitable framework for Antarctica for the benefit of mankind as a whole."[38] Malaysia was not alone in its efforts in the United Nations. Steadfast allies in the First Committee debates have been Sri Lanka,[39] Pakistan,[40] Bangladesh,[41] and Indonesia,[42] with additional support offered on occasion by the governments of Bhutan,[43] Nepal,[44] and the Philippines.[45]

Legal Implications

It is important to appreciate what legal implications would be generated for the United States (and other ATCPs) were the common heritage of mankind concept applied to Antarctica. At least five principal features or conditions would characterize Antarctica under a CHM regime.[46] First, Antarctica would not be subject to appropriation of any kind, either public or private, national or corporate. Under the CHM doctrine, the continent would be regarded legally as a region owned by no one, though hypothetically managed by everyone. Sovereignty would be absent, as would all its attributes and ramifications. Consequently, no jurisdictional privileges, rights, or obligations arising from sovereignty considerations could exist.[47] In sum, Antarctica under a CHM regime could not be owned in whole or in part by the United States or any state or group of states. Legally, the region would be administered by the international community.

A second condition, derived from the first, is that all people would share in the management of Antarctica. The United States, or any other national government, would be precluded from management functions except as the representative agent for all mankind. This requirement purports to expunge national interests from the administration process. Instead, universal popular interests would assume priority and furnish the foundation for administrative decisions affecting the Antarctic.[48]

Third, if natural resources were exploited from Antarctica, the economic benefits derived from those activities would be shared internationally. Under a CHM regime, U.S. agents or corporations engaged in commercial profit or private gain would be deemed inappropriate unless they operated to enhance the common benefit of all mankind. Hard minerals and hydrocarbons on and offshore Antarctica would thus be prohibited from U.S. national and corporate exploitation save under the auspices of a common heritage regime mandate.[49]

Fourth, explicit in an Antarctic CHM regime is the precondition that the area must be used exclusively for peaceful purposes.[50] No American military bases would be permitted, no weapons of any kind could be tested, no maneuvers could be conducted, and no weapons system could be installed anywhere in the region. A CHM regime in effect would demilitarize Antarctica to ensure that it was used for peaceful purposes only.

A final condition arising from the concept's application to Antarctica concerns the conduct of scientific research in the region. Such research would be free and openly permissible as long as the Antarctic environment was neither physically threatened nor ecologically despoiled. National research results would be made available as soon as possible to anyone who genuinely expressed interest in them. Under a CHM regime, American scientific research would be conducted for the benefit of all peoples, not merely for the United States or other government sponsoring the research. Moreover, the scientific fruits of such research would be freely and publicly exchanged to foster greater scientific cooperation and more extensive knowledge about Antarctica.[51]

Interestingly enough, these five attributes for a CHM regime in the Antarctic appear quite similar to conditions already established for the region under the Antarctic Treaty. Indeed, in large part, they are. Where serious concern surfaces, however, is with the application of a more radical form of common heritage—one that draws its legal and philosophical bases from the ideology advocating creation of a New International Economic Order. Under this latter version, three important modifications would be made for Antarctica. First, the world community would obtain full ownership rights—with attendant resource utilization rights—over the continent.[52] This modification would plainly cancel any real or supposed rights of claimant states in the region. As a state that reserves the right to make a claim to Antarctica, based on six decades of extensive financial investment and intensive scientific investigation, American interests would thus be directly affected.

Second, if mineral resources were ever exploited from the Antarctic, any profits derived from those activities would accrue to "all mankind," with preferential distribution to developing countries.[53] This point must not be taken lightly: The NIEO version of the CHM notion advocates exploitation of common space areas—inclusive of Antarctica—to produce revenues for development activities in poor countries. Though perhaps noble, such a policy is expressly exploitative in character. It advocates development of Antarctica's perceived treasure house of living and nonliving natural resources to advance the economic status of needy countries. Such an overtly exploitative policy generates serious concerns about conservation and environmental preservation of the continent and its circumpolar waters.

A third feature of the NIEO ideology is institutional. To carry out the obligations and responsibilities associated with the CHM concept, a special institutional mechanism—an "International Antarctic Authority"— would likely be created, ostensibly under the aegis of the United Nations. Such an agency presumably would be patterned after the International Authority created for the deep seabed mining regime in the 1982 Law of the Sea Convention and would exercise legal jurisdiction over the Antarctic. It would also serve as the trustee for the international community in the region.[54] Given the adversarial experience with the law of the sea, it is highly unlikely that the United States or other ATCPs would ever accept such an arrangement for the polar south. As the government that has invested the most money, manpower, time, equipment, and scientific expertise in Antarctica, the United States (or any other developed ATCP, for that matter) could hardly be expected to hand over all governing responsibility to an unproven international institution. Nor would the United States be enthusiastic about supporting an "Antarctic Authority" financially or technologically, as surely would be the expectation.

Attempts by Malaysia and others during the 1980s to apply the CHM concept to the southernmost continent introduced troubling elements into Antarctic politics for the United States. Why did this movement in the UN General Assembly occur? Three factors explain efforts by developing countries to apply the CHM concept to Antarctica. First, precedents from earlier situations shaped expectations. Developing countries outside the Antarctic Treaty System sought to extend to Antarctica the common heritage of mankind notion that had been inserted into the 1982 Law of the Sea Convention. Second, during the 1980s international perception of Antarctica and its surrounding seas shifted from that of a worthless icy desert to one of a rich storehouse of mineral resources, including superabundant hydrocarbons. Third, the series of closed (which were perceived as "secret") ATCP negotiations on an Antarctic minerals regime that began in 1982 fueled a popular belief that Antarctica possessed tremendous untapped mineral wealth. States outside the Antarctic Treaty framework became suspicious that the ATCPs were about to expropriate the mineral wealth of Antarctica for themselves.[55] The debate was sharpened by new information on the extent of Antarctica's resources, by concern that current mineral supplies were being depleted, by the growing scarcity of world energy supplies, and by improvement of extraction technology for polar regions.[56]

Proponents of transforming Antarctica into a CHM regime contended that the frozen continent, like the deep seabed, contains valuable resources and has no effective legal owners. To allow developed ATCP states to exploit Antarctic resources without compensation to less developed states, it was argued, would perpetuate and exacerbate the disparity between haves and have-nots. Such a situation was viewed as economically unfair and legally unjust.

To summarize, as applied to Antarctica the common heritage notion implies a regime in which all members hold common rights of access, common rights to resources, and common rights and obligations to the protection of the Antarctic environment.[57] Each state is entitled to share in revenues derived from the exploitation of resources on the continent or under its seas.[58] Also implied is the establishment of a global management regime (i.e., an "International Antarctic Authority") to govern Antarctica and to regulate resource exploitation.

U.S. Interests and Common Heritage

Differences over the meaning of the CHM concept have produced disagreement in the Antarctic context. Should the common heritage notion apply to the frozen south? To most treaty members, applying this proposition to the Antarctic denies the political realities of the situation. Although comparison of Antarctica with other areas beyond national jurisdiction reveals similarities, it also shows that vast differences exist as well. Unlike the deep seabed, the electromagnetic spectrum, and outer space, Antarctica can hardly be considered an area outside the sphere of man's juridical activity: For nearly four decades, the region has been governed under a "valid and operative juridical system of advanced maturity."[59] More importantly, advocates of a common heritage regime for the continent conveniently ignore the fact that in Antarctica, unlike other global common spaces, seven states assert sovereignty claims to territory. Although legal opinion disagrees over the validity of these claims, no doubt exists that most claimant states perceive Antarctica to be legitimate slices of their national territory.

A common heritage regime for the Antarctic would be antithetical to

U.S. interests because it would not reflect and protect key American objectives in the region. Application of the CHM notion to Antarctica could unravel the delicate political balance painstakingly achieved over three decades—the very balance that preserves the effectiveness of the Antarctic Treaty System.[60] Disruption of this equilibrium could produce destabilizing results crosscutting the full range of U.S. interests in the region.

Several considerations suggest that a common heritage regime would be contrary to U.S. interests in Antarctica. Two of these relate directly to basic characteristics of the concept; others apply indirectly. The first point concerns the politico-legal realm. A common heritage regime raises a serious possibility that militarization of the continent or acts of outright hostility might break out. In supporting a treaty for Antarctica, the United States was motivated by the desire to transform the continent from an arena of international conflict and discord into a zone of peace and cooperation. Essential to obtaining agreement from other involved states, however, was compromise over the claims issue; that is, the claimant states had to be assured that their juridical position would not be jeopardized. Indeed, cooperation and willingness to observe the treaty's principles have revolved around the claimant states' belief that their Antarctic interests were best protected *within* the treaty framework. This belief has been reinforced by a tacit but long-standing commitment among treaty members that the development of legal instruments to manage Antarctica will respect and preserve that compromise. In keeping with this notion, subsequent legal instruments developed by the treaty members have, like the treaty itself, uniformly and intentionally skirted the sovereignty issue.

Under a common heritage arrangement, Antarctica would become an area legally excluded from national sovereignty. This situation would constitute outright denial of the territorial claims, something the claimant states, at least for the foreseeable future, are hardly prepared to accept.[61] There is no question that intense nationalistic feelings toward Antarctic territory are held, particularly by Argentina and Chile. Nothing indicates that any claimant state is ready to renounce its claims to sovereignty.[62] Such blunt, de facto denial of their claims might push Argentina and Chile—and perhaps even the United Kingdom as well—into defensive postures and precipitate attempts to buttress their territorial claims with military bases, or perhaps even to resort to force.[63]

The use of military force in Antarctica represents the most serious challenge to U.S. interests in the region. Upsetting the geopolitical balance on the continent could also produce other, albeit lesser, effects. Disintegration of the treaty system would clearly damage the spirit of international cooperation in scientific research and complicate the search for collective international solutions to Antarctic problems. It could also foster national competition and rivalry in the region, thereby creating more opportunities for conflict.

Superimposed on the political problems with the common heritage notion in Antarctica is the economic issue. The CHM concept seems prejudicial to states whose economies are based on the private free enterprise system. Two contending philosophies over the exploitation of natural resources in common space areas are clear. The United States, possessing the technological wherewithal, prefers maximum freedom to exploit resources in these areas and produce maximum profits for its nationals. Louis Henkin's observation over two decades ago remains quite pertinent today. As he noted, the United States and other developed states "resist international organizations with substantial authority, surely organizations run by a simple majority. Mankind, they might say, would enjoy the benefits of its heritage, and enjoy them most quickly and cheaply, if the developed states were allowed to exploit it competitively for profit."[64]

The United States concurs that developing states should share in benefits from the economic development of Antarctic resources. It does not agree, however, that such benefits should come through unduly restricting access to those states that have the technology and financial capability to exploit them. If Antarctica is to be exploited, economic benefits should accrue to those that earn them; benefits should not be handed out through a system of international redistributive socialism.

In contrast, the common heritage concept as filtered through the New International Economic Order philosophy maintains that development of resources "to which no single decision-making unit holds exclusive title"[65] must be regulated by strict controls to ensure that the neediest countries benefit most. The overall management of Antarctic resources—including exploration and exploitation—would therefore be carefully managed and regulated by an international equivalent of a government monopoly. This form of international socialism would restrict U.S. economic activity in Antarctica and would represent an unprecedented move toward central management of the international economy, two actions that do not promote U.S. interests.

Environmental considerations under a common heritage regime also concern U.S. policymakers. American policy on Antarctica has consistently reflected deep concern over preservation of the Antarctic environment and conservation of its resources. Treaty-based arrangements developed since 1959 highlight the strong protectionist stance of the United States. There are real doubts, however, that strict environmental protection and resource management could be maintained under a UN-based common heritage regime. Efficient coordination and harmonization of conservationist policies for Antarctica would be extremely unwieldy under the aegis of 185 governments with divergent interests and ideologies. It would be very difficult under a common heritage regime, for example, to proscribe activities that cause pollution or other environmental hazards. Developing countries generally have lackluster conservationist records, and many of these governments are understandably less than enthusiastic about limiting opportunities for economic growth by adherence to restrictive and costly environmental controls.

Another liability to U.S. Antarctic interests from a common heritage regime would be the deterioration of the unique system of cooperation that has been facilitated by ATS arrangements. International collaboration in Antarctica has insulated the region from the tensions of international politics and has contributed to order and stability. As testimony to cooperation in Antarctica, issues that have strained relations between treaty members elsewhere in the world seem to have reverberated little on their relations regarding Antarctica. Two dramatic examples bear this statement out: First, both Argentina and Great Britain sat down together in Wellington, New Zealand, at the special ATCP meeting on minerals in May and June 1982, without political complications from the ongoing Falklands War. The head of the Argentine delegation stressed during the meetings that "we are here to discuss Antarctic matters only."[66] Similarly, during the Afghan crisis of 1979–1980, the United States reduced scientific cooperation with the Soviet Union and boycotted the 1980 Olympics in Moscow as retaliatory measures, but cooperation on Antarctic research remained unaffected.[67]

Proposals for a common heritage regime in Antarctica could generate another unwanted consequence for the United States by placing U.S. negotiators in the difficult position of pursuing U.S. interests in an unsympathetic forum. Since 1970 a politicized Third World majority has often used the UN General Assembly as a platform for criticizing American foreign policy, seeking to diminish U.S. influence in world affairs.

Though such anti-American sentiments have subsided since 1990, doubts remain among American foreign policymakers about the feasibility of obtaining global accords that fully protect U.S. interests. The actions of the Third World majority in the UN General Assembly during the 1970s and 1980s emphatically evidenced a concerted bias against U.S. policies. This bias readily surfaced during the 1980s in the tendency of the Third World majority to assert new political and economic rights for managing the Antarctic.

The UN movement for transforming Antarctica into a common heritage continent faded during the early 1990s with the emergence of an environmental ethic for the polar south. The greening of Antarctica, which in effect dissipated the push by developing governments for the political internationalization of Antarctica, came about with the collapse of the Antarctic minerals treaty in 1989 and the subsequent genesis of a comprehensive environmental protection regime for the region.

Growth in the Number of ATCPs

At the same time common heritage was being debated in the United Nations, another challenge to U.S. national interests was evolving. This second challenge involved the rush by a spate of states to participate as ATCPs in the Antarctic Treaty regime, ostensibly to gain a national stake in the outcome of an Antarctic minerals regime. Between 1980 and 1991, fourteen states sought and gained admission into the consultative party group. For a consensus-based decision-making apparatus, this transformation had to be taken very seriously, especially by those governments with prominent interests and activities in the region. The United States, of course, was among that group.

The ATCPs—the original parties plus other states that have demonstrated serious interest in Antarctica through conduct of substantial scientific research—are the heart of the Antarctic Treaty System. Over the years, consensus rule has proved to be a major strength of the treaty. The requirement that action be subject to unanimous approval by the ATCPs usefully protects vital U.S. interests. The power of veto, with its strong guarantee that basic interests will be protected, has played an essential role in preserving the sense of unity and common purpose on which the success of the treaty largely depends. The original consultative parties— some holding antithetical ideologies and others with conflicting territorial claims in Antarctica—have nevertheless worked in concert amicably and constructively. The 1959 treaty brought the interests of these states into balance. Cooperative work in promoting the goals and objectives of the treaty has contributed to the maintenance of that delicate balance. The extent to which a significantly expanded treaty membership might disturb this equilibrium and impede the treaty's progress over the long term remains to be seen.

The Antarctic Treaty has been sufficiently flexible to permit the expansion of participating states. Expectations of developing Antarctic resources, fueled by the minerals negotiations, stimulated interest in membership by several states. As a result, by 1996 the number of consultative parties had risen to twenty-six and that of acceding states to seventeen. Among more recent members of the consultative group are governments that could exert influence in new and important directions. For example, influential Group of 77 members such as India, China, and Brazil acceded to the treaty and became ATCPs in 1983 for varying political reasons. The question remains whether these states can continue to work within the structure and framework of the treaty system, notwithstanding any lingering aspirations they might have of transforming the international economic order.[68]

An extensive consultative party membership may inevitably complicate the consensus-building process,[69] especially if new ATCPs are unfamiliar with the political composition and operating procedures of the system.[70] The greater the number of actors who must agree before any plan or policy can be implemented, the more cumbersome and difficult the process to achieve consensus. Such conditions also impinge upon the ability to advance any plan.[71]

Political pressures to expand treaty membership have produced new concerns. Perceived impairment of the treaty system's viability is diminished if new ATCPs demonstrate genuine interest in Antarctic science,[72] but in some cases political motives have been key.[73] Some governments have sought ATCP membership solely for political or prestige purposes— to have a stake in Antarctica, so to speak.[74] Such motives add an extra dimension to Antarctic decision making: Obtaining consensus becomes more difficult when certain members in an organization have little expert knowledge of the subject and become preoccupied with political or ideological issues. Consequently, the tendency in recent years has been not to encourage additional governments to seek consultative party membership. Since late 1990, no new state has sought or been admitted into the ATCP group, a pattern that seems likely to persist into the foreseeable future.

An Environmental Ethic for Antarctica

Another salient consideration is whether the Antarctic consultative parties are capable of functioning in Antarctica as the legitimate trustee group for the international community. Much international criticism leveled at the ATS during the 1980s suggested that the regime was operating more along the lines of a condominium arrangement than as a trustworthy agent for the world community. Spurred by the minerals negotiations in the 1980s, doubts focused around the ATCPs' willingness to promote and safeguard global interests in Antarctica, as well as their own collective goals. Those concerns have since abated with the successful negotiation of a new international agreement that promises to bring more comprehensive environmental protection to the Antarctic and its circumpolar waters. When ratified by all ATCPs, this new legally binding instrument for environmental protection will furnish the means through which global governance of Antarctica is secured and the interests of all mankind in the polar south are promoted.

The Madrid Environmental Protocol

On October 4, 1991, the Protocol on Environmental Protection to the Antarctic Treaty was adopted and opened for signature by the consultative parties in Madrid, Spain.[75] Coming after less than twelve months of negotiations, this instrument represents one of the most comprehensive, far-reaching multilateral environmental agreements ever promulgated. It embodies a legal blueprint for preservation of the Antarctic. The Madrid environmental protocol also signaled a profound shift—indeed a reversal in course—in the political and economic aspirations of the United States for the Antarctic. Whereas in late 1988 the direction advocated by U.S. policymakers appeared headed toward potential exploration and possible development of Antarctic minerals and hydrocarbons, by mid-1991 that course had been diverted toward a general commitment toward protection and conservation of the continent and its circumpolar seas.

Negotiation of the Madrid environmental protocol emerged out of a history of considerable debate, tension, and frustration—not only among the environmental, conservationist, and scientific communities but also within the ATCP group itself. A fundamental source of disagreement centered on the intended status of the area: Should Antarctica be susceptible to oversight by a special regime for regulating future mineral resource activities? Or, should the Antarctic be legally declared an international wilderness conservation zone—that is, in essence a world park administered under strict preservation rules and conservation regulations designed to promote environmental protection? The United States, for perceived economic security reasons, had steadfastly supported the former course during the 1980s. By 1991, the U.S. position had radically reversed itself into a more pro-conservationist, environmental protectionist approach.

The Protocol on Environmental Protection and its annexes were not negotiated in isolation. On the contrary, the protocol was designed mindfully to complement and strengthen the entire family of Antarctic Treaty instruments. In fact, it implicitly relies upon these other instruments as institutional pillars for political support and legal enactment.

Significantly, among the designated principles and objectives of the Antarctic Treaty, Article IX calls for the "preservation and conservation of living resources in Antarctica." Towards this aim, the consultative parties have convened since 1961 to discuss and devise means for protecting Antarctica's environment, inclusive of marine areas. Several ATCP acts specifically relate to the Antarctic environment, and all have become integrated into the 1991 Madrid protocol. Formal agreement by the United States on all these measures has been essential for each one's success.

Evolution of the Protocol

The Madrid protocol emerged out of the diplomatic wreckage of the Antarctic Minerals Convention. The ATCPs had negotiated a special regime to regulate prospecting, exploration, and development of mineral resources in the Antarctic, should such activities ever occur there. Agreement came in June 1988 on the text for that minerals treaty, and in November 1988 the Convention on the Regulation of Antarctic Mineral Resource Activities (CRAMRA) was opened for signature in Wellington, New Zealand.[76]

The United States had led the way in negotiating the agreement and in advocating international support for the CRAMRA regime. CRAMRA was not meant to be a mining code, though it did create a regulatory framework for minerals activities. To this end, the United States had worked to ensure that the minerals convention created new institutions, set out impressive environmental restrictions, and established legally binding procedures to permit minerals activities to go forward only with the consensus of all contracting parties. The minerals agreement provided for inspection, monitoring, reporting, compulsory settlement of disputes, access to tribunals, and suspension of activities causing unacceptable damage to the environment.[77]

U.S. government policymakers felt that the Wellington Convention made notable contributions to the Antarctic Treaty System. In their view, the agreement marked completion of the resource protection regime. It furnished a negotiated, consensus approach to deal with the regulation of mineral activities that impinged upon complex questions of national sovereignty and environmental protection in Antarctica. The CRAMRA created formal machinery to address those issues. It established groundbreaking environmental provisions, chief among them the principle that one could not proceed with mineral activities unless sufficient information about exploration were available. Importantly, the sufficient information requirement was a process limited by the requirement of consensus agreement by all parties. For the United States, the promulgation of CRAMRA represented a major accomplishment by the ATCP governments, for which U.S. policymakers could justifiably claim much credit.

Notwithstanding U.S. support, the Wellington convention was stillborn. Despite efforts by U.S. diplomats during 1988–1989 to promote the merits of CRAMRA to other ATCPs, by 1990 international support for the minerals agreement had withered away. Critical environmental concerns had been left hanging about the proposed minerals regime, and these were translated by conservationists into potent political weapons. United States Antarctic policy could do little officially or informally to reverse these concerns.

The demise of the Wellington minerals convention, however, opened the diplomatic door to negotiation of a more far-reaching instrument for environmental protection of the Antarctic environment. Diplomatic tensions among ATCPs were high, especially between the United States/United Kingdom and Australia/France over the latters' minerals policy reversal. These tensions, however, gave way to new aspirations for environmental protection. Prior to the October 1989 fifteenth Antarctic

Treaty Consultative Party meeting in Paris, Australia and France circulated a joint proposal calling for comprehensive measures to protect the Antarctic environment and its dependent and associated ecosystems.[78] Stimulated by this development, the United States—along with Chile, New Zealand, and Sweden—submitted draft proposals for comprehensive protection measures at the Paris ATCP meeting.[79] The result was a decision to convene a special ATCM meeting in Viña del Mar, Chile, from November 19 to December 6, 1990.

Environmental Protection Under the Protocol

Emerging out of the meetings in Viña del Mar and in Madrid, Spain (April to May and June 1991), came the Protocol to the Antarctic Treaty on Environmental Protection. The protocol was negotiated with broad support from the U.S. environmental community but with reluctance by the Bush White House. Even so, by 1995 it had been signed by the United States and the other twenty-five ATCPs, as well as eight non-consultative party states. The instrument obligates parties to consider the Antarctic (i.e., that area south of 60° south latitude, inclusive of ocean space) as a "natural reserve devoted to science" and commits them to comprehensive protection of the region's environment.[80]

Once the protocol has been ratified by the ATCPs and enters into force, the lawfulness of U.S. activities in the Antarctic will be directly affected by its operative provisions. Key for U.S. foreign policy is the principle in the first paragraph of Article 3:

The protection of the Antarctic environment and dependent and associated ecosystems and intrinsic value of Antarctica, including its wilderness and aesthetic values and its value as an area for the conduct of scientific research, . . . shall be fundamental considerations in the planning and conduct of all activities in the Antarctic Treaty area.[81]

To give effect to this principle, paragraph 2 of Article 3 asserts that:

(a) all activities in the Antarctic Treaty area shall be planned and conducted so as to limit adverse impacts on the Antarctic environment and dependent and associated ecosystems;
(b) activities in the Antarctic Treaty area shall be planned and conducted so as to avoid:
    (i) adverse effects on climate or weather patterns;
    (ii) significant adverse effects on air or water quality;
    (iii) significant changes in the atmospheric, terrestrial (including aquatic), glacial or marine environments;

(iv) detrimental changes in the distribution, abundance or productivity of species or populations of species of fauna and flora;
(v) further jeopardy to endangered or threatened species or populations of such species; or
(vi) degradation of, or substantial risk to, areas of biological, scientific, historic, aesthetic or wilderness significance;
(c) activities in the Antarctic Treaty area shall be planned and conducted on the basis of information sufficient to allow prior assessments of, and informed judgments about, their possible impacts on the Antarctic environment and dependent and associated ecosystems and on the value of Antarctica for the conduct of scientific research.[82]

Article 3 furnishes a set of legally binding principles. Specifically, countries must meet certain environmental standards and limit insofar as possible adverse impacts on the environment; they must give priority to scientific research in Antarctica and preserve Antarctica for global research; they must ensure that human activities are planned and carried out on the basis of information sufficient to permit prior assessments of their possible impacts; and they must conduct environmental monitoring.

Article 7 of the Antarctic environmental protocol flatly forbids all mining activity by persons or companies in Antarctica: "Any activity relating to mineral resources, other than scientific research, shall be prohibited."[83] This provision ensures that no mineral development on Antarctica or in its circumpolar waters can lawfully take place within the foreseeable future. Put simply, this prohibition means that no degradation of Antarctica is likely to occur from minerals or hydrocarbon development or transportation activities on or around the continent, nor are natural habitats of Antarctic living marine resources likely to be disrupted or destroyed by these activities for at least fifty years.[84]

Nevertheless, this ban is not necessarily permanent. Although no period for a moratorium is specified, Article 25 allows for modification or amendment of the protocol at any time, provided all ATCPs agree by consensus. In addition, fifty years after the protocol enters into force, the prohibition can be lifted if such a decision is adopted at a Review Conference by a majority of all ATCPs, including three-fourths of the current ATCPs. That decision must then be ratified by three-fourths of all the ATCPs, "including the ratifications of all States that were Consultative Parties at the time of the adoption of this Protocol."[85]

Compliance rests with the governments party to the protocol, who are obligated under Article 13 to take "appropriate measures" to ensure compliance. In addition, the protocol provides that inspections can be carried out "to promote the protection of the Antarctic environment and associated ecosystems, and to ensure compliance with this Protocol" of stations, installations, equipment, ships, and aircraft within the Antarctic Treaty area.[86] There are also procedures for mandatory dispute settlement[87] and for advance environmental impact assessment for proposed activities in the Antarctic.[88]

Attached to the protocol are five annexes dealing with environmental impact assessment, conservation of fauna and flora, waste disposal and waste management, marine pollution, and protected areas, which are to be implemented in furtherance of the environmental protection of Antarctica. These annexes "form an integral part" of the protocol, would be adopted in line with Article IX of the Antarctic Treaty, and would be binding upon the United States as a contracting party.[89]

Annex I represents a significant accomplishment of the protocol, for it sets out procedures for environmental impact assessment.[90] For these purposes, human activities are divided into those having "less than a minor or transitory impact"; those having "a minor or transitory impact"; and those having "more than a minor or transitory impact."[91] Even so, it remains regrettable that these terms are neither precisely defined nor quantifiably explained in either Article 8 of the protocol or in Annex I. Interpretation and implementation of environmental impact assessment procedures for American activities in Antarctica would be left to the discretion and responsibility of the United States government.

Annex II restates the need for conservation of Antarctic fauna and flora and essentially updates the Agreed Measures.[92] Three improvements are made by Annex II to the conservation regime for Antarctica. First, protection is extended to terrestrial and freshwater invertebrates.[93] Second, a prohibition is placed on the presence of dogs in Antarctica after April 1, 1994.[94] And third, significant damage to native terrestrial plants is included within the definition of "harmful interference" to the Antarctic environment.[95] Annex II remains important for integrating the conservation of Antarctic fauna and flora into a more comprehensive, comprehensible protection framework and for reasserting the critical need for wildlife conservation in the Antarctic.

The third annex pertains to waste disposal and waste management,[96] and grew out of the 1975 Code of Conduct for Antarctic Expeditions and Stations Activities[97] and from Recommendation XV-3, which attempted to upgrade the 1975 code.[98] Annex III notably improves the 1975 code, for it places greater emphasis on retrograding waste and other materials from the continent and standardizes collection and circulation of information on waste management. As a major producer of wastes on the continent, the United States will be directly affected by the stipulations in Annex III.

Under Annex III waste is classified into five main groups: Group 1, sewage and domestic liquid wastes; Group 2, other liquids and chemicals, including fuels and lubricants; Group 3, solids to be combusted; Group 4, other solid wastes; and Group 5, radioactive materials.[99] The United States would be required to remove all Group 2, 4, and 5 wastes after entry into force of the annex. Moreover, the United States would be obligated to remove Group 1 wastes "to the maximum extent possible" from the treaty area.

Still, environmentalists complain that Annex III contains certain flaws. For one, provisions contain numerous qualifiers, such as "to the maximum extent practicable" and "as far as practicable." Such indefinite parameters will make it difficult to hold operators accountable for their actions. Conservationists also object to the annex's acceptance of incineration as an environmentally safe form of waste disposal. Incineration pollutes the air and also produces contaminated, toxic ash that must be disposed. Sewage and liquid waste present even more problems. The annex relies on maceration (i.e., softening by soaking in a liquid over time) as a principal means for dealing with such waste products,[100] but this method fails to account for heavy metals, bacteria, viruses, and other chemical contaminants in the waste matter. Moreover, the annex explicitly permits discharge of liquid wastes directly into the sea.[101] Environmentalists prefer that sludge from these waste processes be retrograded from the continent rather than dumped at sea.[102]

The fourth annex, "Prevention of Marine Pollution,"[103] is specifically linked to the International Convention for the Prevention on Pollution from Ships, as amended by its 1978 Protocol (MARPOL 73/78).[104] This annex deals with discharges from ships, in particular oil, noxious liquids, garbage, and sewage. Certain provisions highlight the need for vessel retention capacity, emergency response, and preparedness.

Annex IV prohibits in Article 3 "any discharge of oil or oily mixture," except in circumstances permitted under MARPOL 73/78. In Article 4, the marine pollution annex forbids "[t]he discharge of any noxious liquid substance, and any other chemicals or other substances, in quantities or concentrations that are harmful to the marine environment."

Article 5 goes on to prohibit by name the disposal into the sea of two other categories of substances: (1) plastics, "including but not limited to synthetic ropes, synthetic fishing nets, and plastic garbage bags . . . ;" and (2) all forms of garbage, "including paper products, rags, glass, metal, bottles, crockery, dunnage, incineration ash, lining, and packing materials. . . ." Vessels would be obligated under Article 6 to discontinue any discharge of untreated sewage within twelve nautical miles of land or ice shelves. Any sewage discharge beyond that distance is to be made "at a moderate rate of speed, and where practicable, while the ship is en route at a speed of no less than 4 knots." There is, however, a loophole that could debilitate this fiat; that is, as provided for in paragraph 1 of Article 6, this prohibition applies "[e]xcept where it would unduly impair Antarctic operations." Not only are the terms "unduly," "impair," and "Antarctic operations" left undefined but determination of where and when those conditions exist apparently would be left to the discretion of vessel operators.

Compliance enforcement for Annex IV would be left to the American government to exercise over ships flying its flag or supporting its Antarctic operations.[105] The United States must ensure that all its ships have "sufficient capacity" on board for the retention of garbage while within the Antarctic Treaty area[106] and have "adequate facilities" for the reception of all sludge, dirty ballast, tank washing water, oily residues, and garbage. [107]

The marine pollution annex attempts to close the gaps provided for sovereign immunity that were left in previous instruments, though only with partial success. Article 11 asserts that the annex "shall not apply to any warship, naval auxiliary, or other vessel owned or operated by a State. . . ." Still, the United States would be obligated to "ensure by the adoption of appropriate measures not impairing the operations or operational capabilities of such ships owned or operated by it, that such ships act in a manner consistent, so far as reasonable and practicable," with the annex.[108]

In October 1991 at the sixteenth ATCM in Bonn, the United States and the United Kingdom proposed a fifth annex to the protocol that simplified and significantly expanded the future scope of the Antarctic protected area system.[109] As adopted, this annex supplies an integrated approach to the creation and management of protected areas in the Antarctic.

Annex V consolidates into two the five existing categories of protected areas under the Antarctic Treaty. Antarctic Specially Protected Areas (APSAs) will include areas that are to remain inviolate from human interference, that represent major human ecosystems, or that are of outstanding environmental, scientific, historic, aesthetic, or wilderness value.[110] The second category, Antarctic Specially Managed Areas (ASMAs), allows for coordination of multiple-use activities occurring in the same area, ostensibly to minimize the potential for conflicts.[111] Although permits will not be required, a detailed management plan will be necessary for each area.[112] Importantly, the concept of "sufficient size to protect the values for which the special protection or management is required" is used,[113] which improves upon the previous "minimum size requirement." As if to foreshadow a future annex, management for tourist visits is also included.[114]

The Madrid protocol consolidates environmental measures into a single instrument under the Antarctic Treaty. Prior to this agreement, environmental rules and regulations in the Antarctic Treaty System were negotiated in an ad hoc, piecemeal basis, with little linkage or substantive integration. The Madrid protocol transforms that patchwork of rules into a more comprehensive approach to environmental protection in Antarctica and, in the process, provides a means for revising and improving detailed measures as circumstances evolve.

Diplomatic participation by the United States was fundamental to the successful negotiation of the protocol. No less significant is that the Madrid protocol represents the legal installation of a new environmental ethic in U.S. foreign policy for the Antarctic region. That new ethic does much to assure that environmental considerations will be more purposefully figured into the U.S. national interests that inform American foreign policy objectives in the Antarctic.

At the end of 1995, the environmental protection protocol was still in the process of being ratified by the consultative parties. A total of twenty states have ratified the protocol out of the twenty-six necessary for the instrument to enter into force.[115] Six more governments, including the United States, must still ratify, and nineteen must pass implementing legislation.

The U.S. Senate gave its consent in 1992, but ratification remains contingent upon implementing legislation being adopted by the Congress. The critical obstacle in securing ratification of the protocol is bound up in the administration's insistence that the National Science Foundation, which manages the Antarctic Program, be left to implement the protocol for the United States with little or no oversight. This position is adamantly opposed by the environmental community, which is concerned that such a situation might foster serious abuses of the Antarctic environment for the sake of science, cost effectiveness, or expedience.

Prospects for passage of implementing legislation during the 104th Congress do not appear optimistic. The major unresolved issue is the applicability of the National Environmental Policy Act of 1969 (NEPA)[116] to U.S. activities in Antarctica. Environmental groups have contended that NEPA should apply and that activities by Americans should be regulated accordingly. The legal basis for the environmental community's position is the decision in *Environmental Defense Fund, Inc. v. Walter E. Massey, Director, National Science Foundation*, a case decided on January 29, 1993, by the U.S. District Court of Appeals for the District of Columbia.[117]

In *EDF Massey*, the D.C. Circuit Court had concluded that Antarctica is a "sovereignless continent" and therefore part of the global commons, beyond the jurisdiction of any state. NEPA, which was not designed to regulate the activities of private parties abroad, would therefore not apply. The NSF argued that, even conceding the domestic nature of NEPA and the remote possibility of conflict with foreign law, the potential for NEPA's interference with foreign relations was real. Any future international cooperation on environmental matters in the Antarctic might thus be undermined if the executive branch were made subject to the act.

The Court of Appeals rejected these arguments and concluded that NEPA is consistent with existing and anticipated obligations of the United States, including those in the environmental protection protocol. In short, the court was "not convinced that NSF's ability to cooperate with other nations will be hampered by NEPA's injunctions."[118]

The current administration maintains that American activities in Antarctica should not be subject to NEPA, since it could impede international scientific research in the region. In particular, the administration contends that any joint activity with another government should be exempted from NEPA's legal purview, even if that other government held only a 1 percent share of the project. The contention here is that another government would become subject to U.S. national law, a fundamental breach of the sovereign immunity protection upheld by provisions in the protocol and its annexes.

The environmental community counters that the NEPA process would hold up a project only if that activity threatens to have greater than a minor or a transitory impact on the environment, the very standards used in Article 8 and Annex I of the protocol for determining environmental impact assessment. The concern here focuses on projects such as "building a station or adding new buildings, or ice core drilling where explosives are required,"[119] not the normal, everyday process of scientific research carried out in the Antarctic. The environmental community notes that, in any event, activities must be reviewed by the treaty system at its next annual consultative meeting, and thus the NEPA process would not delay a scientific project any longer than would the protocol process itself. Finally, the environmental community has stipulated that no project should be delayed more than fifteen months by the NEPA process.[120]

The bottom line is that until these differences over the applicability of NEPA are resolved between the administration and environmental interest groups, implementing legislation can not go forward in the U.S. Congress. One gets the sense that U.S. domestic political considerations persist in overshadowing the merits of international environmental protection in the frozen south. That situation seems regrettable. Not only does it undercut the U.S. government's leadership role in negotiating the protocol, but it also delays that instrument's entry into force to protect and conserve the Antarctic polar environment for succeeding generations. The United States should lead policy implementation in the Antarctic by international example rather than lag behind because of the nuances of domestic politics.

The Balance Sheet

U.S. national interests in the Antarctic have been influenced by international ideological considerations since the 1980s. Prominent among these was the movement led by the Malaysian group in the United Nations to apply the common heritage of mankind concept to Antarctica in order to internationalize the region and secure access to and development of its mineral resources. At the same time, several new states rushed to join the treaty regime and obtain decision-making rights to Antarctic policy. This CHM movement and the rush to consultative party membership, however, waned with the rise of a pervasive ATCP environmental ethic in the early 1990s, manifested by a comprehensive environmental protection regime.

The United States applied both domestic and international legal measures in its foreign policy to protect the Antarctic from environmental degradation. Efforts were taken to prohibit degradation of the continent under reinforced national environmental law. The United States and other ATCP governments also sought to negotiate specific instruments for regulating pollution activities that could directly threaten the Antarctic marine ecosystem. The Agreed Measures, Seals Convention, CCAMLR, and the promulgation of a special regulatory regime for mineral resources activities clearly attest to this concern. The recent Madrid environmental protocol brings these regulations into a neater, tighter, more manageable and comprehensive legal package, designed to protect and conserve the Antarctic environment.

The efficacy of international conservation policies, however, must rest with the degree of genuine commitment by national governments, in particular, the United States. Governments have made domestic and international laws protecting the Antarctic environment, and governments must enforce those laws against nationals who violate them.

In the final analysis, blame for degradation of the Antarctic environment will not lie in frail law. Both the domestic and international law is present and plain. Additional environmental law may even be created by the ATCPs as various new needs are perceived, and if the past is prologue, the United States will be a leader in those efforts.

The responsibility for violations will rest in the lack of political will by the United States and other ATCPs to monitor activities, enforce compliance, and compel compensation for liability. The Madrid protocol and attendant law for regulating activities in the polar south can be as strong as the Antarctic Treaty governments are willing to make them. If Antarctica is to be protected and preserved, then raising and sustaining that necessary political will will remain critical.

Conclusion

THIS STUDY HAS examined U.S. Antarctic policy from 1960 through the present. The focus has fallen on interests or goals of the United States, especially as they dictate policy objectives in Antarctica. We have sought to discover what those goals are; whether they can be explained by reference to the national interest; how they are manifested in actual policy decisions; and what factors or influences lie behind them. A number of related questions are consequently posed in the study: How is the U.S. government organized to carry out policy? To what extent have U.S. officials been able to realize objectives through international diplomacy? What constraints inhibit realization of these goals, and what factors facilitate their accomplishment? And, is the pursuit of U.S. national interests in Antarctica compatible or incompatible with global interests, and in what ways?

American Antarctic policy, spurred on by the 1959 Antarctic Treaty, provides a particularly interesting case of how conceived national interest serves as a basis for carrying out foreign policy. The theoretical difficulties with the concept of the national interest cannot be disguised or disputed. Even sophisticated attempts to deal with the concept have met with criticism. Undisputedly—from a policy perspective—the notion of the national interest has, so far, been applied in numerous instances by U.S. foreign policymakers to define, justify, and interpret goals and objectives. As applied to the Antarctic, the notion of the national interest has provided U.S. foreign policymakers with a unifying framework or basis for the articulation and subsequent implementation of a durable and integrated set of policy objectives with respect to the south polar

Conclusion / 189

region. This study shows that, even prior to the signing of the 1959 Antarctic Treaty, U.S. executive branch officials had formulated and implemented a set of goals and objectives toward the Antarctic that were viewed as promoting the national interests of the United States. The Antarctic Treaty has symbolized, promoted, and accommodated these interests for nearly four decades.

The U.S. Antarctic experience over the past half century has revealed that (1) U.S. executive branch officials (i.e., foreign policymakers) have exercised considerable leadership and enjoyed significant automonomy in identifying and pursuing U.S. interests in the Antarctic; (2) the national interests perceived and pursued have been remarkably consistent over time; and (3) these interests, as formulated and applied, have benefited American society as a whole rather than particular individuals or special groups.

The United States does not have any vital national interests at stake in the Antarctic. For vital interests to be at risk, real threats to the physical survival or political and cultural integrity of the country would have to exist. Such threats obviously are not evident. Even so, the United States does pursue a cluster of national interests in its Antarctic policy. The continuation and promotion of these interests are dependent upon the preservation of the Antarctic Treaty and the continued use of Antarctica for peaceful purposes only. These interests have enjoyed determined, long-standing commitment from U.S. foreign policymakers, and they supply the core considerations for determining U.S. policy options for the Antarctic. Theses objectives have not changed since 1959.

The cluster of U.S. national interests in Antarctica include (1) scientific interests, aimed at sustaining opportunities for scientific research and cooperation in the region; (2) environmental interests, defined in terms of resource conservation and pollution prevention, the importance of which have become magnified as proliferating human activities have put greater pressures on the local ecosystem; and, (3) geostrategic interests, defined in terms of economic and security considerations. Economic concerns have increasingly become associated with potential hard mineral and hydrocarbon resource development. Security considerations, in turn, have become construed primarily as the promotion of a stable political and legal order in the region. It is important to realize that, as U.S. Antarctic policy has evolved, geostrategic interests as a policy goal have become overtaken by and subordinated to environmental interests. This shift occurred during the early 1990s as U.S. foreign policymakers were compelled to respond to perceived threats to the survival of the Antarctic Treaty System.

Several other conclusions emerge from the analysis, providing answers to the questions posed earlier in this study. First, the United States is the chief player in events in Antarctica and in south polar diplomacy. Indeed, the United States has a long history of involvement and interest in the south polar region. This interest finds expression in the U.S. government's steadfast determination to establish an effective political/legal framework to manage the Antarctic continent. This management concern has been a constant consideration in U.S. policy toward Antarctica since the late 1940s.

A second conclusion is drawn from the experience of U.S. executive branch officials, who are convinced that the Antarctic Treaty System and the demilitartized status of the region are essential conditions for the successful pursuit of American interests in the polar south. The United States has been an active, consistent, and strong supporter of Antarctic agreements, as well as of the demilitarized status of the region. The pivotal role of the Antarctic Treaty System in U.S. Antarctic policy is amply documented. Since signing the Antarctic Treaty in 1959, U.S. executive branch policymakers have been unequivocal about the importance of preserving that instrument and the attendant demilitarization of the Antarctic. To Department of State and National Science Foundation officials charged with managing Antarctic affairs, the Antarctic Treaty is considered a solid, yet responsive and flexible framework for conducting polar activities in a peaceful, cooperative, and environmentally sound fashion.

The importance of the Antarctic Treaty to American foreign policy was neatly described by a principal U.S. policymaker as "the vehicle, the international vehicle, through which the United States pursues its Antarctic interests."[1] This position was reaffirmed in the "Statement on Foreign Policy" issued by President Reagan on the eve of the Falklands/Malvinas War, as he curtly asserted: "The United States has significant political, security, economic, environmental, and scientific interests in Antarctica."[2] In fact, since signing the treaty in 1959, U.S. policy has generally been to maintain and strengthen the Antarctic Treaty System and to support the U.S. Antarctic Program at a level allowing for an active and influential regional presence.

As a third conclusion, the evidence reveals that American executive branch officials have followed a rather independent course in Antarctic policymaking. The direction and character of the shift in U.S. Antarctic interests in the early 1990s—with economic interests being made subordinated to environmental considerations—were not primarily the result of congressional or interest group pressures; rather, the shift was motivated by perceptions among U.S. executive branch officials that a breakdown in consensus was occurring among the ATCP group, precipitated by the Australian-French refusal to sign the 1988 minerals agreement. When the situation appeared to threaten the very survival of the Antarctic Treaty System, American foreign policymakers adjusted their priorities. Greater long-term gains for the U.S. national interest were seen as lying in negotiations for a comprehensive Antarctic environmentalism. As a result, stubborn insistence by the United States upon the need for a minerals agreement was compromised in favor of an environmental protocol.

Despite executive branch independence, a fourth significant conclusion cannot be ignored, namely, that American policy activities affecting Antarctica are closely regulated by U.S. domestic legislation. The domestic authority over Antarctic activities is a complex matter, involving several statutes and agencies. For example, the National Science Foundation, an executive branch agency, has full authority over the United States Antarctic Program.[3] The NSF, funded to support basic science, to finance research projects, and to promote scientific education, is bound in its Antarctic activities by several pieces of U.S. legislation.

The importance of U.S. legislation naturally leads to a fifth conclusion regarding the role of the U.S. Congress in Antarctic policymaking. The influence of Congress has been a mixed bag. Congressional actions have constrained as well as promoted executive branch policy goals for the south polar region. As this study shows, Congress has not been without influence in the formulation and implementation of U.S. Antarctic policy. Legislation passed by the Congress provides legal guidelines that regulate the behavior of and activities by U.S. nationals in Antarctica. Still, the congressional committees with oversight over U.S. Antarctic policy did not fully succeed in steering executive branch officials away from their support for CRAMRA.

A sixth conclusion indicates that the U.S. pluralistic political system—providing for interest group access to governmental decision making—has served environmentalists well. These groups' interest in Antarctica policymaking has grown considerably since the mid-1970s, as has the influence of nongovernmental environmental groups in the formulation and implementation of U.S. policy. Environmentalists have used the legislature and the courts to enhance their objectives, while executive branch officials have tended to exercise restraining influence. Actions by environmentalists in their watchdog capacity have contributed to increased awareness about the need to protect the Antarctic environment. As this study shows, however, the rise of environmental activism predictably collided at times with the development of executive branch strategies to regulate Antarctic mineral activities. Still, growing pressures applied through Congress and environmental groups was not decisive in the executive branch decision to abandon CRAMRA.

American Antarctic scientific research holds significance for the world community at large, a seventh conclusion from the study that undermines the popular assumption that national interests and general global interests are endemically conflictual. The United States has been for several decades the preeminent contributor to Antarctic science and has devoted considerable resources to that purpose. Scientists involved in the major research areas pursued in Antarctica recognize that U.S. Antarctic research has made substantial contributions to science and to understanding the planet as a whole. In fact, a key to unlocking secrets about the earth's present and future geophysical problems may lie in the Antarctic.[4] U.S. treaty-based proposals for the management of Antarctic resources have sought to deter pollution and to generate more knowledge about how the destruction of resources might directly impact the entire world community.

An eighth conclusion from this study suggests that international events external to the Antarctic Treaty System can affect decisions taken by the ATCPs. Clearly nongovernmental organizations affected the outcome of the 1988 minerals convention. Also, criticisms by Third World governments led to greater international awareness about Antarctica and to the rush by nonparty states to join the consultative group. In this respect, the U.S. strategy was plain and purposeful: Give ATCP membership to potential Third World critics who qualify so that they will acquire a vested stake in the Antarctic Treaty System regime. In this way, the political, scientific, and conservation interests of the United States in Antarctica can be protected under the legal umbrella of the Antarctic Treaty System.

The evidence also supports a final conclusion, namely, that there is significant congruency among U.S. national interests in Antarctica and general global interests. The Antarctic Treaty System works efficiently and effectively, not only to preserve the interests of the United States but also to the benefit of all mankind. The cooperative efforts of all the consultative parties serve the interests of the entire international community. Moreover, no other treaty promotes the purposes and principles of the United Nations Charter to the extent that the Antarctic Treaty does. However difficult to quantify, the level of order and acceptance of rules of conduct promoted by the Antarctic Treaty System have provided international security and stability in a world full of conflict.

The Antarctic Treaty has held a prominent place in international diplomacy for more than thirty-five years. It stands as a cohesive, yet dynamic force for sociopolitical interaction among states with divergent world views and interests. More importantly, the Antarctic Treaty has achieved a political balance that has served and continues to serve U.S. national interests well. Not only does the treaty support and facilitate U.S. scientific activities in the region, it also works to offset latent conflict over territorial claims within a climate of peaceful cooperation and orderly change. Equally notable is that U.S. Antarctic policy gives meaningful content to U.S. national interests in the region and to general global interests there as well. That realization bodes well for the future of Antarctica and for the successful pursuit of U.S. national interests there over the long term.

Appendixes

A: The Antarctic Treaty*

Signed at Washington December 1, 1959; Ratification advised by the Senate August 10, 1960; Ratified by the President August 18, 1960; Ratification deposited at Washington August 18, 1960; Proclaimed by the President June 23, 1961; Entered into force June 23, 1961.

BY THE PRESIDENT OF THE UNITED STATES OF AMERICA

A PROCLAMATION

WHEREAS the Antarctic Treaty was signed at Washington on December 1, 1959 by the respective plenipotentiaries of the United States of America, Argentina, Australia, Belgium, Chile, the French Republic, Japan, New Zealand, Norway, the Union of South Africa, the Union of Soviet Socialist Republics, and the United Kingdom of Great Britain and Northern Ireland;

WHEREAS the text of the said treaty, in the English, * * * languages, is word for word as follows:

The Antarctic Treaty

The Governments of Argentina, Australia, Belgium, Chile, the French Republic, Japan, New Zealand, Norway, the Union of South Africa, the Union of Soviet Socialist Republics, the United Kingdom of Great Britain and Northern Ireland, and the United States of America,

Recognizing that it is in the interest of all mankind that Antarctica shall continue forever to be used exclusively for peaceful purposes and shall not become the scene or object of international discord;

Acknowledging the substantial contributions to scientific knowledge resulting from international cooperation in scientific investigation in Antarctica;

Convinced that the establishment of a firm foundation for the continuation and development of such cooperation on the basis of freedom of scientific investigation in Antarctica as applied during the International Geophysical Year accords with the interests of science and the progress of all mankind;

Convinced also that a treaty ensuring the use of Antarctica for peaceful purposes only and the continuance of international harmony in Antarctica will further the purposes and principles embodied in the Charter of the United Nations.

*Citation: 12 UST 794: TIAS 4780.

Have agreed as follows:

Article I

1. Antarctica shall be used for peaceful purposes only. There shall be prohibited, *inter alia*, any measures of a military nature, such as the establishment of military bases and fortifications, the carrying out of military maneuvers, as well as the testing of any type of weapons.

2. The present Treaty shall not prevent the use of military personnel or equipment for scientific research or for any other peaceful purpose.

Article II

Freedom of scientific investigation in Antarctica and cooperation toward that end, as applied during the International Geophysical Year, shall continue, subject to the provisions of the present Treaty.

Article III

1. In order to promote international cooperation in scientific investigation in Antarctica, as provided for in Article II of the present Treaty, the Contracting Parties agree that, to the greatest extent feasible and practicable:

(a) information regarding plans for scientific programs in Antarctica shall be exchanged to permit maximum economy and efficiency of operations;

(b) scientific personnel shall be exchanged in Antarctica between expeditions and stations;

(c) scientific observations and results from Antarctica shall be exchanged and made freely available.

2. In implementing this Article, every encouragement shall be given to the establishment of cooperative working relations with those Specialized Agencies of the United Nations and other international organizations having a scientific or technical interest in Antarctica.

Article IV

1. Nothing contained in the present Treaty shall be interpreted as:

(a) a renunciation by any Contracting Party of previously asserted rights of or claims to territorial sovereignty of Antarctica;

(b) a renunciation or diminution by any Contracting Party of any basis of claim to territorial sovereignty in Antarctica which it may have whether as a result of its activities or those of its nationals in Antarctica, or otherwise;

(c) prejudicing the position of any Contracting Party as regards its recognition or non-recognition of any other State's right of or claim or basis of claim to territorial sovereignty in Antarctica.

2. No acts or activities taking place while the present Treaty is in force shall constitute a basis for asserting, supporting or denying a claim to territorial sovereignty in Antarctica or create any rights of sovereignty in Antarctica. No new claim, or enlargement of an existing claim, to territorial sovereignty in Antarctica shall be asserted while the present Treaty is in force.

Article V

1. Any nuclear explosions in Antarctica and the disposal there of radioactive waste material shall be prohibited.

2. In the event of the conclusion of international agreements concerning the use of nuclear energy, including nuclear explosions and the disposal of radioactive waste material, to which all of the Contracting Parties whose representatives are entitled to participate in the meetings provided for under Article IX are parties, the rules established under such agreements shall apply in Antarctica.

Article VI

The provisions of the present Treaty shall apply to the area south of 60° South Latitude, including all ice shelves, but nothing in the present Treaty shall prejudice or in any way affect the rights, or the exercise of the rights, of any State under international law with regard to the high seas within that area.

Article VII

1. In order to promote the objectives and ensure the observance of the provisions of the present Treaty, each Contracting Party whose representatives are entitled to participate in the meetings referred to in Article IX of the Treaty shall have the right to designate observers to carry out any inspection provided for by the present Article. Observers shall be nationals of the Contracting Parties which designate them. The names of observers shall be communicated to every other Contracting Party having the right to designate observers, and like notice shall be given of the termination of their appointment.

2. Each observer designated in accordance with the provisions of paragraph 1 of this Article shall have complete freedom of access at any time to any or all areas of Antarctica.

3. All areas of Antarctica, including all stations, installations and equipment within those areas, and all ships and aircraft at points of discharging or embarking cargoes or personnel in Antarctica, shall be open at all times to inspection by any observers designated in accordance with paragraph 1 of this Article.

4. Aerial observation may be carried out at any time over any or all areas of Antarctica by any of the Contracting Parties having the right to designate observers.

5. Each Contracting Party shall, at the time when the present Treaty enters into force for it, inform the other Contracting Parties, and thereafter shall give them notice in advance of

   (a) all expeditions to and within Antarctica, on the part of its ships or nationals, and all expeditions to Antarctica organized in or proceeding from its territory;
   (b) all stations in Antarctica occupied by its nationals; and
   (c) any military personnel or equipment intended to be introduced by it into Antarctica subject to the conditions prescribed in paragraph 2 of Article I of the present Treaty.

Article VIII

1. In order to facilitate the exercise of their functions under the present Treaty, and without prejudice to the respective positions of the Contracting Parties relating to jurisdiction over all other persons in Antarctica, observers designated under paragraph 1 of Article VII and scientific personnel exchanged under subparagraph 1(b) of Article III of the Treaty, and members of the staffs accompanying any such persons, shall be subject only to the jurisdiction of the Contracting Party of which they are nationals in respect of all acts or omissions occurring while they are in Antarctica for the purpose of exercising their functions.

2 Without prejudice to the provisions of paragraph 1 of this Article, and pending the adoption of measures in pursuance of subparagraph 1 (e) of Article IX, the Contracting Parties concerned in any case of dispute with regard to the exercise of jurisdiction in Antarctica shall immediately consult together with a view to reaching a mutually acceptable solution.

Article IX

1. Representatives of the Contracting Parties named in the preamble to the present Treaty shall meet at the City of Canberra within two months after the date of entry into force of the Treaty, and thereafter at suitable intervals and places, for the purpose of exchanging information, consulting together on matters of common interest pertaining to Antarctica, and formulating and considering, and recommending to their Governments, measures in furtherance of the principles and objectives of the Treaty, including measures regarding:

   (a) use of Antarctica for peaceful purposes only;
   (b) facilitation of scientific research in Antarctica;
   (c) facilitation of international scientific cooperation in Antarctica;
   (d) facilitation of the exercise of the rights of inspection provided for in Article VII of the Treaty;
   (e) questions relating to the exercise of jurisdiction in Antarctica;
   (f) preservation and conservation of living resources in Antarctica.

2. Each Contracting Party which has become a party to the present Treaty by accession under Article XIII shall be entitled to appoint representatives to participate in the meetings referred to in paragraph 1 of the present Article, during such time as that Contracting Party demonstrates its interest in Antarctica by conducting substantial scientific research activity there, such as the establishment of a scientific station or the dispatch of a scientific expedition.

3. Reports from the observers referred to in Article VII of the present Treaty shall be transmitted to the representatives of the Contracting Parties participating in the meetings referred to in paragraph 1 of the present Article.

4. The measures referred to in paragraph 1 of this Article shall become effective when approved by all the Contracting Parties whose representatives were entitled to participate in the meetings held to consider those measures.

5. Any or all of the rights established in the present Treaty may be exercised as from the date of entry into force of the Treaty whether or not any measures

facilitating the exercise of such rights have been proposed, considered or approved as provided in this Article.

Article X

Each of the Contracting Parties undertakes to exert appropriate efforts, consistent with the Charter of the United Nations, to the end that no one engages in any activity in Antarctica contrary to the principles or purposes of the present Treaty.

Article XI

1. If any dispute arises between two or more of the Contracting Parties concerning the interpretation or application of the present Treaty, those Contracting Parties shall consult among themselves with a view to having the dispute resolved by negotiation, inquiry, mediation, conciliation, arbitration, judicial settlement or other peaceful means of their own choice.

2. Any dispute of this character not so resolved shall, with the consent, in each case, of all parties to the dispute, be referred to the International Court of Justice for settlement; but failure to reach agreement on reference to the International Court shall not absolve parties to the dispute from the responsibility of continuing to seek to resolve it by any of the various peaceful means referred to in paragraph 1 of this Article.

Article XII

1. (a) The present Treaty may be modified or amended at any time by unanimous agreement of the Contracting Parties whose representatives are entitled to participate in the meetings provided for under Article IX. Any such modification or amendment shall enter into force when the depositary Government has received notice from all such Contracting Parties that they have ratified it.

(b) Such modification or amendment shall thereafter enter into force as to any other Contracting Party when notice of ratification by it has been received by the depositary Government. Any such Contracting Party from which no notice of ratification is received within a period of two years from the date of entry into force of the modification or amendment in accordance with the provisions of subparagraph 1(a) of this Article shall be deemed to have withdrawn from the present Treaty on the date of the expiration of such period.

2. (a) If after the expiration of thirty years from the date of entry into force of the present Treaty, any of the Contracting Parties whose representatives are entitled to participate in the meetings provided for under Article IX so requests by a communication addressed to the depositary Government, a Conference of all the Contracting Parties shall be held as soon as practicable to review the operation of the Treaty.

(b) Any modification or amendment to the present Treaty which is approved at such a Conference by a majority of the Contracting Parties there represented, including a majority of those whose representatives are entitled to participate in the meetings provided for under Article IX, shall be communicated by the depositary Government to all the Contracting Parties immediately after the termination of the Conference and shall enter into force in accordance with the provisions of paragraph 1 of the present Article.

(c) If any such modification or amendment has not entered into force in accordance with the provisions of subparagraph 1(a) of this Article within a period of two years after the date of its communication to all the Contracting Parties, any Contracting Party may at any time after the expiration of that period give notice to the depositary Government of its withdrawal from the present Treaty; and such withdrawal shall take effect two years after the receipt of the notice by the depositary Government.

Article XIII

1. The present Treaty shall be subject to ratification by the signatory States. It shall be open for accession by any State which is a Member of the United Nations, or by any other State which may be invited to accede to the Treaty with the consent of all the Contracting Parties whose representatives are entitled to participate in the meetings provided for under Article IX of the Treaty.

2. Ratification of or accession to the present Treaty shall be effected by each State in accordance with its constitutional processes.

3. Instruments of ratification and instruments of accession shall be deposited with the Government of the United States of America, hereby designated as the depositary Government.

4. The depositary Government shall inform all signatory and acceding States of the date of each deposit of an instrument of ratification or accession, and the date of entry into force of the Treaty and of any modification or amendment thereto.

5. Upon the deposit of instruments of ratification by all the signatory States, the present Treaty shall enter into force for those States and for States which have deposited instruments of accession. Thereafter the Treaty shall enter into force for any acceding State upon the deposit of its instrument of accession.

6. The present Treaty shall be registered by the depositary Government pursuant to Article 102 of the Charter of the United Nations.

Article XIV

The present Treaty, done in the English, French, Russian and Spanish languages, each version being equally authentic, shall be deposited in the archives of the Government of the United States of America, which shall transmit duly certified copies thereof to the Governments of the signatory and acceding States.

* * * *

IN WITNESS WHEREOF, the undersigned Plenipotentiaries, duly authorized, have signed the present Treaty.

DONE at Washington this first day of December, one thousand nine hundred and fifty-nine.

B: Interview Questions

1. What would you say are the principal national interests of the United States in Antarctica?
2. How would you rank them in order of importance?
3. How do these national interests translate into policy?
4. Do you find that American foreign policymakers have pursued these national interests actively and consistently since 1960 to the present, or would you say that these have been pursued in a haphazard manner?
5. How would you characterize the outcome of U.S. Antarctic policy? That is, would you say that it comes about as a series of compromises among the positions of executive branch agencies having jurisdiction in the area, or that Antarctic policy is set primarily by the State Department?
6. As it concerns domestic interest groups (e.g., environmental groups), would you say that they have (a) constrained the ability of government officials to pursue U.S. national interests in Antarctica; (b) that they have promoted U.S. national interests in the region; or (c) that they have not exercised any influential role in U.S. Antarctic policymaking?
7. In terms of international forces, do you find that the other ATCPs have promoted, inhibited, or had no discernible impact on the pursuit of U.S. interests in Antarctica? What about with respect to non-ATCPs?
8. Would you say that the U.S. pursuit of its national interests in Antarctica is compatible or incompatible with general global interests?
10. Do you think that U.S. interests could be protected if there were no Antarctic Treaty System? If not, why not? If yes, how could that be done?
11. What do you see as threats to the survival of the Antarctic regime?
12. What are your views on the notion of reserving Antarctica as an international park?

C: Government and Nongovernment Officials Interviewed for this Study

Name	Position agency affiliation	Date of interview
Raymond Arnaudo	Chief, Division of Polar Programs, Office of Oceans, Department of State	11/20/91
E. U. Curtis Bohlen	Assistant Secretary, Bureau of Oceans, Department of State	11/10/92
David A. Colson	Deputy Assistant Secretary, Bureau of Oceans, Department of State	10/15/91
Ted E. De Laca	Section Head, Polar Science Section, National Science Foundation	11/18/91
Capt. Dwight Fisher	Aviations Project Manager, DOD Operations in Antarctica, Attention: National Science Foundation	02/28/92
Guy G. Guthridge	Manager, Polar Information Program, Polar Coordination and Information Section, National Science Foundation	01/17/92
Anton Inderbitzen	Deputy Assistant Director for Research, U.S. Geological Survey	11/22/91
Lee Kimball	Senior Associate, World Resources Institute	10/22/91
Thomas Laughlin	Senior Policy Analyst, Office of Policy and Planning, Department of Commerce	11/05/92
Erice Melby	National Security Council	01/23/92
Lawrence Rudolph	Assistant General Counsel, National Science Foundation	03/12/92
Wesley Scholz	Director, Office of International Commodities, Bureau of Economic and Business Affairs, Department of State	01/25/92
R. Tucker Scully	Director, Office of Oceans and Polar Affairs, Department of State	11/05/91
John B. Talmadge	Section Head, Polar Program, Polar Coordination and Information Section, National Science Foundation	02/03/92
Peter E. Wilkniss	Director, Division of Polar Programs, National Science Foundation	03/05/92

D: Antarctic Treaty Consultative Meetings

Meetings	Location and dates
First Consultative Meeting	Canberra July 10–24, 1961
Second Consultative Meeting	Buenos Aires July 18–28, 1962
Third Consultative Meeting	Brussels June 2–13, 1964
Fourth Consultative Meeting	Santiago November 3–18, 1966
Fifth Consultative Meeting	Paris November 18–29, 1968
Sixth Consultative Meeting	Tokyo October 19–31, 1970
Seventh Consultative Meeting	Wellington October 30–November 10, 1972
Eigth Consultative Meeting	Oslo June 9–20, 1975
Ninth Consultative Meeting	London September 19–October 7, 1977
Tenth Consultative Meeting	Washington, D.C. September 17–October 5, 1979
Eleventh Consultative Meeting	Buenos Aires June 23–July 7, 1981
Twelve Consultative Meeting	Canberra September 13–27, 1983
Thirteen Consultative Meeting	Brussels October 8–18, 1985
Fourteenth Consultative Meeting	Rio de Janeiro October 5–15, 1987
Fifteenth Consultative Meeting	Paris October 9–20, 1989
Sixteenth Consultative Meeting	Bonn October 7–18, 1991
Seventeenth Consultative Meeting	Milan November 11–20, 1992
Eighteenth Consultative Meeting	Kyoto April 11–22, 1994
Nineteenth Consultative Meeting	Seoul May 8–19, 1995
Twentieth Consultative Meeting	Utrecht April 29–May 10, 1996

E: Antarctic Treaty Consultative Parties and Acceding States

Original Signatories (1959)

Argentina	New Zealand
Australia	Norway
Belgium	Republic of South Africa
Chile	Union of Soviet Socialist Republics*
France	United Kingdom
Japan	United States

Consultative Parties That Gained Status After 1959

Country	Year of accession	Year granted consultative status
Brazil	1975	1983
Ecuador	1987	1990
Finland	1984	1989
Germany	1979/1974†	1981/1987†
India	1983	1983
Italy	1981	1981
The Netherlands	1967	1990
People's Republic of China	1983	1985
Peru	1981	1981
Poland	1961	1977
Republic of Korea	1986	1987
Spain	1982	1982
Sweden	1984	1984
Uruguay	1980	1980

*Known as Russia since December 1990.
†The first date listed applies to the former Federal Republic of Germany; the second to the former German Democratic Republic.

Acceding Parties

Country	Year of accession
Austria	1987
Bulgaria	1978
Canada	1988
Colombia	1989
Cuba	1984
Czechoslovakia	1962[‡]
Democratic Peoples Rupublic of Korea	1986
Denmark	1965
Greece	1987
Guatemala	1991
Hungary	1984
Papua New Guinea[§]	1981
Romania	1971
Switzerland	1990
Turkey	1996
Ukraine	1992

[‡]In October 1993, the new Czech and Slovak Republics acceded to the treaty effective January 1, 1993.
[§]Succeeded to the treaty after independence from Australia.

F: List of Selected Congressional Hearings/Reports on Antarctic Issues 1939–1994

List of Selected Congressional Hearings/Reports on Antarctic Issues 1939–1992

House of Representatives committee/subcommittee	Issue and year of hearing/report
Appropriations	
Deficiencies	Expedition to the Antarctic Region (1939)
Independent Office	NSF Appropriations (Annual)
	Report on the International Geophysical year (1957)
	Review of the First Eleven Months of the IGY (1958)
	Report on the IGY (1959)
Foreign Affairs*	
National Security Policy and Scientific Development	The Political Legacy of the IGY: A Report (1973)
Human Rights and International Organizations	Preserving Antarctica's Ecosystem (1990)
Economic Policy, Trade and Environment; Merchant Marine Fisheries	Implementing Legislation for the Protocol on Environmental Protection to the Antarctic Treaty (1993)
Interior and Insular Affairs	
Territorial and Insular Affairs	Antarctica Legislation (1960)
	Antarctica Legislation (1961)
	Deep Freeze Operations (1962)
	Deep Freeze Operations (1964)
	Deep Freeze Operations (1965)
Insular and International Affairs	Establish an Antarctica World Park (1990)
Interstate and Foreign Commerce	
Committee of the Whole	International Geophysical Year: the Arctic and Antarctic (1958)
Merchant Marine and Fisheries†	
Fisheries and Wildlife Conservation and the Environment	Antarctic Marine Living Resources (1983)
Oversight and Investigations	Convention on the Regulation of Antarctic Mineral Resource Activities (1990)
Fisheries and Wildlife Conservation and the Environment; Oceanography and Great Lakes Coast Guard and Navigation	Antarctic Environmental Protection (1990)
Oceanography, Great Lakes, and the Outer Continental Shelf; Coast Guard and Navigation; and Fisheries and Wildlife Conservation and the Environment	Antarctic Treaty Protocol on Environmental Protection (1992)

Congressional Hearings/Reports on Antarctica / 209

(*continued*)

House of Representatives committee/subcommittee	Issue and year of hearing/report
Science, Space and Technology[‡]	
Science, Research and Technology	U.S. Antarctic Program (1979)
Science, Space and Technology	Review of the Results of the Antarctic Ozone Expedition (1987)
	The Antarctic Environmental Protection Act (1994)
Science, Space and Technology and Transportation, Aviation and Materials	Antarctic Minerals Policy (1990)
Science	NSF Antarctic Environment Act of 1991 (1991)
	The Antarctic Environmental Protection Act (1993)

Senate Committee/subcommittee	Issue and year of hearing/reports
Armed Services	
	Antarctic Expedition (1954)
Commerce, Science and Transportation	The Antarctic Scientific Research, Tourism and Marine Resources Act of 1993 to Implement the Protocol on Environmental Protection to the Antarctic Treaty (1993)
	Antarctic Scientific Research, Tourism and Marine Resources Act of 1993: A Report on S. 1427 (1994)
National Ocean Policy Study	Antarctic Marine Living Resources Negotiations (1978)
Science, Space and Technology	Antarctica (1984)
	Protecting Antarctica's Environment (1989)
	Monitoring the Arctic and Antarctic Environments (1991)
Energy and Natural Resources	Oversight of U.S. Activities in Antarctica (1979)
Foreign Relations	The Antarctic Treaty (1960)
	Protocol on Environmental Protection to the Antarctic Treaty (1992)
Oceans and International Environment	U.S. Policy with respect to Mineral Exploration and Exploitation in Antarctica (1975)
Arms Control, Oceans and International Environment	U.S. Policy with Respect to Exploitation of Antarctic Resources (1978)
	Antarctica Legislation (1990)

*Renamed Foreign Relations in the 104th Congress
[†]Renamed Resources in the 104th Congress.
[‡]Renamed Science in the 104th Congress.

G: U.S. Antarctic Program Funding

U.S. Antarctic Program Funding, 1955–1971 (thousands of dollars)

Fiscal year	Science (USARP) (US$)	Logistics (US$)	Total (US$)
1955	73	4,716	4,789
1956	2,195	17,768	19,963
1957	712	30,049	30,761
1958	1,892	17,268	19,160
1959	2,306	26,724	29,030
1960	6,180	16,278	22,458
1961	5,461	25,026	30,487
1962	7,188	20,946	28,134
1963	6,359	21,555	27,914
1964	7,175	21,645	28,820
1965	7,608	20,984	28,592
1966	8,363	22,591	30,954
1967	7,577	26,907	34,484
1968	7,844	28,106	35,950
1969	6,859	25,625	32,484
1970	7,000	24,802	31,802
1971	7,407	24,752	32,159

SOURCE: National Science Foundation, "U.S. Antarctic Program Information Sheet," no. 10 (March 30, 1992).

U.S. Antarctic Program Funding, 1972–1995 (thousands of dollars)

Fiscal year	Science projects	Direct science support	Base level support	Major construction and procurement	Total
1972	4,285	4,410	14,240	4,065	27,000
1973	3,275	3,059	14,860	22,810	44,004
1974	3,047	5,044	14,250	2,394	24,735
1975	4,467	3,900	16,109	1,422	25,898
1976	4,308	4,130	20,147	20,000	48,585
1977	6,349	6,100	25,546	7,300	45,295
1978	7,002	7,100	26,270	8,075	48,447
1979	7,282	7,574	28,013	8,222	51,091
1980	7,375	9,011	32,819	6,630	55,835
1981	9,041	8,960	37,280	12,174	67,455
1982	8,498	10,100	42,807	7,100	68,505
1983	8,996	10,800	55,300	8,100	83,196
1984	10,433	12,054	60,400	19,600	102,487
1985	11,155	14,000	67,600	18,085	110,840
1986	11,016	16,000	70,045	13,100	110,161
1987	12,617	17,879	59,080	27,700	117,276
1988	13,520	19,100	66,140	25,900	124,660
1989	15,570	19,300	69,200	27,000	131,070
1990	17,020	24,280	75,090	32,290	151,680
1991	19,000	25,200	96,800	34,000	175,000
1992	22,000	32,200	101,800	37,000	193,000
1993	26,740	33,200	161,420*		221,360
1994	28,600	43,729	124,910		197,240
1995	29,020	42,080	125,190		196,290

SOURCE: National Science Foundation, "U.S. Antarctic Program Information Sheet," no. 10 (July 15, 1996).
*For FY 1993, 1994, and 1995, items under columns 4 and 5 were merged.

Notes

1. Introduction

1. House Committee on Foreign Affairs, Subcommittee on National Security Policy and Scientific Developments, *The Political Legacy of the International Geophysical Year*, report prepared by Harold Bullis, Congressional Research Service (Washington, D.C.: GPO), November 1973, 57.

2. Interests are "those things that a government regards as valuable." Interests may be material things to be possessed (like territory), opportunities for activity (such as access to mineral resources), or environmental conditions to be enjoyed (such as the maintenance of world order). See M. J. Peterson, *Managing the Frozen South: The Creation and Evolution of the Antarctic Treaty System* (Berkeley: University of California Press, 1988), 11.

3. Inis L. Claude, Jr., "Global Goals and Foreign Policy," in *National Interest and Global Goals*, ed. George C. McGhee et al. (Lanham: University Press of America, 1989), 93.

4. See, for example, Hans Morgenthau, *Politics Among Nations* (New York: Knopf, 1967); E. H. Carr, *The Twenty Years' Crisis: 1919–1939: An Introduction to the Study of International Relations* (London: Macmillan, 1939; New York: Harper and Row Torch Books, 1964); Nicholas Spykman, *America's Strategy in World Politics* (New York: Hartcourt Brace Jovanovich, 1942); George Kennan, *American Diplomacy: 1900–1950* (New York: North American Library, 1951); Arnold Wolfers, ed., *Discord and Cooperation* (Baltimore: Johns Hopkins University Press, 1962); and Robert E. Osgood, *Ideals and Self-Interest in America's Foreign Relations* (Chicago: University of Chicago Press, 1953).

5. "Vital" or "core" national interests concern the preservation of the state, its territorial integrity, and regime survival. Obviously, these are essential to the pursuit of non-vital national interests such as creating and strengthening regional arrangements and global institutions.

6. See George Modelski, *A Theory of Foreign Policy* (New York: Praeger Publishers, 1962), 70–72; Robert C. Johansen, *The National Interest and the Human Interest: An Analysis of U.S. Foreign Policy* (Princeton: Princeton University Press, 1980), 20–23. For an excellent recent, spirited defense of the utility and relevance of the concept of the national interest in the context of U.S. foreign policy, see W. David Clinton, *The Two Faces of National Interest* (Baton Rouge: Louisiana State University Press, 1994).

7. See Alexander L. George and Robert O. Keohane, "The Concept of National Interest: Uses and Limitations," in *Presidential Decision-making in Foreign Policy: The Effective Use of Information and Advice*, ed. Alexander L. George (Boulder: Westview Press, 1980), 217–37.

8. See Robert C. Good, "National Interest and Moral Theory: The 'Debate' Among Contemporary Political Realists," in *Foreign Policy in the Sixties: The Issues and Instruments*, ed. Roger Hilsman and Robert C. Good (Baltimore: The Johns Hopkins University Press, 1965), 271.

9. Seyom Brown, *The Faces of Power: Constancy and Change in United States Foreign Policy from Truman to Johnson* (New York: Columbia University Press, 1968), 8.
10. Ibid., 9.
11. Cecil B. Crabb, Jr., *Policy-Makers and Critics: Conflicting Theories of American Foreign Policy* (New York: Praeger Publishers, 1976), 167.
12. Stephen D. Krasner, "Approaches to the State: Alternative Conceptions and Historical Dynamics," *Comparative Politics* (January 1984): 224.
13. See Stephen D. Krasner, *Defending the U.S. National Interest: Raw Materials Investments and U.S. Foreign Policy* (Princeton: Princeton University Press, 1978), 5–6, 10–12.
14. Graham T. Allison, *Essence of Decision: Explaining the Cuban Missile Crisis* (Boston: Little Brown, 1971), 29.
15. See Sidney Verba, "Assumptions of Rationality and Non-Rationality in Models of the International System," in *International Politics and Foreign Policy: A Reader in Research and Theory*, rev. ed., ed. James N. Rosenau (New York: The Free Press, 1969), 226.
16. Ibid., 218–26.

2. The Antarctic Setting

1. The name Antarctic is derived from the Greek words "anti" and "arktos," which mean "opposite the bear," or the Arctic. See Paul A. Siple, *90° South: The Story of the American South Polar Conquest* (New York: G.P. Putnam's Sons, 1959), 23.
2. Dr. Leland J. Haworth's statement is in the House Committee on Interior and Insular Affairs, *Antarctica Report—1965: Hearing before the Subcommittee on Territorial and Insular Affairs*, 89th Cong., 1st sess., 12, 13 April; 6, 7 May; and 15 June 1965, 112.
3. The South Pole lies approximately 9,500 feet above sea level. Antarctica, the highest continent, averages 6,000 feet compared to 2,000 feet for the average height of all continents. See Central Intelligence Agency, *Polar Regions Atlas* (Washington, D.C.: GPO, May 1978, reprinted November 1981), 4.
4. National Academy of Sciences, National Research Council, Committee on Polar Research, *Polar Research: A Survey* (Washington, D.C., 1970), 1; and CIA, *Polar Regions Atlas*, 35.
5. Library of Congress, Office of Technology Assessment, *Polar Prospects: A Minerals Treaty for Antarctica*, OTA-O-428 (Washington, D.C.: GPO, September 1989), 7. Estimates of the size of the Antarctic vary. For example, in Senate Committee on Energy and Natural Resources, *U.S. Activities in Antarctica: Hearing before the Senate Committee on Energy and Natural Resources*, 96th Cong., 1st sess., 23 April 1979, 4, the Antarctic area is estimated at 5.5 million square miles; and in House Committee on Science and Technology, Subcommittee on Energy Research, Development and Demonstration, *Polar Energy Resources Potential*, report prepared by the Congressional Research Service, 94th Cong., 2d sess., September 1976, 3, it is estimated at 5.1 million square miles.
6. David Sugden finds the terms "East" and "West" to refer to Antarctica misleading since the continent is centered on the South Pole. See his *Arctic and Antarctic: A Modern Geographical Synthesis* (Totowa: Barnes and Noble, 1982), 29.
7. The Drake Passage is reputed to be the most dangerous stretch of the

Southern Ocean, where waves build up to unprecedented heights. Fifty-foot waves are not uncommon, and others reaching a hundred feet are not unknown. See Eliot Porter, *Antarctica* (New York: E.P. Dutton, 1978), 44; and Barney Brewster, *Antarctica: Wilderness at Risk* (San Francisco: Friends of the Earth, 1982), 1.

8. Edwin Mickleburg, *Beyond the Frozen Sea: Visions of Antarctica* (London: The Bodley Head, 1987), 48.

9. Gordon de Q. Robin, "The Ice of the Antarctic," *Scientific American* 27 (September 1962): 138.

10. D. W. H. Walton, ed., *Antarctic Science* (New York: Cambridge University Press, 1987), 153.

11. See G. E. Fogg, and David Smith, *The Explorations of Antarctica: The Last Unspoilt Continent* (New York: Sterling Publishing, 1990), 193. The ice sheet welds the continent together, for if the ice were removed, West Antarctica would appear as an archipelago of scattered mountainous islands. CIA, *Polar Regions Atlas*, 35.

12. In October 1987, a massive iceberg of about twice the size of Rhode Island and with an estimated thickness of 750 feet broke free of the Ross Ice Shelf. The break occurred near the site of several "Little America" camps operated by Admiral Richard E. Byrd. See *Antarctic Journal of the United States (AJUS)* 22 (December 1987): 1.

13. CIA, "National Intelligence Survey—Antarctica" (NIS) 69, January 1956, chap. 2, sec. 23, 29.

14. See J. Peter Bernhardt, "Sovereignty in Antarctica," *California Western International Law Journal* 5 (spring 1975): 302. For an extended discussion of the features and significance of Antarctic ice, see Stephen Pyne, *The Ice: A Journey to Antarctica* (Iowa City: University of Iowa Press, 1986).

15. Bernhardt, "Sovereignty in Antarctica," 302.

16. Ibid.

17. Richard S. Lewis, *A Continent for Science: The Antarctic Adventure* (New York: The Viking Press, 1965), 194–95.

18. Lewis, *Continent for Science*, 194.

19. Paul A. Carter, *Little America: Town at the End of the World* (New York: Columbia University Press, 1979), 2.

20. Ibid.

21. Lawson W. Brigham, "The Soviet Antarctic Program," *Oceanus* 31 (summer 1988): 87.

22. J. R. Dudeney, "The Antarctic Atmosphere" in Walton, ed., *Antarctic Science*, 193. Ian Cameron is particularly graphic when he says that, in this sort of cold, "if you drop a steel bar it is likely to shatter like glass, tin disintegrates into loose granules, mercury freezes into a solid metal, and if you haul up a fish through a hole in the ice within five seconds it is frozen so solid it has to be cut with a saw." See Ian Cameron, *Antarctica: The Last Continent* (Boston: Little, Brown, 1974), 14.

23. Paul A. Siple, *90° South*, 19; and John May, *Greenpeace Book of Antarctica: A New View of the Seventh Continent* (New York: Doubleday, 1989), 52. Winds in excess of 125 miles per hour are not uncommon. Brewster, *Wilderness at Risk*, 5. In 1984 the Soviet Russkaya station recorded wind speeds of over 135 miles per hour. Peter Beck, *The International Politics of Antarctica* (London: Croom Helm, 1986), 10.

24. Dudeney, "The Antarctic Atmosphere," 193.

25. Ibid.

26. Cameron, *Antarctica: The Last Continent*, 14.
27. Brewster, *Wilderness at Risk*, 6.
28. Richard E. Byrd, *Discovery: The Story of the Second Byrd Expedition* (New York: G.P. Putnam's Sons, 1935), 340.
29. Ibid.
30. Ibid.
31. Ibid.
32. James H. Zumberge, "Potential Mineral Resource Availability and Possible Environmental Problems in Antarctica," in *The New Nationalism and the Use of Common Spaces*, ed. Jonathan Charney (Totowa: Allanheld, Osmun, 1982), 119–20.
33. Ibid.
34. Jack Child, *Antarctica and South American Geopolitics: Frozen Lebensraum* (New York: Praeger Publishers, 1988), 7.
35. Siple, *90° South*, 290.
36. See Appendix A, Antarctic Treaty, Article VI.
37. Louis J. Halle, *The Sea and the Ice: A Naturalist in Antarctica* (Boston: Houghton Mifflin Co., 1973), 9.
38. Harm J. de Blij, "Physical Aspects: Resources, Environment and Ecology, A Regional Geography of Antarctica and the Southern Ocean," *University of Miami Law Review* 33 (December 1978): 300.
39. Paul Quigg, *A Pole Apart: The Emerging Issue of Antarctica* (New York: McGraw Hill, 1983), 37.
40. de Blij, "Physical Aspects of Antarctica," 312.
41. Convention on the Conservation of Antarctic Marine Living Resources, 20 May 1980, TIAS No. 10240 (entered into force 7 April 1982), Article I(1).
42. See Appendix A, Antarctic Treaty, Article VI.
43. Siple, *90° South*, 337.
44. Ibid.
45. Ibid.
46. Sayed Z. El-Sayed, "Biology of the Southern Ocean," *Oceanus* 18 (1975): 40–41.
47. CIA, *Polar Regions Atlas*, 54.
48. Ibid.
49. Michael Tobias, "The Next Wasteland: Can the Spoiling of Antarctica Be Stopped?," *The Sciences* (March/April 1989): 20.
50. Ibid., 20–21.
51. Ibid.
52. May, *Greenpeace Book of Antarctica*, 80.
53. For fascinating and vivid accounts of the hazards and the rewards of life in the Antarctic, see the following works of Admiral Richard E. Byrd: *Little America* (New York: G.P. Putnam's Sons, 1930); *Discovery*; and *Alone* (New York: G.P. Putnam's Sons, 1938). See also Siple, *90° South*; Adm. G. J. Dufek, *Operation Deepfreeze* (New York: Harcourt Brace, 1957); Norman D. Vaughan, *With Byrd at the Bottom of the World: The South Pole Expedition of 1928–1930* (Harrisburg: Stackpole Books, 1990); and Harry Adams, *Beyond the Barrier with Byrd: An Authentic Story of the Byrd Antarctic Exploring Expedition* (New York: M.A. Donohue & Co., 1932).
54. Originally costing fourteen cents a gallon, the diesel fuel delivered to the South Pole in 1956 "came to $3.60 by the time Globemasters had dropped it by

parachute at the Pole. This made a total of a quarter of a million dollars just to heat and power one Antarctic base!" Rear Adm. George J. Dufek, "What We Have Accomplished in Antarctica," *Geographic Magazine* 116 (October 1959): 531.

55. Ian Cameron observed that "each knot of wind has an effect on life commensurate to a drop of one degree in temperature. Thus whereas a man can live quite happily at -20° C in the still air, when the temperature is -20 and the wind 60 knots he will very quickly die." Cameron, *Antarctica: The Last Continent*, 14. Also see Dufek, *Operation Deepfreeze*, 62.

56. Byrd, *Discovery*, 282.

57. Dufek, *Operation Deepfreeze*, 44.

58. Byrd, *Discovery*, 175.

59. Admiral Byrd wrote that against such drifts there is no protection. "Propelled by high winds, drift moves down with the solid rush and substance of a mountain river. . . . The boxes of stores cached at Retreat Camp, . . . the planes anchored at the mouth of Ver-Sur-Mer Inlet, even Little America itself, had become monstrous heaps of hard-packed drift. . . . Even at the end of May, with the sun more than a month below the horizon, a company of men was still shoveling . . . in search of machine shop equipment." Byrd, *Discovery*, 175.

60. CIA, *Polar Regions Atlas*, 37.

61. National Science Foundation, *Facts About the U.S. Antarctic Program*, NSF 88–66 (July 1978), 2.

62. Ibid., 3. Byrd Station, which operated from 1957 to 1972, had to be closed because the increasing snow load was crushing it.

63. H. G. R. King, *The Antarctic* (London: Blandford Press, 1969), 183.

64. Walton, "Geography, Politics and Science," in Walton, ed., *Antarctic Science*, 53.

65. Amundsen-Scott Station was established entirely by air. Globemaster aircraft dropped the materials and supplies required. There was no other way the station could have been built. Planes that did land had great difficulty taking off again in the thin air. Fogg and Smith, *The Explorations of Antarctica*, 190.

66. Sir Vivian Fuchs, "Polar Travel," in *Antarctic Research: A Review of British Scientific Achievement in Antarctica*, ed. Sir Raymond Priestley, Raymond J. Adie, and G. de Q. Robin (London: Butterworths, 1964), 20.

67. See Byrd, *Discovery*, 353, 355.

68. See Vaughan, *With Byrd at the Bottom of the World*, 100.

69. Richard E. Byrd, "Exploring the Ice Age in Antarctica," *National Geographic Magazine* 68 (October 1935): 456.

70. Francis M. Auburn, *Antarctic Law and Politics* (Bloomington: University of Indiana Press, 1982), 1–2.

71. During Admiral Byrd's second Antarctic expedition, the crew counted two hundred icebergs in view at one time, the largest about four miles long and three miles wide. Later, in the "Devil's graveyard," they steered past mountains of ice that dwarfed those seen earlier. He writes that "on some we saw the whole of Manhattan could have been disposed, even to its subways." Byrd, *Discovery*, 36.

72. CIA, *Polar Region Atlas*, 38.

73. Antarctic icebergs may reach immense sizes, with the largest ones clustering near the shore. One of the greatest, tracked in 1965, was 143 kilometers long and had a surface area of 7,000 square kilometers. Pyne, *The Ice*, 11.

74. See George Deacon, *The Antarctic Circumpolar Ocean* (New York: Cambridge University Press, 1984), 133.

75. Robert D. Hayton, "The Antarctic Settlement of 1959," *American Journal of International Law* 54 (1960): 351.
76. Cameron, *Antarctica: The Last Continent*, 15.
77. See Siple, *90° South*, 114.
78. Byrd, *Alone*, 19.
79. Beck, *International Politics of Antarctica*, 14.
80. For example, the U.S. icebreaker *Polar Star* alone cost $53 million. Accidents can cost millions of dollars. In one season, 1964–65, the United States had seven air crashes, including two total losses. Auburn, *Antarctic Law and Politics*, 3.
81. Diesel fuel in 1960 cost the navy twelve cents per gallon, but it was worth forty cents a gallon by the time they had transported it to McMurdo and six dollars a gallon by the time it got to the South Pole. Carter, *Little America*, 255.
82. CIA, *Polar Regions Atlas*, 49.
83. Dufek, "What We Have Accomplished in Antarctica," 528.
84. Ibid. The high cost of establishing and maintaining inland stations is also exorbitant—estimated by the United States at around one million dollars per man for Amundsen-Scott base in 1957. See Walton, ed., *Antarctic Science*, 55.

3. The Antarctic Treaty

1. Statement by Richard C. Atkinson, director of the National Science Foundation, House Committee on Science and Technology, *U.S. Antarctic Program: Hearings before the Subcommittee on Science, Research and Technology*, 96th Cong., 1st sess., 1, 3 May 1979, 81.
2. The actual discoverer of mainland Antarctica will likely never be known. The British claimed it was Captain Edward Bransfield; the Americans have credited Nathaniel Palmer. The Soviets, in similar fashion, credit the Russian Navy Officer Thaddeus von Bellingshausen with the discovery. Carter, *Little America*, 11.
3. The expedition was authorized only after Congress had been subjected for many years to urgent pleas by American whalers, sealers, and shipowners for better charts of the waters into which their ships sailed. Charles E. Fowler and Harry H. Moore, "U.S. Naval Oceanographic Office: Charting of the Antarctic," *Antarctic Journal of the United States (AJUS)* 1, 2 (March/April 1966): 51.
4. Kenneth J. Bertrand, *Americans in Antarctica: 1775–1948* (New York: American Geographic Society, 1971), Special Publication 39, 160.
5. Exploration of the Antarctic was only one phase of the broad program of the expedition, which included surveys and scientific observations extending from the Atlantic across the far reaches of the Pacific. Bertrand, *Americans in Antarctica*, 159. For an excellent treatment of the expedition, see William Stanton, *The Great United States Exploring Expedition of 1838–1842* (Berkeley: University of California Press, 1975).
6. The land that Wilkes claimed to have sighted is now included within the Australian Antarctic Territory and Adelie Land. W. M. Bush, ed., *Antarctica and International Law: A Collection of Interstate and National Documents*, vol. 3 (New York: Oceana Publications, 1982), 432.
7. Wilkes was not the first to sight the coast of Antarctica, and no member of the expedition set foot on the mainland. Wilkes, however, was the first to conclude, on the basis of actual field observation, that an Antarctic continent existed beyond the icy barrier of the pack. He vigorously maintained his stand in the face

of contradictions by others, including Sir James Clark Ross. Bertrand, *Americans in Antarctica*, 188.

8. Ibid., 160.

9. Ibid., 161.

10. The best sources for Byrd's expeditions are his own writings. See *Little America* (New York: G.P. Putnam's Sons, 1930); *Discovery: The Story of the Second Byrd Expedition* (New York: G.P. Putnam's Sons, 1935); and *Alone* (New York: G.P. Putnam's Sons, 1938).

11. Joseph J. Sisco, "The United States Program for Antarctica," *AJUS*, 1, no. 1 (January/February 1966): 2.

12. Since Byrd's proposed program included extensive scientific research, he obtained financial support from the National Geographical Society. Although generous donations came from the public, his principal supporters were businessmen, particularly John D. Rockefeller Jr. and Edsel Ford. B. M. Plott, "The Development of U.S. Antarctic Policy" (Ph.D. diss., Tufts University, 1969), 31.

13. Byrd, *Little America*, 10–11.

14. Plott, "Development of U.S. Antarctic Policy," 38

15. Unlike those bases of previous Antarctic expeditions, Little America was equipped with electricity and telephones and remained in almost constant radio contact with the outside world. Ibid., 39.

16. Bertrand, *Americans in Antarctica*, 290.

17. Byrd, *Little America*, 341. Byrd did not land at the South Pole. After Robert Scott, the next person to set foot at the South Pole was George J. Dufek on October 31, 1956. Guy G. Guthridge, "A New Research Station at the South Pole," *AJUS* 10, no. 2 (March/April 1975): 37.

18. Byrd, *Little America*, 357, 359, 408.

19. *Congressional Record*, 71st Cong., 1st sess., 1930, 72, pt. 2: 12179–12180.

20. Ibid.

21. If the resolution had passed, it would have had no force other than as an expression of the wishes of the Senate. For an insightful discussion of U.S. territorial rights to Antarctic territory, see Todd J. Parriott, "Territorial Claims in Antarctica: Will the United States be Left in the Cold?" *Stanford Journal of International Law* 22 (spring 1986).

22. Plott, "Development of U.S. Antarctic Policy," 53.

23. President to Byrd, September 7, 1933, in Samuel Rosenman, ed., *The Years of Crisis: The Private Papers and Addresses of Franklin Delano Roosevelt*, 5 vols. (New York: Random House, 1938), 2: 355. This moral support did not, however, help raise the million dollars that the expedition would cost. Siple, *90° South*, 51.

24. In *Alone*, Byrd narrates his experiences manning the Bolling Advance Weather Station.

25. Lincoln Ellsworth, "My Flight Across Antarctica," *National Geographic Magazine* 70, no. 1 (July 1936), 13. The land claimed by Ellsworth is adjacent to Marie Byrd Land in the Antarctic Peninsula.

26. Lincoln Ellsworth, *Beyond Horizons* (New York: Book League of America, 1938), 316–45.

27. Letter from President Roosevelt to the Secretary of State dated July 13, 1939 reprinted in *Foreign Relations of the United States* 1939 (FRUS), 2, 7.

28. Siple, *90° South*, 61.

29. Congressional funding for the USAS was provided by the Second Defi-

ciency Appropriation Act, fiscal year 1939, approved May 2, 1939, and by the Urgent Deficiency and Supplemental Appropriation Act, fiscal years 1939 and 1940, approved June 30, 1939. See, *FRUS* 1939, 2: 11. The $350,000 appropriation from Congress was far from adequate. Financing the expedition was possible only through a combination of private funds from Byrd and his supporters and the Congressional appropriation. See Gordon W. Baldwin, "The Dependence of Science on Law and Government: The International Geophysical Year, A Case Study," *Wisconsin Law Review* (January 1964): 104.

30. Memorandum by the Department of State's Division of European Affairs dated July 18, 1939, in *FRUS* 1939, 2, 7–8.

31. Bertrand, *Americans in Antarctica*, 407.

32. Ibid., 408.

33. Reprinted in *FRUS* 1939, 2, 11–14.

34. Baldwin, "Dependence of Science on Law and Government," 104.

35. Letter from President Roosevelt to Admiral Byrd dated November 25, 1939, in *FRUS* 1939, 2, 11–14.

36. Secretary of State Hull believed that this agency would both strengthen U.S. rights and increase the rewards from scientific investigation. Cordell Hull, *The Memoirs of Cordell Hull*, 2 vols. (New York: The Macmillan Co., 1948), 2: 758–59.

37. Letter from Roosevelt to Byrd, July 12, 1939, in Elliott Roosevelt, ed., *F.D.R. His Personal Letters* (New York: Duell, Sloan, and Pearce, 1950), 2: 906–907.

38. Plott, "Development of U.S. Antarctic Policy," 108.

39. Ibid., 109.

40. Ibid.

41. Ibid.

42. Operation Highjump included more than 4,400 naval and marine personnel, over 40 civilians, and an assortment of equipment and materials. Kenneth J. Bertrand, "Operation Highjump Twenty Years Later," *AJUS* 2: no. 1 (January/February 1967): 5.

43. Plott, "Development of U.S. Antarctic Policy," 114.

44. Bertrand, "Operation Highjump," 5.

45. U.S. Department of the Navy, Press Release, November 12, 1946, "United States Navy Department Expedition to the Antarctic (Operation Highjump), 1946–47" in *Polar Record*, 4 (July 1946), 399–401.

46. The navy's directive for Operation Highjump listed the extension and consolidation of "United States sovereignty over the largest practicable area of the Antarctic continent" as one of its objectives. The Second Antarctic Development Project (1947–1948), generally known as Operation Windmill, a reference to the extensive use of helicopters, was similarly instructed. Siple, *90° South*, 77, 81.

47. Lewis, *Continent for Science*, 62.

48. Siple, *90° South*, 81.

49. "Antarctica," (PPS-31) is reprinted in *(FRUS)* 1948, 1, 977–83.

50. Office of Management and Budget, *The U.S. Antarctic Program*, report to the Committees on Appropriations of the U.S. Senate and House of Representatives, May 1983, Appendix A, 1.

51. Ibid.

52. *United States Policy and International Cooperation in Antarctica*, H.R. Doc. 358,

88th Cong., 2d sess., May 1964 (message from the president of the United States transmitting Special Report on U.S. Policy and International Cooperation in Antarctica), 3.

53. Henry M. Dater, "Organizational Developments in the United States Antarctic Program, 1954–1965," *AJUS* 1, no. 1 (January/February 1966): 23.

54. Ibid.

55. In his last year, President Eisenhower made his special assistant for national security affairs chairman of the OCB. James P. Pfiffner and R. Gordon Hoxie, *The Presidency in Transition* (New York: Center for the Study of the Presidency, 1989), 199.

56. John Prados, *Keepers of the Keys: A History of the National Security Council from Truman to Bush* (New York: William Morrow & Co., 1991), 75.

57. Bush, ed., *Antarctica and International Law*, 3: 484. In 1953, the National Science Board endorsed the concept of a third IGY and the National Research Council recommended that NSF administer the program. Congress appropriated funds specifying that NSF serve as the focal point for coordinating all government agencies involved in IGY activities. Robert H. Rutford, "United States Antarctic Program," in *Antarctic Politics and Marine Resources: Critical Choices for the 1980s*, ed. Lewis M. Alexander and Lynne Carter Hanson (Kingston: University of Rhode Island Press, 1985), 56–57.

58. *U.S Policy and International Cooperation in Antarctica*, H.R. Doc. 358, 3–4.

59. Dater, "Organizational Developments," 23.

60. The bill was referred to the Committee on Public Works, where it received no further consideration. *Congressional Record*, 85th Cong., 1st sess.,1957 103, 8130–8135.

61. Bush, ed., *Antarctica and International Law*, 484.

62. See the statements of the under secretary of the interior, 5–6; the assistant secretary of state, 6–7; the assistant general manager of the Atomic Energy Commission, 7; and the general counsel, Department of Defense, 13, in House Committee on Interior and Insular Affairs, *Antarctica Legislation—1960: Hearings before the Subcommittee on Territorial and Insular Affairs*, 86th Cong., 2d sess., 13, 14 June 1960.

63. Letter from the general counsel, Department of Defense, to the chairman, House Interior and Insular Affairs Committee, dated June 13, 1960. Ibid., 13.

64. Quigg, *A Pole Apart*, 49.

65. Henry M. Dater has appropriately noted that the establishment of the U.S. bases had interesting political implications. Except for Byrd Station, the other six bases were located on territory claimed by another state, and the South Pole station would lie where most of the claims converged. Dater, "Organizational Developments," 22.

66. *U.S. Policy and International Cooperation in Antarctica*, H.R. Doc. 358, 6.

67. Rutford, "United States Antarctic Program," 57.

68. By 1960 the staff had grown to seven professionals. Ibid., 58.

69. The Bureau of the Budget favored the NSF since the Committee on Polar Research of the National Academy of Sciences and the Antarctic Working Group of the OCB—the two key bodies involved in administering the U.S. IGY Antarctic program—did not seem well suited to a continuing program. Quigg, *A Pole Apart*, 59.

70. Ibid., 58.

71. Ibid., 59.
72. Ibid.
73. Ibid.
74. Department of State, "Statement of U.S. Policy on Antarctica, NSC 5804/1" in *FRUS* 1958–1960, vol. 2: (Washington, D.C., GPO, 1991), 485.
75. Several excellent accounts are available of the scientific discoveries made during the IGY and the spirit of international cooperation it generated. See, for example, Walter Sullivan, *Assault on the Unknown* (New York: McGraw-Hill, 1961); Sydney Chapman, *IGY: Year of Discovery, The Story of the International Geophysical Year* (Ann Arbor: University of Michigan Press, 1959); Sir Harold Spencer Jones, "The Inception and Development of the International Geophysical Year," *Annals of the International Geophysical Year*, vol. 1 (London: Pergammon Press, 1959); House Committee on Appropriations, *Report on the International Geophysical Year: Hearing before the Subcommittee of the Committee on Appropriations*, 85th Cong., 1st sess., 1 May 1957; and House, Committee on Appropriations, *Report on the International Geophysical Year: Hearing before the Subcommittee of the Committee on Appropriations*, 86th Cong., 1st sess., February 1959.
76. It was an American, Dr. L. V. Berkner, who introduced and steered through its early years the idea for the IGY of 1957–1958. He made the suggestion in 1950 to a group of fellow scientists. That same year he and other scientists submitted the proposal to the Joint Commission on the Ionosphere, which recommended the project to the International Council of Scientific Unions. See Albert P. Crary, "International Geophysical Year: Its Evolution and U.S. Participation," *AJUS*, 17, no. 4 (December 1982), 1–2.
77. *FRUS 1958–1960*, 2: 481. See also "Discussions Asked on Territorial Problem of Antarctica," *Department of State Bulletin*, 19 (September 5, 1948): 301.
78. For a discussion of the U.S. proposal and the responses it drew from various claimant states, see John Hanessian, "The Antarctic Treaty of 1959," *International and Comparative Law Quarterly* 9 (July 1960).
79. For a draft agreement prepared by the Department of State establishing an international trusteeship for the administration of Antarctica, see *FRUS 1948* vol. 1, part 2 (Washington, D.C.: GPO, 1976), 984–87.
80. For a discussion of the advantages and disadvantages of establishing a condominium over Antarctica, see *FRUS 1958–1960* 2, 488–90. As with a trusteeship, this arrangement was tantamount to renunciation of exclusive sovereignty claims, which explains why Chile and Argentina rejected the proposal outright. See Hanessian, "The Antarctic Treaty of 1959," 438.
81. Press statement, Foreign Office, January 18, 1949. Reprinted in *Polar Record* 5 (January/July 1949): 361.
82. See Hanessian, "The Antarctic Treaty of 1959," 436–44.
83. Memorandum from the embassy of the Soviet Union to the Department of State, in *FRUS 1950*, vol. 1 (Washington, D.C.: GPO, 1977), 911–13.
84. See *FRUS 1958–1960*, 2: 481.
85. Hanessian, "The Antarctic Treaty of 1959," 419.
86. The IGY brought together over 25,000 scientists and technicians working at 2,500 stations around the world with a total expenditure of over $2 billion. See Crary, "International Geophysical Year," 4. In Antarctica, the United States had five stations: Little America, Byrd, Amundsen-Scott (Pole), Ellsworth, and Wilkes. It also operated Hallet station jointly with New Zealand. Interestingly, most states operated stations within their claimed sectors, but the United States had stations

in the Argentine, Australian, British, New Zealand, and unclaimed areas as well as the one at the South Pole, the apex where all claims met. The Congress appropriated a total of $43 million in support of the scientific aspects of the United States IGY Antarctic program. See Wallace Atwood Jr., "The International Geophysical Year in Retrospect," *Department of State Bulletin* 40 (May 11, 1959): 684.

87. Frank C. Alexander Jr., "Legal Aspects: Exploitation of Antarctic Resources: A Recommended Approach to the Antarctic Resource Problem," *University of Miami Law Review* 33 (December 1978): 378.

88. See *United States Policy and International Cooperation in Antarctica*, H.R. Doc. 358, and House Committee on Foreign Affairs, *The Political Legacy of the International Geophysical Year*.

89. The text of the resolution is in Marjorie M. Whiteman, ed., *Digest of International Law*, 13 vols. (Washington D.C.: GPO, 1963–73), 1:500. See also Peterson, *Managing the Frozen South*, 39.

90. Among the more serious threats to peace in the south polar region were the increasing disputes among Argentina, Chile, and Great Britain. For an insightful analysis into the controversy, see Christopher C. Joyner, "Anglo-Argentine Rivalry After the Falklands/Malvinas War: Laws, Geopolitics and the Antarctic Connection," *Lawyer of the Americas*, 15 (winter 1984).

91. See for example, Hugh Odishaw, "The International Geophysical Year and World Politics," *Journal of International Affairs*, 13, no. 1 (1959); Walter Sullivan, "The International Geophysical Year," *International Conciliation* 521 (January 1959); and Howard J. Taubenfeld, "A Treaty for Antarctica," *International Conciliation*, 531 (January 1961).

92. It has been argued that the true motivation of the United States in initiating treaty negotiations was to gain access to the whole continent. See Auburn, *Antarctic Law and Politics*, 100.

93. For reactions to the proposal, see Department of State, "Statements by Americans Favoring an International Solution to the Antarctic Question," Historical Division, Bureau of Public Affairs, Research Project no. 425 (June 1959).

94. For the statement by President Eisenhower and text of the United States note addressed to the foreign ministers of the eleven countries, see *Department of State Bulletin* 38 (June 2, 1958): 910.

95. Ibid., 911.

96. See Peter J. Beck, "Preparatory Meetings for the Antarctic Treaty, 1958–59," *Polar Record* 22, no. 141 (September 1985), which discloses interesting sidelights of the discussions during the first twenty-six preparatory meetings.

97. The chairman of the Argentine delegation made it unequivocally clear in his opening statement that it was not the mission of the conference to "change or alter anything. Nothing that is done here will give rise to, affect, or disregard rights. . . . With that clear understanding, Argentina will take part in the work with a sincere desire to cooperate." See *The Conference on Antarctica: Washington October 15 December 1, 1959 (Conference Documents, Antarctic Treaty and Related Papers)*, Department of State Publication 7060 (September 1960), 31.

98. Beck, "Preparatory Meetings for the Antarctic Treaty, 1958–59," 663.

99. Ibid.

100. Statement by Ambassador Herman Phleger, head of the U.S. delegation and permanent chairman of the conference, during congressional hearings to consider the Antarctic Treaty. See Senate Committee on Foreign Relations, *The*

Antarctic Treaty: Hearings before the Committee on Foreign Relations, 86th Cong., 2d sess., June 14, 1960, 44–45.

101. Ambassador Phleger presided through more than fifty formal and informal sessions between June 1958 and October 1959. See *U.S. Policy and International Cooperation in Antarctica, H.R. Doc.* 358, 10. Throughout this period Chile and Argentina remained adamant about their claims.

102. "Twelve Nations Sign Treaty Guaranteeing Nonmilitarization of Antarctica and Freedom of Scientific Investigation," *Department of State Bulletin* 41 (December 21, 1959), 911. Texts of the Final Act and the treaty appear in *The Conference on Antarctica Documents* 56–67. The Text of the Treaty is also in United States Arms Control and Disarmament Agency, *Arms Control and Disarmament Agreements: Texts and Histories of Negotiations* (Washington, D.C., 1982 ed.), 22–27.

103. The United States ratified the treaty on August 18, 1960. *Arms Control and Disarmament Agreements*, 22.

104. Senate Committee on Foreign Relations, *Antarctic Treaty Hearings*, 45.

105. Ibid.

106. These charges were made by Elizabeth Kendall of Washington, D.C., who testified as a "taxpayer and a citizen." Ibid., 78, 81. It appears from reading Ms. Kendall's statement that her strong opposition to the ratification of the treaty was based on her perception that the Soviets planned "to take over Antarctica," and that the treaty denied to the United States "control of what may be the most important area in the space race." Ibid., 80.

107. Ibid., 2.

108. Antarctic Treaty, Article XIII(3).

109. *U.S. Policy and International Cooperation in Antarctica, H.R. Doc.* 358, 12.

110. Antarctic Treaty, Article VI. The text of the treaty is reprinted in Appendix A.

111. In his welcoming address at the opening session of the conference, Secretary of State Herter said that the United States government "is dedicated to the principle that the continuation of this cooperation should be assured, and that Antarctica should be used for peaceful purposes only, should not become an object of political conflict, and should be open for the conduct of scientific investigations." *The Conference on Antarctica Documents*, 1.

112. Ibid., 77–78.

113. This view is consistent with that expressed by the U.S. government officials interviewed for this study. Also see Office of Technology Assessment, *Polar Prospects: A Minerals Treaty for Antarctica*, OTA–O–428 (Washington, D.C.: GPO, 1989); and David A. Colson, "The United States Position on Antarctica," *Cornell International Law Journal* 19, no. 2 (summer 1986), 296.

114. See Henry S. Francis, "The Antarctic Treaty: A Reality Before its Time," in Alexander and Hanson, eds., *Antarctic Politics and Marine Resources*, 87–98. For a thorough discussion of the treaty, see Sir Arthur Watts, *International Law and the Antarctic Treaty System* (Cambridge: Grotius Publications, 1992); Peter Beck, *The International Politics of Antarctica* (London: Croom Helm, 1986); Jeffrey D. Myhre, *The Antarctic Treaty System: Politics, Law, and Diplomacy* (Boulder: Westview Press, 1986); Paul C. Daniels, "The Antarctic Treaty," *Science and Public Affairs*, 26, no. 10 (1970); K. D. Suter, *World Law and the Last Wilderness* (Sydney: Friends of the Earth, 1980). For background commentary, see E. W. Hunter Christie, *The Antarctic Problem: An Historical and Political Study* (London: Allen and Unwin, 1951).

115. Under the treaty, Antarctica was the first part of the world to be declared a demilitarized and nuclear weapons–free zone.

116. The inclusion of a ban on all types of nuclear explosions appealed particularly to both the Australians and New Zealanders. There had been enough talk about Antarctica as a test site to make radiation danger seem real. See M. J. Peterson, *Managing the Frozen South*, 73.

117. The contracting parties have the right to designate nationals as observers and such observers, "have complete freedom of access at any time to any or all areas of Antarctica." It was the United States that insisted on unilateral inspection in opposition to the British and the French suggestions of an international machinery. Senate Committee on Foreign Relations, *Antarctic Treaty: Hearings*, 69–70. The Department of State attached great significance to the inspection provision, partly because the Antarctic Treaty was seen as a valuable source of practical experience. More importantly, the inspection provision provided a "weighty argument for ratification," since this was the first time that the Soviet Union had accepted free aerial inspection by treaty. Auburn, *Antarctic Law and Politics*, 10.

118. Barry Plott cogently argues that this is hardly "a valid criticism of the Treaty itself, or of the United States delegation at the Conference." It would have been "unrealistic to expect the claimant nations to freely relinquish their claims to sovereign rights over a territory whose value is not known and will probably remain unknown for years." Plott, "Development of U.S. Antarctic Policy," 215.

119. In 1959 during preparatory talks for the treaty conference, U.S. foreign policymakers found the resource issue too contentious to be addressed. At the same time they recognized the possible resource potential of Antarctica and referred to it during the Senate debate preceding the ratification of the treaty. See the statement by Dixy Lee Ray, former assistant secretary of state and chairman, Antarctic Policy Group, before the Senate Committee on Foreign Relations, *U.S. Policy with Respect to Mineral Exploration and Exploitation in the Antarctic: Hearing before the Subcommittee on Oceans and International Environment*, 94th Cong., 1st sess., 15 May 1975, 5.

120. Peterson, *Managing the Frozen South*, 49.

121. Some commentators opposed the treaty because they believed the United States was foregoing a strategic advantage by allowing Soviet participation in Antarctic affairs. See Senate Committee on Foreign Relations, *Antarctic Treaty Hearings*, 69–70.

122. For an examination of this issue, see Gerald S. Schatz, ed., *Science, Technology and Sovereignty in the Polar Regions* (Washington, D.C.: D. C. Heath, *1974*); and Elizabeth K. Hook, "Criminal Jurisdiction in Antarctica," *University of Miami Law Review* 33 (December 1978).

123. Myhre, *The Antarctic Treaty System*, 39.

124. Article XII (2)(b) and (c). If the amendment does not become effective, then any contracting party may opt to withdraw from the Treaty within two years after receiving notice of the amendment. Such withdrawal is not effective until two years after the withdrawing party gives notice of its withdrawal.

125. For a discussion of theories on acquisition of territory, see R. Y. Jennings, *The Acquisition of Territory in International Law* (Manchester: Manchester University Press, 1963); G. Smedal, *Acquisition of Sovereignty over Polar Areas* (Oslo: J. Dybwad, 1931); and Georg Schwarzenberger, "Title to Territory: Response to a Challenge," *American Journal of International Law* 51 (1957), 2.

126. Those seven countries making territorial claims and the years of their initial claims are: United Kingdom, 1908; New Zealand, 1923; Australia, 1933; France, 1924 (defined in 1938); Norway, 1939; Chile, 1940; and Argentina, 1942. These countries are known as the claimant states. The claims of Argentina, Chile, and Great Britain overlap substantially in the Antarctica Peninsula region of the continent.

127. Under international law, discovery alone is insufficient to establish sovereignty over *terra nullius* (the territory of no one). Inchoate title to land claimed by discovery must be perfected by effective occupation within a reasonable period. Effective occupation requires both the intent to exercise sovereignty and the actual exercise of sovereignty. See J. L. Brierly, *The Law of Nations: An Introduction to the International Law of Peace*, 6th ed. by Sir Humphrey Waldock (London: Oxford University Press, 1963), 163.

128. In resorting to the sector principle, the claimant states "are relying on the legal validity of their claims under the occupation principle because such a legitimate claim is necessary before the sector principle may be utilized." See Rudy Cerone, "Survival of the Antarctic Treaty: Economic Self-Interest versus Enlightened International Cooperation," *Boston College International and Comparative Law Journal* 2:1 (1978–1979): 119.

129. *Uti possidetis juris* refers to the rights Argentina and Chile claim they inherited from Spain. In this case, inheritance of Antarctica would be based on the Bull of Pope Alexander VII, issued in 1493, in which he divided the world between Spain and Portugal. See Joyner, "Anglo-Argentine Rivalry After the Falklands/Malvinas War," 477.

130. Paul Siple found it ironic that the United States, whose Antarctic explorations had uncovered most of the continent, had not raised its voice to demand a single foot of Antarctica. Siple, *90° South*, 82. Senator Ernest Gruening of Alaska echoed similar sentiments when he remarked that Americans had seen approximately 80 percent of the continent yet had not claimed any portion of it. Senate Comittee on Foreign Relations, *Antarctic Treaty Hearings*, 19.

131. See Arthur D. Watts, "The Antarctic Treaty as a Conflict Resolution Mechanism," in *Antarctic Treaty System, An Assessment: Proceedings of a Workshop held at Beardmore South Field Camp, Antarctica, January 7–13, 1985*, Polar Research Board (Washington, D.C., National Academy Press, 1986), 65–75.

132. Bernard H. Oxman, "The Antarctic Regime: An Introduction," *University of Miami Law Review* 33 (December 1978): 290.

133. Deborah Shapley, *The Seventh Continent: Antarctica in a Resource Age* (Washington, D.C. Resources for the Future, 1985), 19.

134. The long American history of exploration and discovery in Antarctica commenced in the early part of the nineteenth century. See *FRUS* 1958–1960, 2: 482. In fact, from the late 1920s to the time of the signing of the Antarctic Treaty, the United States flew more planes, mapped and photographed more territory, and sent more expeditions to Antarctica than any other state. Siple, *90° South*, 83.

135. In 1986, David A. Colson, deputy assistant secretary, Bureau of Oceans and International Environmental and Scientific Affairs, Department of State, listed among U.S. national interests in Antarctica that of preserving "any basis for a United States claim to territorial sovereignty in Antarctica. . . ." See Colson, "The United States Position on Antarctica," 296.

136. Bush, ed., *Antarctica and International Law*, 3: 420.

137. See *FRUS 1958–1960*, 2:483.

138. Letter from Secretary Hughes to A. W. Prescott, Republican Publicity Assn., in Bush, ed., *Antarctica and International Law*, 3:430.

139. See "Report of Comments by the Department of State on Areas Claimed by Admiral Byrd," dated June 8, 1936, reprinted in Bush, ibid., 435.

140. The area claimed by Ellsworth, between 80°W and 120°W, embraces part of an extended Marie Byrd Land and what is now known as Ellsworth Land. See "Account of Claim by Lincoln Ellsworth on Behalf of the United States," reprinted in Bush, ibid.

141. A. C. Miller, "Antarctica: White Continent of Promise," *U.S. Naval Institute Proceedings* 88 (August 1962): 52.

142. Auburn, *Antarctic Law and Politics*, 55.

143. See *FRUS 1947*, vol. 1 (Washington, D.C.: GPO, 1973), 1047. Paul Siple corroborates this when he writes that "our primary activity was to be exploration and execution of a scientific program, yet the official aims of the United States Antarctic Service were nationalistic. The idea proposed by the State Department was that if the U.S. maintained continuing bases in the Antarctic, our claim on Antarctic territories if and when we made one would be upheld by any international court, or by international law." Siple, *90° South*, 65.

144. Siple, *90° South*, 63.

145. Robert D. Hayton, "Polar Problems and International Law," *American Journal of International Law* 52 (1958): 746, 762–63.

146. Siple, *90° South*, 76–77.

147. Miller, "Antarctica: White Continent of Promise," 52.

148. Ibid.

149. Letter from Acheson to the secretary of the navy dated December 14, 1946, in *Foreign Relations of the United States (FRUS) 1946*, vol. 1 (Washington, D.C.: GPO, 1972), 1497–98.

150. Admiral Byrd was a strong supporter of the formal assertion of American territorial claims to Antarctica. During Congressional hearings on the activities of the U.S. Antarctic Service, he criticized the government for not perfecting a claim. House Committee on Appropriations, *Expedition to the Antarctic Regions: Hearing before the Subcommittee on Deficiencies*, 76th Cong., 1st sess., 2 June 1939, 12–13.

151. Auburn, *Antarctic Law and Politics*, 64.

152. Ibid., 83.

153. *FRUS 1958–1960*, 2:484.

154. Ibid., 483.

155. Siple, *90° South*, 83.

156. Quigg, *A Pole Apart*, 14.

157. M. J. Peterson shares the view that in 1959 the United States was not in a position to make a wise choice of areas to claim, partly because of the paucity of information about the whole continent. As she points out, behind the U.S. decision lay a calculation that United States interests in exploration and scientific research were best served by maintaining access to the whole continent. Asserting a claim would make free access impossible. See her *Managing the Frozen South*, 53.

158. Quigg, *A Pole Apart*, 14.

159. Ibid.

160. Bertrand, *Americans in Antarctica*, 18.

161. Laurence M. Gould, "Antarctica in World Affairs," *Foreign Policy Association*, Headline Series, 128 (March/April 1958): 26. For discussion of other states'

attitudes toward American interest in Marie Byrd Land, see Hayton, "The Antarctic Settlement of 1959," 351.
162. Taubenfeld, "A Treaty for Antarctica," 255.
163. Auburn, *Antarctic Law and Politics*, 64–65.
164. Although the treaty does not establish a secretariat or other permanent headquarters, Article IX provides for continuing consultation in matters of coimmon interest pertaining to Antarctica.
165. Essentially, a secretariat is needed to support the increased and more complex operations of the consultative mechanism, to circulate information, and to implement the protocol. The United States has been disappointed at the lack of progress toward establishing a secretariat at each meeting convened since 1989. See *AJUS* 25, no. 1 (March 1990): 3; and *AJUS* 27, NO. 1 (March 1992): 5.
166. Quigg, *A Pole Apart*, 158.
167. Although the treaty does not stipulate how frequently meetings should be held, they have been convened approximately every two years. At the 16th meeting, the member states reviewed the need to increase the frequency of consultative meetings. With entry into force of the Environmental Protocol, consultative parties will need to meet more frequently. They agreed that, starting with the 17th meeting, consultative meetings should be held annually. See "Bonn Meeting Stresses Environmental Stewardship and International Cooperation," *AJUS*, 27, no. 1 (March 1992): 5.
168. For ATCP meetings held to date, see Appendix D.
169. Auburn, *Antarctic Law and Politics*, 284.
170. Peterson, *Managing the Frozen South*, 94.
171. Ibid.
172. Antarctic Treaty, Article XI. Establishing a scientific station and sending a scientific expedition to Antarctica are examples of "substantial scientific research."
173. Although fifteen states have joined the original twelve signatories as consultative parties during the last thirty years, the unification of Germany in 1991 rendered 26 rather than 27 consultative parties. *AJUS* 25, no. 4 (December 1990): 11; and *AJUS* 27, no. 2 (June 1992): 17.
174. In January 1, 1993, the Czech Republic and Slovak Republic succeeded to the treaty as parts of former Czechoslovakia. Turkey joined as an acceding party in 1996. Appendix E lists the consultative parties and the acceding states.
175. The Rules of Procedure of Antarctic Treaty Consultative Meetings, adopted July 10, 1961, are reprinted in John Heap, ed., *Handbook of the Antarctic Treaty System*, 8th ed. (Washington, D.C.: U.S. Department of State, April 1994), Annex C.
176. Antarctic Treaty, Article IX, para 1.
177. Enumeration of the 186 recommendations adopted at the first seventeen consultative meetings are in the 1994 (Heap, ed.) *Handbook of the Antarctic Treaty System*, 1–6.
178. David A. Colson, "The Antarctic Treaty System: The Mineral Issue," *Law and Policy in International Business* 12 (1980): 875.
179. Antarctic Treaty, Article IX(4).
180. Antarctic Treaty System, Rules of Procedure, Rule 23 (Heap, ed., *Handbook of the Antarctic Treaty System*).
181. See ibid., 1–6, and Draft Final Report of the XVIII Antarctic Treaty Meeting, AT Doc XVIII ATCM/WP37 (22 April 1994), 30.

182. Normally, an acceptance is communicated to the depository government, which notifies the consultative parties when all have accepted a recommendation. Antarctic Treaty, Article IX(4), and Article XII(1) (a).

183. The Rules of Procedure provide that the final report contain a brief account of the proceedings of the meeting, that it be approved by a majority of the representatives present, and that it be transmitted to the governments entitled to consultative status for their consideration. See 1994 (Heap, ed.) *Handbook of the Antarctic Treaty System*, Annex C2.

184. Special consultative meetings should be viewed as distinctive events, supportive of the operations and evolution of the Antarctic Treaty System. See Beck, *International Politics of Antarctica*, 160.

185. Ibid.

186. SCAR is a scientific committee of UNESCO's International Council of Scientific Unions. It was established in 1958 to plan and coordinate post-IGY Antarctic research projects. SCAR, including the composition of its membership, is discussed in more detail in chapter 6.

187. Beck, *International Politics of Antarctica*, 163.

188. Agreed Measures for the Conservation of Antarctic Flora and Fauna, June 2–13, 1964, 17 UST 996, TIAS no. 6058, modified in 24 UST 1802, TIAS no. 7692 (1973) (hereinafter cited as Agreed Measures).

189. Convention for the Conservation of Antarctic Seals, adopted June 1, 1972, 29 UST 441 no. 8826 (entered into force March 11, 1978) (hereinafter cited as Seals Convention).

190. Convention on the Conservation of Antarctic Marine Living Resources, done May 20, 1980, TIAS no. 10240 (entered into force April 7, 1982) (hereinafter cited as CCAMLR).

191. Convention on the Regulation of Antarctic Mineral Resources Activities, done June 2, 1988, Doc. AMR/SCM/88/78 (hereinafter cited as Minerals Convention). The Minerals Convention and the reasons it did not come into force was discussed in chapters 8 and 9.

4. Making U.S. Antarctic Policy

1. Alfred N. Fowler, "Antarctic Logistics," *Oceanus* 31, no. 2 (summer 1988), 82. Mr. Fowler is former deputy director, Division of Polar Programs, National Science Foundation.

2. Statement of R. Tucker Scully, in House Committee on Interior and Insular Affairs, *Establish an Antarctica World Park: Hearing before the Subcommittee on Insular and International Affairs*, 101st Cong., 2d Sess., 18 September 1990, 61. In interviews with the authors, Mr. Scully has reiterated the importance of the treaty system for the pursuit of U.S. Antarctic interests. In interview with Ambassador David Colson, the Treaty was characterized as "the cornerstone of U.S. Antarctic policy." Interview by Ethel R. Theis with David A. Colson, deputy assistant secretary for oceans and international environmental and scientific affairs, Department of State, October 15, 1991.

3. F. M. Auburn disagrees strongly with the notion that U.S. Antarctic policy has been generally consistent, even prior to the inception of the Antarctic Treaty. See *Antarctic Law and Politics*, 77.

4. See, House Committee on Interior and Insular Affairs, *Antarctica Report—1965: Hearings*, 30.

5. See Senate Committee on Foreign Affairs, *U.S. Policy with Respect to Mineral Exploration and Exploitation*, 5.
6. See House Committee on Science and Technology, *U.S. Antarctic Program: Hearings*, 96.
7. F. M. Auburn suggests that the fragmentation of effort resulting from the participation of several executive branch agencies in Antarctic policymaking has had a negative impact on policy coherence and program continuity. See Auburn, *Antarctic Law and Politics*, 77.
8. House Committee on Interior and Insular Affairs, *Antarctica Legislation— 1960: Hearings*, 13.
9. Ibid., 98.
10. Ibid., 13.
11. Ibid.
12. Patrick Anderson, *The President's Men* (New York: Doubleday, 1969), 210.
13. Ibid., 317. Circular A-51 was slightly modified as a result of the president's termination of the OCB.
14. *U.S. Policy and International Cooperation in Antarctica*, H.R. Doc. 358, 13.
15. Ibid.
16. Statement by Ambassador Cleveland, in House Committee on Interior and Insular Affairs, *Antarctica Report—1965 Hearings*, 34.
17. Ibid.
18. Initially, in 1958 NSF created an "Antarctic Research Program" in its Office of Special International Programs. In 1961, the "Antarctic Research Program" was upgraded as an "office" of the foundation. In 1963, the Office of Antarctic Programs became part of the International Activities Division, and in 1965 it joined the newly formed Division of Environmental Sciences. Rutford, "United States Antarctic Program," 57–58. As a result of a broad reorganization of the NSF announced on July 10, 1975, the Office of Polar Programs became part of the newly created Directorate for Astronomical, Earth, and Ocean Sciences. *AJUS* 10, no. 4 (July/August 1975): 202.
19. *U.S. Policy and International Cooperation in Antarctica*, H.R. Doc. 358, 13.
20. Office of Technology Assessment, *Polar Prospects*, 81.
21. Lyndon B. Johnson, "Statement by the President in Response to a Progress Report by the Antarctic Policy Group." *Public Papers of the Presidents of the United States: Lyndon B. Johnson, 1965*, vol. 1 (Washington, D.C.: GPO, 1966), 468–69.
22. Ibid., 564.
23. Quoted by Dr. Ray in Senate Committee on Foreign Relations, *U.S. Policy with Respect to Mineral Exploration and Exploitation*, 3.
24. For example, position papers drafted primarily by the Department of State that provide guidance to U.S. delegations attending Antarctic Treaty Consultative Meetings are coordinated with and cleared by not only the NSF but also by other agencies with interests in Antarctic affairs. Ethel R. Theis's interview with Lawrence Rudolph, deputy general counsel, National Science Foundation, March 12, 1992.
25. See House Committee on Interior and Insular Affairs, *Antarctica Report— 1965: Hearings*, 34–35.
26. See *AJUS* 6, no. 3 (May/June 1971): 66.
27. Philip Quigg has strongly criticized the organization and personnel within the Department of State on Antarctic affairs. According to Quigg, "No

other nation with interests in the Antarctic manages its affairs in so haphazard a manner." He finds the OES "a grab bag of leftover missions that have not found a home elsewhere," and notes that in "its first five years, the OES was headed by a succession of three inexperienced assistant secretaries, and for thirty-one months the post was vacant." See Quigg, *A Pole Apart*, 214–15.

28. See Statement from the Office of the White House Press Secretary in Senate Committee on Foreign Relations, *U.S. Policy with Respect to Mineral Exploration and Exploitation*, 30.

29. Ibid.

30. NSDM 71 is reprinted in *United States Antarctic Program: Personnel Manual*, National Science Foundation, 1990 ed., 49.

31. OMB Circular No. A-51 Revised, is reprinted in *U.S. Antarctic Policy Hearing*, 27–28.

32. See House Committee on Science and Technology, *U.S. Antarctic Program: Hearings*, 77.

33. Ibid., 2.

34. Ibid., 122.

35. Commenting on the role of the NSF, Philip Quigg criticized that some of the weaknesses of the Division of Polar Programs "all come together in the division's inability or unwillingness to adapt to changing circumstances. . . . NSF has resisted every effort to involve it in technical and scientific questions concerning the resources of Antarctica. Despite the directive of the National Security Council to extend scientific research in Antarctica to resource appraisal, . . . the NSF has only made a pretense of complying." See Quigg, *A Pole Apart*, 217.

36. On October 3, 1985, the NSF and the Department of Defense signed a Memorandum of Agreement, which set forth operational and logistic support responsibilities for the U.S. Antarctic Program. The memorandum is reprinted in *Safety in Antarctica: Report of the U.S. Antarctic Program Safety Review Panel*, NSF Pub. 88–78 (June 30, 1988), Appendix 5.

37. See *AJUS* 6, no. 3 (May/June 1971): 66.

38. Antarctic Services, a subsidiary of the Federal Electric Corp. (a subsidiary of the International Telephone and Telegraph Corp.), was selected as the new support contractor effective April 1, 1980. It replaced Holmes & Narver which had been providing this service since 1968. *AJUS* 14, no. 1 (March 1980): 8.

39. On April 19, 1976, the Office of Polar Programs became the Division of Polar Programs. The organizational structure of the new division remained unchanged. *AJUS* 11, no. 2 (June 1976): 112. The Division of Polar Programs is part of the Geosciences Directorate, one of seven directorates that assist NSF's director and the twenty-four-member National Science Board in fulfilling the foundation's mission. *AJUS* 22, no. 3 (September 1987): 20.

40. U.S. National Science Foundation, Division of Polar Programs, *Facts about the United States Antarctic Research Program* (July 1986), 6.

41. Ibid., 70.

42. Ibid.

43. "Statement of Rear Admiral Edward P. Travers, Director of Budget and Reports, Office of the Comptroller of the Navy," in House Committee on Science and Technology, *U.S. Antarctic Program: Hearings*, 117.

44. Ibid.

45. NSDM 318 is reprinted in NSF's *Antarctic Program: Personnel Manual*, 1990 ed., 50.

46. With the issuance of NSDM 318, the status of Circular A-51 Revised

became moot. Bush, ed., *Antarctica and International Law*, 3:485. The OMB determined, in the light of presidential directives since 1971, that there was no longer a need for the circular and cancelled it December 12, 1985. *Federal Register* 51, no. 76 (April 21, 1986): 14760–61. "Description of the Agency Responsibilities for the United States Antarctic Research Program," which appears in NSF's *Facts about the U.S. Antarctic Program* (July 1985), 5–6.

47. Memorandum 6646 is reprinted in National Science Foundation, *Facts about the U.S. Antarctic Program* (October 1994), 26.

48. "U.S. Antarctic Program Receives Presidential Support," *AJUS* 17, no. 1 (March 1982): 4.

49. See Senate Committee on Commerce, Science, and Transportation, *Antarctica: Hearing before the Subcommittee on Science, Technology, and Space*, 98th Cong., 2d sess., 24 September 1984, 2, 3.

50. Interview by Ethel R. Theis with Eric Melby, National Security Council official responsible for Antarctic affairs, Washington, D.C., January 23, 1992.

51. As indicated in an unclassified version of NSD-1, the Principals Committee, chaired by the assistant to the president for national security affairs, is the senior interagency forum for consideration of policy issues affecting national security. The Deputies Committee is chaired by the deputy national security adviser. Beneath this apex is a network of eight Policy Coordinating Committees that have regional or functional specialties.

52. Interview by Ethel R. Theis with Raymond Arnaudo, chief of Division of Polar Affairs, Department of State, November 20, 1991.

53. In separate interviews with both the present authors, David Colson, R. Tucker Scully, and Raymond Arnaudo all emphasized this fact. See also Frank G. Klotz, *America on the Ice* (Washington, D.C.: National Defense University Press, 1990), 138.

54. The assistant secretary for oceans and international environmental and scientific affairs is the current chair of the Policy Coordinating Committee for Oceans, Environment and Science (former APG with expanded responsibilities covering Arctic policy as well).

55. Interview by Ethel R. Theis with R. Tucker Scully, November 5, 1991; Raymond Arnaudo, November 20, 1991; and Thomas Laughlin, senior policy analyst, Office of Policy and Planning, National Oceanic Atmospheric Administration, November 5, 1992.

56. *AJUS* 26, no. 1 (March 1991): 11.

57. The National Oceanic and Atmospheric Administration is not an independent agency but was established in 1970 under the Department of Commerce.

58. Comment by Dr. Ray in Senate Committee on Foreign Relations, *U.S. Policy with Respect to Mineral Exploration and Exploitation*, 3.

59. "U.S. Antarctic Program Receives Presidential Support," *AJUS* 19, no. 1 (March 1984): 4.

60. *AJUS* 24, no. 4 (December 1989), 11.

61. Ibid.

62. Ibid.

63. Interview by Ethel R. Theis with David A. Colson, October 15, 1991.

64. The APG routinely meets once or twice a year. Ethel R. Theis interview with Dr. Anton Inderbitzen, former Head, Antarctic Staff, Division of Polar Programs, on November 22, 1991.

65. Interview by Ethel R. Theis with Raymond Arnaudo, November 20, 1991.

232 / Notes to Pages 54–57

66. The chairman of the APG in 1996 was R. Tucker Scully, who required special dispensation to remain in Washington beyond the prescribed eight-year limit. He resigned the Foreign Service to retain his post as director of the Office of Oceans and Polar Affairs. See Quigg, *A Pole Apart*, 215, 288. A normal requirement of Foreign Service Officers (FSOs), which comprise most of the State Department's professional staff, is overseas service. It is not uncommon to require that FSOs spend two years overseas out of every eight years of service. Refusal to accept overseas assignments can ruin an FSO's opportunities for advancement and possibly lead to termination of employment.

67. Ethel R. Theis interview with Raymond Arnaudo, November 20, 1991.

68. Ibid.

69. During an interview with Ethel R. Theis on November 10, 1992, Ambassador E. U. Curtis Bohlen, assistant secretary for ocean and international environmental and scientific affairs and chairman of the Antarctic Policy Coordinating Committee (former APG), observed that the role played by agencies participating in Antarctic decision making varies according to the degree of their involvement in Antarctic activities. To illustrate, Ambassador Bohlen mentioned that, though EPA's role is limited, its staff has been very helpful in assisting with the development of standards for environmental impact statements applicable to Antarctica. As for NOAA, although its role in marine resources issues is large, its involvement in activities on the "ice" is relatively small. The Department of Interior plays a major role through the U.S. Coast Guard. Interior, although initially opposed to a moratorium on mining activities, participated actively on policy discussions leading to the environmental protocol.

70. Klotz, *America on the Ice*, 137.

71. Ethel R. Theis interview with Eric Melby, January 23, 1992.

72. In an interview with Ethel R. Theis on March 5, 1991, Peter Wilkniss, director, Division of Polar Programs, NSF, remarked that the consistency and coherence evident in U.S. goals and objectives in Antarctica derives largely from specific presidential statements on the nature of these interests.

73. Interview by Christopher C. Joyner with R. Tucker Scully, April 30, 1992.

74. Done at Wellington June 2, 1988, opened for signature November 25, 1988. Document AMR/SCM/88/78 (June 2, 1988), reprinted in *International Legal Materials* 27 (July 1988): 859–900. For discussion of the negotiations that produced this agreement, see Christopher C. Joyner, "The Antarctic Minerals Negotiating Process," *American Journal of International Law* 81 (October 1987): 888–905; and Franciso Orrego Vicuna, *Antarctic Mineral Exploitation: The Emerging Framework* (Cambridge: Cambridge University Press, 1988).

75. For discussion of the principal objections to CRAMRA, see C. Joyner, "CRAMRA: The Ugly Duckling of the Antarctic Treaty System?" in *The Antarctic Treaty System in World Politics*, ed. A. Jørgensen-Dahl and W. Østreng (New York: Macmillan, 1991), 161, 170–73.

76. "Press Release from the Prime Minister for Australia: Joint Statement with the Minister for Foreign Affairs and Trade, Senator Gareth Evans QC, and the Minister for Arts, Sport, the Environment, Tourism & Territories, Senator the Hon. Graham Richardson," May 22, 1989. See David Scott, "Australia Advocates Wilderness Status for Antarctica," *Christian Science Monitor*, May 24, 1989, 4; and Malcolm W. Browne, "France and Australia Kill Pact on Limited Antarctic Mining and Oil Drilling," *New York Times*, September 25, 1989, A10.

Under Article 62 of the Wellington convention, all seven states that have claims to the continent would have to sign and ratify the convention for it to enter into force. Both Australia and France are claimant states; thus, by their refusal even to sign the treaty, those governments effectively precluded the possibility of the agreement's entry into force.

77. See "Implications of Alaskan Oil Spill for the Antarctic," Antarctic & Southern Ocean Coalition (ASOC) Information Paper No. 1, PREP ATCM XV/ASOC INF. 1 (May 9, 1989).

78. See XV ATCM/WP/2 and XV ATCM/WP/3, in *Final Report of Fifteenth Antarctic Treaty Consultative Meeting* (1989), 202–13.

79. See Docs. XV ATCM/WP/7 (Chile), in *Ibid.* at 227; XV ATCM/WP/4 (New Zealand), in *Ibid.* at 214; XV ATCM/WP/8 (United States), in *Ibid.* at 237; XV ATCM/WP/14 (Sweden), in *Ibid.* at 243.

80. For discussion of policy positions and negotiations during the Viña del Mar meeting, see C. Joyner and B. Ewing, "Antarctica and the Latin American States: The Interplay of Law, Geopolitics and Environmental Priorities," *Georgetown International Environmental Law Review* 4, no. 1 (1991): 33–41.

81. "U.S. Raises Objections to Antarctica Pact," *New York Times,* June 18, 1991.

82. See "U.S. Opposes Antarctic Mining Ban Now," *New York Times,* June 23, 1991, 3.

83. The White House, Office of the Press Secretary, "Statement by the President," July 3, 1991.

84. Christopher C. Joyner interview with R. Tucker Scully, April 30, 1992.

85. Christopher C. Joyner interviews with R. Tucker Scully, April 30, 1992, and Thomas Laughlin, May 1, 1992.

86. Christopher C. Joyner interview with R. Tucker Scully, April 30, 1992.

87. David R. Mayhew, *Congress: The Electoral Connection* (New Haven: Yale University Press, 1974), 5.

88. Senate Committee on Commerce, Science, and Transportation, *Protecting Antarctica's Environment: Hearing before the Subcommittee on Science, Technology and Space,* 101st Cong., 1st sess., 8 September 1989, 5.

89. House Committee on Interior and Insular Affairs, *Antarctica Report—1965 Hearings,* 37.

90. For an insightful treatment of these congressional actions, see Joan Bondareff, "The Congress Acts to Protect Antarctica," *Territorial Sea Journal* 1, no. 2 (1991): 223–44, on which portions of the following analysis rely.

91. Senate Joint Resolution, *Congressional Record,* 101st Cong., 2d sess., 1990, 136:180043.

92. House of Representatives Joint Resolution 418, *Congressional Record,* 101st Cong. 2d sess., 1990, 135: 46746.

93. U.S. Constitution, Article II, section 2, clause 2.

94. See Congressional Research Service, *Treaties and Other International Agreements: The Role of the United States Senate,* study prepared for the Committee on Foreign Relations, United States Senate, 98th Cong., 2d sess., 1984, 10.

95. H.R. 3977, 101st Cong., 2d sess., 1990, section 4.

96. Ibid., sections 9–110.

97. S.R. 2575, 101st Cong., 2d sess., 1990.

98. Ibid., sections 4–5.

99. See *Convention on the Regulation of Antarctic Mineral Resource Activities: Hearing before the Subcommittee on Oversight and Investigations* of the House Committee on Merchant Marine and Fisheries, 101st Cong., 2d sess., 1990.

234 / Notes to Pages 65–71

100. Ibid., 104.
101. Ibid., 2.
102. *Antarctica Briefing with Jacques-Yves Cousteau: Hearing before the Subcommittee on Oceanography and Great Lakes* of the House Committee on Merchant Marine and Fisheries, 101st Cong., 2d sess., 1990.
103. Senate Foreign Relations Committee, *Hearing on Legislation to Protect the Environment of Antarctica*, 101st Cong., 2d sess., 1990, statement of Ambassador Bohlen.
104. Antarctica Treaty—Global Ecological Commons, Public Law 101–620, *U.S. Statutes at Large* 104 (1990): 3340.
105. *U.S. Code*, vol. 42, sections 4321–47 (1985).
106. *Antarctic Protection Act of 1990*, Public Law 101–594, *U.S. Statutes at Large* 104 (1990): 2975.
107. H.R. 4210, 101st Cong., 2d sess., 1990.
108. S.R. 2571, 101st Cong., 2d sess., 1990.
109. A list of relevant congressional hearings and reports on Antarctic issues since 1939 is found in Appendix C.
110. Christopher C. Joyner's interviews with Susan Sabella, Antarctic Campaigner, Greenpeace, May 26, 1992; Raymond Arnaudo, April 30, 1992; Thomas Laughlin, May 1, 1992.
111. In fact, since 1977 a membership slot on the U.S. Antarctic delegation has been explicitly "reserved" for a nongovernmental organization representative.
112. This point is made by M. J. Peterson. See her *Managing the Frozen South*, 165.
113. For an excellent treatment on the national and international activities of Antarctic interest groups, see Lee Kimball, "The Role of Non-Governmental Organizations in Antarctic Affairs," in *The Antarctic Legal Regime*, Christopher C. Joyner and Sudhir K. Chopra, ed. (Hague: Martinus Nijhoff, 1988), 33–63.
114. Christopher C. Joyner interview with Susan Sabella, chief Antarctic Campaigner, Greenpeace USA, May 26, 1992.
115. Christopher C. Joyner interview with Susan Sabella, May 26, 1992.
116. In referring to Greenpeace's plans, Sir Anthony Parsons highlights a point on which there is substantial agreement. He points out that the primary objective of Greenpeace members is not science, "[i]t is political. By maintaining a non-governmental presence they hope to highlight the alternatives open to Treaty governments. It is, in short, classical pressure group politics and although extremely ambitious, not to say risky, is no different in concept from the protester chained to the proverbial railings." See Sir Anthony Parsons, *Antarctica: The Next Decade* (New York: Cambridge University Press, 1987), 40.
117. Christopher C. Joyner interview with Susan Sabella, May 26, 1992. See also Kimball, "Non-Governmental Organizations in Antarctic Affairs," 48–49.
118. Christopher C. Joyner interview with R. Tucker Scully, April 30, 1992.
119. Christopher C. Joyner interview with Thomas Laughlin, May 1, 1992.
120. Kimball, "Non-Governmental Organizations in Antarctic Affairs," 49–50.
121. See Peterson, *Managing the Frozen South*, 165.
122. Ibid.
123. Kimball, "Non-Governmental Organizations and Antarctic Affairs," 51–52.

124. Christopher C. Joyner interviews with R. Tucker Scully, April 30, 1992; and Raymond Arnaudo, April 30, 1992.
125. Christopher C. Joyner interview with Raymond Arnaudo, April 30, 1992.
126. Christopher C. Joyner interview with Thomas Laughlin, May 1, 1992.
127. Christopher C. Joyner interviews with Raymond Arnaudo, April 30, 1992; and Thomas Laughlin, May 1, 1992.
128. Christopher C. Joyner interview with R. Tucker Scully, April 30, 1992. Not surprisingly, interest groups that use a media strategy to develop constituencies among both the public and the press on Antarctic issues tend to be blamed for slanted media coverage. Joyner interview with Susan Sabella, May 26, 1992.
129. Christopher C. Joyner interviews with John B. Talmadge, May 14, 1992; R. Tucker Scully, April 30, 1992; and Thomas Laughlin, May 1, 1992.
130. *AJUS* 12, no. 4 (December 1977): 6.
131. Ibid.
132. Christopher C. Joyner interview with John B. Talmadge, May 14, 1992.
133. Ibid.
134. Christopher C. Joyner interview with Thomas Laughlin, May 1, 1992.
135. Christopher C. Joyner interview with R. Tucker Scully, April 30, 1992.
136. Christopher C. Joyner interview with Raymond Arnaudo on April 30, 1992, and interview with Thomas Laughlin on May 1, 1992.
137. Christopher C. Joyner interview with Thomas Laughlin, May 1, 1992.
138. House Committee on Science and Technology, *U.S. Antarctic Program: Hearings*, 34.

5. Freedom of Scientific Research

1. Ambassador Phleger's statement is in Senate Committee on Foreign Relations, *The Antarctic Treaty: Hearings*, 44.
2. Dr. Zumberge's statement is in Senate Committee on Commerce, Science, and Transportation, *Antarctica: Hearing*, 44.
3. See letter from the president to Admiral Byrd, September 7, 1933, in Samuel Rosenman, ed. *The Years of Crisis*, 2:354–55.
4. For instructive discussions into the scientific achievements of the IGY, see House Committee on Appropriations, *Report on the International Geophysical Year: Hearing before the Subcommittee on Independent Offices*, 86th Cong., 1st sess., February 1959.
5. See Laurence M. Gould, "The U.S.-IGY Program in the Antarctic," in *Geophysics and the IGY: Proceeding of the Symposium at the Opening of the International Geophysical Year*, ed. Hugh Odishaw and Stanley Ruttenberg (Washington, D.C.: American Geophysical Union, 1958), 203.
6. See House Committee on Interstate and Foreign Commerce, *International Geophysical Year, the Arctic and Antarctica: Report submitted by the Committee on Interstate and Foreign Commerce to the Committee of the Whole House*, 85th Cong., 2d sess., 17 February 1958, 11. The United States established and maintained six IGY stations in the Antarctic. Ibid., 97.
7. The U.S.-IGY funds expended in Antarctica (salaries and instruments) were about $4.6 million, but this figure does not include shipboard oceanography, the meteorology at three IGY stations and McMurdo Naval Air Facility, the biology and medical sciences, and the deep drill holes at Byrd and Little America Stations. All of these were funded by offices in the navy, army, or air force. The

total science budget, still excluding the main logistics items such as ships and aircrafts, was about $10 million. See Crary, "International Geophysical Year," 5.

8. House Committee on Interstate and Foreign Commerce, *International Geophysical Year, the Arctic and Antarctica: Report*, 12.

9. Ibid.

10. The statement of R. Tucker Scully, director, Office of Oceans and Polar Affairs, is in Senate Committee on Commerce, Science and Transportation, *Antarctica: Hearing*, 2.

11. Memorandum 6646 is reprinted in National Science Foundation, *Facts about the United States Antarctic Program*," NSF 92-134 (October 1994), 26. Also see "U.S. Antarctic Program Receives Presidential Support," *AJUS* 17, no. 1 (March 1982): 4.

12. Memorandum 6646, reprinted in *Facts about the United States Antarctic Program*," 26.

13. Statement of Robert H. Rutford, in House Committee on Science, Space, and Technology, *NSF Antarctic Environment Act of 1991: Hearing before the Subcommittee on Science*, 102d Cong., 1st sess., 14 May 1991, 134.

14. See statement of John T. McNaughton in House Committee on Interior and Insular Affairs, *Antarctica Report—1965: Hearings*, 78.

15. Statement of Peter Wilkniss, in House Committee on Science, Space and Technology, *NSF Antarctic Environment Act of 1991: Hearing*, 28.

16. Statement of A. N. Fowler, deputy and acting director, Division of Polar Programs, National Science Foundation, in Senate Committee on Commerce, Science and Transportation, *Antarctica: Hearing*, 37.

17. Navy helicopters in use in Antarctica are transportable in the Hercules—a combination that allows Americans to operate in many different places in Antarctica.

18. NSF 92-134, *Facts about the U.S. Antarctic Program*, 1.

19. Public Law 81-507; *U.S. Statutes at Large* 64:149. See House Committee on Science and Technology, *A History of Science Policy in the United States, 1940–1985*, background report no. 1, prepared for the Task Force on Science Policy, 99th Cong., 2d sess., September 1986, 94. This study provides absorbing reading of the events that led to the creation of NSF.

20. A former NSF director opined, "In this way our Nation can be assured that the most useful and valuable program will be developed and supported." Statement of Dr. Leland J. Haworth is in House Committee on Interior and Insular Affairs, *Antarctica Report—1965: Hearings*, 105.

21. The division, one of the four divisions of the Geosciences Directorate, is the only component of NSF with responsibility for a geographic region rather than a scientific area. National Science Foundation, "A Long-Range Science Plan for the Division of Polar Programs" (April 1990), i.

22. Enumeration of these functions is drawn from NSF 92-134, *Facts about the U.S. Antarctic Program*, 9.

23. See statement of Leland J. Haworth in House Committee on Interior and Insular Affairs, *Antarctica Report—1965*, 106.

24. Ibid., 11.

25. Because NSF is organized according to scientific disciplines, the DPP may draw on expertise available in other sections whenever more specific scientific advice is needed. Ibid., 12.

26. Ibid.

27. "United States Antarctic Activities: Long-Range Projection, 1965–1970," *AJUS* 1, no. 3. (May/June 1966): 79.

28. See "Plans and Events of the 1965–1966 Summer Season," *AJUS* 1, no. 1 (January/February 1966): 5; and statement by Ambassador Harlan Cleveland, assistant secretary of state for international organization affairs in House Committee on Interior and Insular Affairs, *Antarctica Report—1965: Hearings*, 32. Leland Haworth, former director of the NSF, has indicated that the coordination between the division and NSFA is essential to reconcile differences concerning the provision of logistic support capabilities. Statement of Leland J. Haworth, ibid., 106.

29. The statement of Adm. James S. Gracey, Commandant, U.S. Coast Guard, Department of Transportation, in Senate Committee on Commerce, Science and Transportation, *Antarctica: Hearing*, 83.

30. The logistics support requested by the NSF is provided, on a reimbursable basis, under a "Memorandum of Agreement" between NSF and the Coast Guard dated October 26, 1982. See Gracey, ibid., 84.

31. Statement of Leland J. Haworth in House Committee on Interior and Insular Affairs, *Antarctica Report—1965: Hearings*, 106.

32. A. P. Crary, "Long Range Planning," *AJUS* 3, no. 1 (January/February 1968): 12. Crary was the chief scientist of the Office of Antarctic Programs (now the Division of Polar Programs) from 1960 to 1967.

33. For the role of NOAA in U.S. Antarctic science, see Thomas L. Laughlin, "Antarctic Science Programs: National Atmospheric Administration's Antarctic Activities," in Alexander and Hanson, eds., *Antarctic Politics and Marine Resources*, 65–71; and U.S. Congress, Office of Technology Assessment, *Polar Prospects: A Minerals Treaty for Antarctica*, OTA–O–428 (Washington, D.C.: GPO, September 1989), 27–28.

34. Terry W. Offield, "An Antarctic-Marine Geology Program: Possibilities," in Alexander and Hanson, eds., *Antarctic Politics and Marine Resources*, 68.

35. "Plans and Events of the 1965–1966 Summer Season," *AJUS* 1, no. 1 (January/February 1966): 5.

36. Statement of Robert M. Mangan, deputy undersecretary, Department of the Interior, in House Committee on Interior and Insular Affairs, *Antarctica Report—1965: Hearings*, 136.

37. Dater, "Organizational Developments", 31.

38. Robert Bell, *Impure Science: Fraud, Compromise and Political Influence in Scientific Research* (New York: John Wiley and Son, 1992), 3, 4. As Dr. Bell notes, the peer review system is not without critics. A National Institute of Health scientist whom he interviewed gave perhaps the most scathing comment on the practice. He said: "Your fate is decided by your competitors." Ibid.

39. Ibid.

40. Some projects are completed in several weeks; others are long-term efforts that may continue for years. Investigators can propose to perform research and analysis individually, in small teams, or in large interdisciplinary groups. Prior discussion with a program manager in the Division of Polar Programs can help define research objctives that match the logistic and operational capabilities at any given time.

41. Dr. Haworth has pointed out that introducing into the program the first full-time Antarctic research vessel has given the United States a unique capability to explore the rich ocean regions surrounding Antarctica. At that time, he also

238 / Notes to Pages 85–87

noted that "the United States is the only country engaged in such a program of research and exploration of the southern oceans." See House Committee on Interior and Insular Affairs, *Antarctica Report—1965: Hearings*, 107.

42. The publication of AJUS seems to be chronically delayed. As of March 1996, no 1995 issues had been published.

43. See *AJUS* 10, no. 4 (July/August 1975): 203.

44. The National Academy of Sciences, established in 1863, was designated by President Lincoln as a private, nonprofit, self-governing corporation.

45. Its establishment reflected the academy's recognition of the growing national importance of research in the polar regions. Polar Research Board, "Annual Report 1978 and Future Plans" (issued February 1979), i.

46. Ibid.

47. W. Timothy Hushen, "Antarctic Science—The Role of the Polar Research Board," in Alexander and Hanson, eds., *Antarctic Politics and Marine Resources*, 71. Hushen is former executive secretary of the board.

48. In 1916, the academy established the National Research Council. The council is the principal operating agency of the National Academy of Sciences, National Academy of Engineering, and Institute of Medicine. Its mission is to join the broad community of science and technology and to advise the federal government.

49. Polar Research Board, "Annual Report 1978," i.

50. Board members are appointed for four-year terms by the chairman of the Assembly of Mathematical and Physical Sciences with the approval of the chairman of the National Research Council. Appointments are staggered so that approximately one-fourth of the membership is rotated each year. Board members are appointed on the basis of their personal professional qualifications and do not represent any group or institution. Polar Research Board, *Annual Report 1988 and Future Plans*, 2.

51. Hushen, "Antarctic Science: The Role of the Polar Research Board," 71.

52. The discussion on the Advisory Committee draws from: *AJUS* 13, no. 4 (December 1978): 9; and *AJUS* 12, no. 4 (December 1977): 6–7.

53. The Advisory Committee was created under the provisions of the Federal Advisory Committee Act (P.L. 92-643). See Library of Congress, Congressional Research Service, *The Role of Advisory Committees in U.S. Foreign Policy*, report prepared for the Senate Committee on Foreign Relations and the House Committee on International Relations, 94th Cong., 1st sess., April 1975.

54. The charter of the committee places membership at about fifty-five persons, selected from the polar research community. Membership terms run for one year but can be extended by two additional one-year terms.

55. The ICSU is an international, nongovernmental umbrella organization, whose principal objective is to encourage international scientific activity. ICSU membership is composed of twenty Scientific Unions, representing various scientific disciplines, and national bodies such as academies or research councils in approximately seventy countries. Since its creation in 1931, the ICSU fulfills its objectives by initiating, designing, and coordinating international interdisciplinary research.

56. All original signatories to the treaty are also members of SCAR. Each member state is represented in SCAR by one scientific delegate.

57. Polar Research Board, "Annual Report 1978," 5.

58. Ibid. SCAR's headquarters are at the Scott Polar Research Institute in

Cambridge, England, but the place of its biennial meetings rotates among member states.
59. Ibid.
60. Statement of Dr. James Zumberge in Senate Committee on Commerce, Science and Transportation, *Antarctica: Hearing*, 46.
61. Scientific matters for SCAR are handled in nine of ten working groups: Biology, Geodesy and Cartography, Geology, Glaciology, Human Biology and Medicine, Meteorology, Oceanography, Solid Earth Geophysics, and Upper Atmosphere Physics. A tenth working group, Logistics, is an exception to the generalization that SCAR deals only in scientific affairs. Ibid.
62. Charles R. Bentley, "International Science Programs in Antarctica," in Alexander and Hanson, eds., *Antarctic Politics and Marine Resources*, 45.
63. Ibid., 46. Bentley states that at present there are five groups of specialists: (1) Antarctic Climate Research; (2) Antarctic Environmental Implications of Possible Mineral Exploration and Exploitation; (3) Seals; (4) Southern Ocean Ecosystems and Their Living Resources; and (5) Antarctic Sea Ice.
64. Ibid., 45.
65. The discussion that follows is based on: Council of Managers of National Antarctic Programs (COMNAPS), "Report to the XVI Antarctic Treaty Consultative Meeting," held in Bonn, October 1991; and Ethel R. Theis's interview with Al Fowler, executive secretary of the council, April 14, 1992.
66. Peter Wilkniss was instrumental in establishing COMNAPS. Ethel R. Theis's interview with Al Fowler, April 14, 1992. Following establishment of COMNAPS, SCAR disbanded the Working Group on Logistics.
67. COMNAPS, "Report to the XVI Antarctic Treaty Consultative Meeting," 1.
68. Ibid., 19.
69. Roger H. Davidson and Walter J. Oleszek, *Congress and Its Members*, 3d ed. (Washington, D.C.: Congressional Quarterly Inc., 1990), 377.
70. Becoming effective in 1975, the Budget and Impoundment Control Act of 1974 requires executive branch agencies to submit to Congress a "current services budget" by November 10 for the new fiscal year that starts the following October 1. David J. Ott and Attiat F. Ott, *Federal Budget Policy* (Washington, D.C.: The Brookings Institution, 1977), 25.
71. Frank G. Klotz, *America on the Ice: Antarctic Policy Issues* (Washington, D.C.: National Defense University Press, 1990), 143.
72. Ibid.
73. Even so, unexpected events such as emergencies, miscalculations in estimates, and inflationary costs for goods and services can generate substantial increases in operating expenses.
74. Statement of Richard Atkinson in House Committee on Science and Technology, *U.S. Antarctic Program: Hearings*, 98.
75. Ibid.
76. The USAP budget has tended to increase every year since the late 1970s. The fiscal year 1990 budget amounted to $151.47 million; $71.76 million was for research activities and $79.71 million for logistic support. *National Science Foundation Annual Report 1990*, 53. Appendix G provides tables listing USAP funding since 1955.
77. June Wakefield and Susan Bucci Mockler, "Small Expectations from Congress for U.S. Science Enterprise," *EOS, Transactions, American Geophysical Union* (January 17, 1995): 27.

78. For starters, the U.S. Geological Survey, the U.S. Bureau of Mines, and the National Biological Survey top the ax list. Ibid.

79. Geophysical Science in the 1996 Budget: NSF Gets Qualified Good News," *EOS, Transactions, American Geophysical Union* (February 21, 1995): 77.

80. Ibid.

81. Ibid.

82. Ibid.

83. House Committee on Interior and Insular Affairs, *Antarctica Report— 1965: Hearings*, 27.

84. Dr. Atkinson's remarks are in House Committee on Science and Technology, *U.S. Antarctic Program: Hearings*, 76.

85. Ibid., 77.

86. Ibid., 77–78.

87. Quigg has been generally critical of NSF's approach to Antarctic research. He notes that those who administer USAP "seem skeptical of the value of multinational projects or simply find them a nuisance." He goes on to assert that a number of such projects "often have a kind of hands-across-the-seas artificiality or they fulfill exchange obligations. Rarer now is the selection of a foreign scientist simply because he is the most qualified person to fill a particular area of expertise on a U.S. research team. No doubt there are exceptions, but this tendency toward pro forma cooperation in international science represents a loss." See *A Pole Apart*, 217.

88. *Role of the National Science Foundation in Polar Regions: A Report to the National Science Board*, NSB-87-128, June 1987, 7.

89. Statement of Adm. James S. Gracey, Commandant, U.S. Coast Guard, Department of Transportation, in Senate Committee on Commerce, Science and Transportation, *Antarctica: Hearing*, 85.

90. *Role of the National Science Foundation in Polar Regions*, 32.

91. Ibid., 31.

92. Ibid.

93. Glaciology, the science of all major forms of naturally occurring ice (floating ice, glaciers and ice sheets, seasonal ice, and ice in the ground) is particularly appropriate for Antarctic programs. The general discussion on glaciology is based on: *Role of the National Science Foundation in Polar Regions*, 27–31; and statement of Al Fowler, in Senate Committee on Commerce, Science and Transportation, *Antarctica: Hearing*, 19, 27.

94. Fowler, ibid., 19, 28.

95. The discussion of meteorology draws from *Role of the National Science Foundation in Polar Regions*, 15–19.

96. Statement of Representative Boucher, in House Committee on Science, Space, and Technology, *NSF Antarctic Environment Act of 1991: Hearing*, 3.

97. *Role of the National Science Foundation in Polar Regions*, 19.

98. The discussion on oceanography is based on: *Role of the National Science Foundation in Polar Regions*, 19–23; and comments of A. Fowler in Senate Committee on Commerce, Science and Transportation, *Antarctica: Hearing*, 18, 30, 31.

99. The discussion of polar marine biology is based on: *Role of the National Science Foundation in Polar Regions*, 34–38; statement of Representative Rich Boucher of Virginia, in House Committee on Science, Space, and Technology, *NSF Antarctic Environment Act of 1991: Hearing*, 4; and comments of A. Fowler in Senate Committee on Commerce, Science and Transportation, *Antarctica: Hearing*, 24.

100. Statement of John B. Byrne, administrator, National Oceanic Atmospheric Administration in Senate Committee on Commerce, Science and Transporation, *Antarctica: Hearing*, 44.

101. American scientists actively participated in the BIOMASS program, giving it great stimulus and adding much to the successful completion of two expeditions. See statements of R. Tucker Scully and Robert J. Hofman, in Senate Committee on Commerce, Science and Transportation, *Antarctica: Hearing*, 79 and 16–17, respectively; and Bentley, "International Science Programs in Antarctica," 47.

102. Bentley, ibid.

103. NOAA participated in Phase I in cooperation with researchers sponsored by National Science Foundation. Ibid. 43, 44.

104. Thomas L. Laughlin, "National Oceanic Atmospheric Administration's Antarctic Activities," 66.

105. The discussion on U.S. Antarctic stations draws from: NSF, *Facts about the U.S. Antarctic Program*, (October 1994) 1–4; NSF, *The United States Antarctic Program*, NSF 90–97, 20–21; and *AJUS*, 1, no. 3 (May/June 1966): 79–86.

106. Statement of Leland J. Haworth, in House Committee on Interior and Insular Affairs, *Antarctica Report–1965: Hearings*, 108.

107. Ibid.

108. The South Pole Station is situated at an elevation of 9,301 feet, having an ice thickness in excess of 9,000 feet at that location. According to Guy G. Guthridge, the remoteness of the site and the extreme environment made the design and construction of the South Pole Station difficult and costly. The total direct cost of re-building the station was about $6 million. See "A New Research Station at the South Pole," *AJUS* 10, no. 2 (March/April 1975): 37–44.

109. The South Pole is an ideal location for research in astronomy because the sun circles above the horizon for weeks and the cold, clean, dry atmosphere permits continuous observations for days at a time. Telescopic observations from the South Pole have confirmed internal characteristics of the sun that previously had been predicted only by mathematical models.

110. The comments about NOAA's scientific activities are drawn substantially from Laughlin, "Antarctic Science Programs," 65–68.

111. Measurements since 1974 have shown gradual long-term increases of climatologically active gases.

112. An excellent discussion of the scientific research conducted at Palmer Station is in Guy G. Guthridge's, "Palmer: An Antarctic Peninsula Research Station," *AJUS* 16, no. 4 (December 1981): 1–6.

113. See NSF 92–134, *Facts about the U.S. Antarctic Program*, 1–4; and NSF, *U.S. Antarctic Program*, 22.

114. Statement of R. Tucker Scully, director, Oceans and Polar Affairs, Department of State, in Senate Committee on Commerce, Science and Transportation, *Antarctica: Hearing*, 3.

115. Article III ensures this cooperation by calling for the transfer of scientific personnel between expeditions and stations and the exchange and free availability of scientific observations and results.

116. R. Tucker Scully, "The Antarctic Treaty System: Overview and Analysis," in Alexander and Hanson, eds., *Antarctic Politics and Marine Resources*, 6.

117. The Polar Research Board endorsed the scientific objectives of the international BIOMASS program and found that the studies proposed were essential to effective guidance in addressing the problems certain to arise from exploi-

tation of living marine resources of the Southern Ocean. See their "Annual Report 1978," 12.

118. CIA, *Polar Regions Atlas*, 46.
119. Scully, "The Antarctic Treaty System," 8, 9.
120. Grahame Cook, *The Future of Antarctica: Exploitation versus Preservation* (Manchester: Manchester University Press, 1990), vii.
121. Sisco, "The United States Program for Antarctica," 4.
122. Protocol on Environmental Protection to the Antarctic Treaty, XI Special Consultative Meeting in Madrid, Doc. XI ATSCM/2, June 21, 1991, adopted 4 October 1991, Article 12, para. 3.

6. Environmental Interests

1. *U.S. Code*, vol. 42, secs. 4321–61 (1976), Public Law 91–190 (1970). NEPA has been called "an environmental bill of rights." See Comment, "NEPA's Role in Protecting the World Environment," *University of Pennsylvania Law Review* 131, no. 2 (December 1982): 354.
2. The Marine Mammal Protection Act, *U.S. Code*, vol. 16, secs. 1361–1407 (1976); and the Endangered Species Act, *U.S. Code*, vol. 16, secs. 1531–43 (1976) represent U.S. strategies to preserve creatures whose existence is threatened by human activities.
3. David A. Colson, "The United States Position on Antarctica," 292. Ambassador Colson reiterated this position during an interview with Ethel R. Theis on October 15, 1991.
4. Peterson, *Managing the Frozen South*, 74.
5. Ibid.
6. The first step was taken at the first consultative meeting in 1961. With recommendation I-VIII, the ATCPs recognized the need for conservation in the treaty area and agreed, as interim measures, to (1) place limits on killing of indigenous animals and plants; (2) prevent introduction of alien forms of fauna and flora; and, (3) prevent harm to wildlife from the conduct of operations. See David Anderson, "The Conservation of Wildlife Under the Antarctic Treaty," *Polar Record* 14, no. 88 (January 1968): 25. Conservation issues took center stage at the fourth ATCP meeting (1966). Of the twenty-eight recommendations adopted, twenty-two dealt with conservation of Antarctic animal and plant life. See Peter Roberts, "Fourth Consultative Meeting Under the Antarctic Treaty," *Antarctic Journal of the United States (AJUS)* 2, no. 1 (January/February 1967): 13.
7. The ATCPs adopted initial measures to control tourism at the 1966 ATCP meeting. Roberts, "Fourth Consultative Meeting," 14. For an insightful look into man's impact on the Antarctic environment, see W. S. Benninghoff and W. N. Bonner, *Man's Impact on the Antarctic Environment: A Procedure for Evaluating Impacts from Scientific and Logistic Activities* (Cambridge: SCAR, 1985).
8. See Shapley, *The Seventh Continent*, 105.
9. At that time, only three other countries had ratified all of the recommendations: Belgium, New Zealand, and South Africa. "United States Now Has Ratified All Antarctic Treaty Recommendations," *AJUS* 14, no. 3 (September 1979): 20.
10. Statement of R. Tucker Scully, director, Office of Oceans and Polar Affairs, Department of State, in House Committee on Interior and Insular Affairs, *Establish an Antarctica World Park: Hearing*, 5. The list of recommendations

adopted at the first seventeen ATCP meetings is in Heap, ed., *Handbook of the Antarctic Treaty System*, 1-6.

11. "NSF Reports Focus on Improving USAP Environmental Practices," *AJUS* 25, no. 2 (June 1990): 1.

12. Testimony of Susan Sabella (Greenpeace) in House Committee on Foreign Affairs, *Preserving Antarctica's Ecosystem: Hearing before the Subcommittee on Human Rights and International Organization*, 101st Cong., 2d sess., 2 May and 19 July, 1990, 103.

13. Ibid.

14. See "Bonn Meeting Stresses Environmental Stewardship and International Cooperation," *AJUS* 27, no. 1 (March 1992): 4.

15. Shapley, *The Seventh Continent*, 105.

16. House Committee on Foreign Affairs, *Preserving Antarctica's Ecosystem: Hearing*, 169.

17. R. M. Laws, ed., *Antarctica: The Last Frontier* (London: Boxtree, 1989), 198-99.

18. The text of the Agreed Measures is Appendix A of the National Science Foundation, *Antarctic Conservation Act of 1978 (Public Law 95-541) with Regulations, Maps of Special Areas, and Application Forms*, NSF 89-59 (June 1989).

19. In presenting its views on H.R. 7749, the Department of State mentioned this fact. See House Committee on Science and Technology, *Antarctic Conservation Act of 1978*, report to accompany H.R. 7749, 95th Cong., 2d sess., 18 May 1978, 11.

20. National Science Foundation, Division of Polar Programs, Office of Safety, Environment and Health, *Final Supplemental Environmental Impact Statement for the U.S. Antarctic Program (Final SEIS)* (October 1991), 4-20.

21. See Heap, ed., *Handbook of the Antarctic Treaty System*, 2131-83.

22. W. N. Bonner, "Recent Developments in Antarctic Conservation," in *The Antarctic Treaty Regime: Law, Environment and Resources*, ed. Gillian D. Triggs (Cambridge: Cambridge University Press, 1987), 144.

23. Heap, ed., *Handbook of the Antarctic Treaty System*, 2184-2242.

24. Christopher C. Joyner and Ethel R. Theis, "The United States and Antarctica: Rethinking the Interplay of Law and Interests," *Cornell International Law Journal* 20 (winter 1987): 65, 87.

25. Disturbing for some was that the Agreed Measures had not become effective sixteen years after their unanimous recommendation by the ATCPs. No particular urgency attached to the approval of the Agreed Measures because they were implemented in practice on a voluntary basis. See Auburn, *Antarctic Law and Politics*, 235-36.

26. T. O. Jones, "U.S. Antarctic Research Program," *AJUS* 3, no. 4 (July/August 1968): 79.

27. Ibid. Quigg notes that the Agreed Measures are "so sweeping and place such severe constraints on the citizens of democracies that most of the participating nations had to enact special legislation to bring themselves into legal compliance. The United States, while effectually complying with the Agreed Measures, dawdled for fourteen years before the Antarctic Conservation Act of 1978 was adopted." See Quigg, *A Pole Apart*, 159.

28. Jones, "U.S. Antarctic Research Program," 80.

29. Quigg, *A Pole Apart*, 160.

30. The text of the convention is in Heap, ed., *Handbook of the Antarctic Treaty System*, 156-61.

31. As M. J. Peterson points out, the Seals Convention was the first of the ATCPs' ventures into regulating the exploitation of Antarctic marine resources. See her *Managing the Frozen South*, 103–4.

32. Ibid., 104.

33. Ibid.

34. For parties to the convention, see U.S. Department of State, *Treaties in Force*, January 1, 1994, 311. Also see Patsy T. Mink, "Oceans: Antarctic Resource and Environmental Concerns," *Department of State Bulletin* 78, no. 2013 (April 1978): 53; and "U.S. Ratifies Seal Convention," *AJUS* 12, nos. 1 & 2 (March/June 1977): 50.

35. Mink, ibid. Of the original twelve states party to the Antarctic Treaty, only New Zealand is not a party to the Seals Convention.

36. Cook, *The Future of Antarctica*, 3.

37. Gregory M. Travalio and Rebecca J. Clement, "International Protection of Marine Mammals," *Columbia Journal of Environmental Law* 5, no. 2 (spring 1979): 220.

38. Donald B. Siniff, "Antarctic Marine Living Resources: Seals," *Oceanus* 31, no. 2 (summer 1988): 72. One of the present authors has suggested that such a precautionary strategy under the Antarctic Treaty System may be designated "preclusive restoration." See Christopher C. Joyner, "Fragile Ecosystems: Preclusive Restoration in the Antarctic," *Natural Resources Journal* 34 (fall 1994): 879–904.

39. Joyner, "Fragile Ecosystems," 74.

40. On February 6, 1978, Patsy Mink, former chairwoman of the APG, said that American interests would be best served by negotiation of an international convention to establish a conservation regime for Antarctic marine living resources. See *American Foreign Policy Basic Documents, 1977–1980*, Department of State Publication 9330, Document 152 (1983), 384.

41. Joseph Macknis, "United States Policy in Antarctica," *Marine Policy Reports* 2, no. 2 (May 1979). (The reproduction of the article does not show page numbers.)

42. Department of State, Press Release 224 of September 14, 1979, in *Department of State Bulletin* 79, no. 2032 (November 1979): 21.

43. Ibid., 22.

44. Mink, *American Foreign Policy Basic Documents, 1977–1980*, 384.

45. Christopher C. Joyner, "Maritime Zones in the Southern Ocean: Problems Concerning the Correspondence of Natural and Maritime Zones," *Applied Geography* 10, no. 4 (October 1990): 316.

46. Martin H. Belsky, "Marine Ecosystem Model: The Law of the Sea's Mandate for Comprehensive Management," in *New Developments in Marine Science and Technology: Economic, Legal and Political Aspects of Change*, ed. Lewis M. Alexander and Lynne Carter Hanson (Narrangansett, R.I.: Proceedings of the 22d Annual Conference of the Law of the Sea Institute, June 12–16, 1988), 115. An ecosystem can be defined as the pattern of relationships between all biotic (living) and abiotic (nonliving) entities within a defined boundary of space and time. Ibid.

47. "Rational use" occurs only when harvesting does not reduce populations below levels necessary for ensuring sustainable yields. See Robert Friedheim and Tuseneo Akaha, "Antarctic Resources and International Law: Japan, the United States and the Future of Antarctica," *Ecology Law Quarterly* 16, no. 1 (1989): 130. The ecosystem standard represents a major contribution to conservation. It re-

quires a long-term commitment to support a long-term scientific program and a willingness to curtail harvesting until sufficient scientific data becomes available to reach sound scientific decisions. See Belsky, "Marine Ecosystem Model," 115.

48. Ibid.

49. Ibid.

50. In 1996, members of CCAMLR's governing Commission include Argentina, Australia, Belgium, Brazil, Chile, European Economic Community, France, Germany, India, Italy, Japan, Republic of Korea, New Zealand, Norway, Poland, Russian Federation, South Africa, Spain, Sweden, United Kingdom, and the United States. States that have acceded to the Convention, but who are not members of the Commission, include Bulgaria, Canada, Finland, Greece, Netherlands, Peru, Ukraine, and Uruguay.

51. In December 1981, the U.S. Senate gave its consent to ratification, and President Reagan signed the instrument of ratification on February 2, 1982. That instrument was conveyed to the government of Australia, the depository government, on February 18, 1982. See "Statement by Press Secretary Speakes on U.S. Antarctic Policy" (March 29, 1982), in *Public Papers of the Presidents of the United States: Ronald Reagan*, 392; and "April 1982 Marks Beginning of Antarctic Marine Conservation Regime," *AJUS* 17, no. 1 (March 1982): 4. The decisive eighth ratification was made by New Zealand in March 1982.

52. Heap, ed., *Handbook of the Antarctic Treaty System*, 178. The full text of the CCAMLR is reprinted in the *Handbook*, 178–88.

53. Not surprisingly, the CCAMLR is considered a model agreement in its approach to conservation and management of marine resources. Kenneth Sherman and Alan F. Ryan, "Antarctic Marine Living Resources, *Oceanus* 31, no. 2 (summer 1988): 59, 60.

54. Auburn, *Antarctic Law and Politics*, 239.

55. Ibid.

56. Convention on the Regulation of Antarctic Mineral Resource Activities, *opened for signature* November 25, 1988, 27 I.L.M. 859 [hereinafter CRAMRA]. The United States signed the convention on November 30, 1988. See *ILM* (*International Legal Materials*) 27 (1988), 868–900. For treatments of negotiating the convention, see the following articles by Christopher C. Joyner: "The Evolving Antarctic Minerals Regime," *Ocean Development and International Law* 19 (1988): 73–95; "1988 Antarctic Minerals Convention," *Marine Policy Reports* 1 (1989): 81–98; and "The Antarctic Minerals Negotiating Process," 888–905.

57. "U.S. Signature of the Antarctic Mineral Resource Convention," statement read by the Department of State Acting Spokesman (Oakley), December 2, 1988, reprinted in Department of State, *American Foreign Policy Current Documents 1988* (Washington, D.C., 1989), Doc. 79, 197.

58. Cook, *The Future of Antarctica*, 6.

59. Browne, "France and Australia Kill Pact."

60. Even so, France's withdrawal of support for CRAMRA appears disingenuous, considering that its airstrip project in the Antarctic remained a prime example of environmental protection taking a back seat to logistical considerations. See statement of Susan Sabella, Greenpeace, in House Committee on Foreign Affairs, *Preserving Antarctica's Ecosystem: Hearing*, 103.

61. Quigg, *A Pole Apart*, 196.

62. Ibid.

63. Ibid.

64. This sentiment was expressed by all of the Department of State and NSF officials interviewed by the authors.

65. See R. Tucker Scully, "Institutionalization of the Antarctic Treaty Regime," in *Antarctic Challenge II*, ed. Rudiger Wolfrum (Berlin: Duncker and Humblot, 1986), 283–96.

66. See, for example, Quigg, *A Pole Apart*, 197.

67. Ibid.

68. "Preserving Antarctica: Treaty Nations Agree on New Environmental Protocol," *AJUS* 26, no. 4 (December 1991): 1. The provisions of the protocol are covered in twenty-seven articles; it also includes four annexes and three appendices. The full text of the protocol and its annexes (but not the appendices) appears in the December 1991 issue of *AJUS*. A draft for a fifth annex to the protocol was introduced at the sixteenth ATCP meeting. See "Bonn Meeting Stresses Environmental Stewardship and International Cooperation," *AJUS* 27, no. 1 (March 1992): 4.

69. "Preserving Antarctica," 1. See also the discussion here in chapter 8.

70. See, for example, House Committee on Merchant Marine and Fisheries, *Antarctic Treaty Protocol on Environmental Protection: Hearing before the Subcommittees on Oceanography, Great Lakes, and the Outer Continental Shelf, Coast Guard and Navigation, and Fisheries and Wildlife Conservation & the Environment*, 102d Cong., 2d sess., 30 June 1992; House Committee on Science, Space and Technology, *The Antarctic Environmental Protection Act of 1993: Hearing before the Subcommittee on Science*, 103d Cong., 1st sess., 23 Feb. 1993; House Committee on Foreign Affairs, *Implementing Legislation for the Protocol on Environmental Protection to the Antarctic Treaty: Joint Hearings before the Subcommittee on Economic Policy, Trade and Environment and the Committee on Merchant Marine and Fisheries*, 103d Cong., 1st sess., 16 November 1993; House Committee on Science, Technology and Space, *H.R. 3532—The Antarctic Environmental Protection Act: Hearing before the Subcommittee on Science, Space, and Technology*, 103d Cong., 2d sess., 8 February 1994; Senate Committee on Foreign Relations, *Protocol on Environmental Protection to the Antarctic Treaty (Treaty Doc. 102–22): Hearing before the Committee on Foreign Relations*, 102d Cong., 2d sess., 4 May 1992; Senate Committee on Commerce, Science and Transportation, *S. 1427, The Antarctic Scientific Research, Tourism, and Marine Resources Act of 1993, to Implement the Protocol on Environmental Protection to the Antarctic Treaty: Hearing before the Committee on Commerce, Science and Transportation*, 103d Cong., 1st sess., 20 October 1993.

71. See Mr. Scully's statement in Senate Committee on Commerce, Science, and Transportation, *Antarctica: Hearing*, 2.

72. "Remarks following a meeting with members of the Antarctic Policy Group," (May 20, 1965) in *Public Papers of the Presidents of the United States—Lyndon B. Johnson*, vol. 1 (1965), 564.

73. White House Press Release, October 13, 1970, reproduced in Senate Committee on Foreign Relations, *U.S. Antarctic Policy: Hearing before the Subcommittee on Oceans and International Environment*, 94th Cong., 1st sess., 15 May 1975, 30.

74. Mink, *American Foreign Policy Basic Documents, 1977–1980*, 382.

75. "Parties to Antarctic Treaty Meet in London," address by Robert C. Brewster, former deputy assistant secretary for oceans and international environmental and scientific affairs, Department of State, at the opening session of the ninth ATCP meeting in London on September 19, 1979. *Department of State Bulletin* 79, no. 2032 (November 1979): 738.

76. See R. Tucker Scully's statement in Senate Committee on Commerce, Science, and Transportation, *Antarctica: Hearing*, 2.

77. "Successes of the Antarctic Treaty," address by John Negroponte, former assistant secretary for oceans and international and scientific affairs, Department of State, before the eleventh annual seminar for the Center for Oceans Law and Policy, March 27, 1987, reprinted in Department of State, *American Foreign Policy: Current Documents 1987* (1988), Document 92, 189.

78. Ms. Mullin was then assistant secretary for legislative affairs, Department of State. She made this statement on behalf of President Bush in reply to a letter from Congressman Morris K. Udall, cosigned by five of his colleagues, to the president on the future of Antarctica. The letter is reprinted in House Committee on Merchant Marine and Fisheries, *Convention on Regulation of Antarctic Mineral Resource Activities: Hearing*, 48.

79. In February 1993 President Clinton announced his intention to abolish the council.

80. *Federal Register* 44, 32699 (June 7, 1979).

81. Ibid.

82. See "NSF Issues a Directive on Pollution Control Provisions of the Antarctic Conservation Act," *AJUS* 18, no. 2 (June 1983): 8.

83. Directives, unlike regulations, are not required to be published in the *Federal Register* following notice and public comment. For this reason, NSF reprinted the directive, along with a brief explanation, in the *Antarctic Journal of the United States*, an NSF periodical. See Bruce S. Manheim, Jr., "On Thin Ice: The Failure of the National Science Foundation to Protect Antarctica," (Washington, D.C.: Environmental Defense Fund, 1988).

84. Greenpeace, for example, has persistently criticized NSF's performance as the lead agency on Antarctic environmental matters and has questioned whether NSF is the agency best suited to address such issues. To Greenpeace, NSF does not have the tools nor the expertise to discharge environmental protection and pollution control responsibilities under the 1978 Antarctic Conservation Act. See House Committee on Science, Space and Technology, *NSF Antarctic Environment Act of 1991: Hearing*, 115.

85. Ibid., statement of Robert H. Rutherford, 138.

86. As noted earlier, in recognition of the vulnerability of the Antarctic marine ecosystem to unregulated commercial harvesting of living resources, executive branch officials through the Antarctic Policy Group advocated during the late 1970s negotiation of an international agreement establishing a conservation regime. Macknis, "United States Policy in Antarctica."

87. President's Reorganization Act No. 4, July 9, 1970. The creation of NOAA was intended to provide centralized ocean planning. NOAA's mission includes responsibility for directed and applied research to support marine resource management. Its areas of direct responsibility include fisheries, marine mammals (in conjunction with the Fish and Wildlife Service), marine and estuarine pollution.

88. House Committee on Government Operations, *Reorganization Plan No. 4: Hearings before the Subcommittee on Government Operations*, 91st Cong., 2d sess., 1970, 8.

89. On November 8, 1984, President Reagan signed Public Law 98-623, Title III, dealing with the Antarctic Marine Living Resources Convention. See "Marine Living Resources Act Becomes Law" *AJUS* 19, no. 4 (December 1984): 11.

90. Ibid., 12.

91. M. J. Peterson has pointed out that the lack of interest in Southern Ocean fisheries among U.S. fishing fleets gave "the conservationist-oriented Marine Mammals Commission more influence than it might otherwise have." This situation is likely to change as the more "use-oriented" NOAA takes over implementation of the CCAMLR. Peterson notes that though NOAA supports fairly strong management, "it has never supported a ban on fishing before the survival of a particular stock appeared to be in immediate danger." Peterson, *Managing the Frozen South*, 179.

92. On international issues, the EPA coordinates with State, Treasury, Commerce, as well as the White House and various technical agencies like NOAA and USGS. See Marc K. Landy, Marc J. Roberts, and Stephen R. Thomas, *The Environmental Protection Agency: Asking the Wrong Questions—From Nixon to Clinton*, expanded ed. (New York: Oxford University Press, 1994), 328.

93. Norman J. Vig and Michael E. Kraft, *Environmental Policy in the 1990s: Reagan's New Agenda* (Washington, D.C.: Congressional Quarterly Press, 1990), 17.

94. See Glenn Pincus, "The 'NEPA-Abroad' Controversy: Unresolved by an Executive Order," *Buffalo Law Review* 30, no. 3 (summer 1981): 612.

95. NEPA, sec. 102(2)(c), 42 *U.S. Code*, vol. 42, sec. 4332(2)(c) (1976). The EIS was required to provide sufficient evidence and analysis of the impact of a proposed action and evaluate environmental effects of alternatives to the proposed action. See Thomas O. McGarity, *Reinventing Rationality: The Role of Regulatory Analysis in the Federal Bureaucracy* (New York: Cambridge University Press, 1991), 17.

96. Charles N. Brower, "Is NEPA Exportable?" *Albany Law Review* 43, no. 3 (spring 1979): 517. The Department of State was the first foreign affairs agency to declare the EIS procedures inapplicable to its activities in foreign nations. See J. D. Head, "Federal Agency Responsibility to Assess Extraterritorial Environmental Impacts," *Texas International Law Journal* 14, no. 3 (summer 1979): 432.

97. Testimony of Robert Corell, assistant director for geosciences, NSF, in House Committee on Science, Space, and Technology, *NSF Antarctic Environment Act of 1991: Hearings*, 18.

98. Another major objection pointed out by Peter Wilkniss, former director of the Office of Polar Programs, is the inhibiting effect on international cooperation in Antarctic scientific research that lawsuits brought under NEPA could have. Ibid., 29.

99. Brower, "Is NEPA Exportable?" 518.

100. Executive Order No. 12114, "Environmental Effects Abroad of Major Federal Actions," 3 C.F.R. 356 (1980), reprinted in *U.S Code*, vol. 42, sec. 4321 app., at 1604.

101. *U.S. Code*, vol. 42, sec. 1–1.

102. Ibid., secs. 2–3(a), 2–4(a)(i), 2–4(b)(i).

103. Ibid., secs. 2–3(d), 2–4(a)(i), (ii), (iii), 2–4(b)(iv), 3 C.F.R. 357–58.

104. Ibid.

105. Ibid.

106. "Final Environmental Impact Statement for USARP Published," *AJUS* 15 (June 1980): 4.

107. *Final SEIS*, 1–1.

108. Ibid., 1–10 (also in *Federal Register* 56, 2777). The NSF solicited comments from the public, environmental organizations, and federal agencies. It also

circulated the draft SEIS to individuals, organizations, and agencies for comment and provided copies to ATCPs as part of the annual treaty information exchange.

109. *Final SEIS*, 1–1.

110. Ibid., 1–4.

111. The NSF Office of General Counsel reviewed the applicability of U.S. environmental laws to USAP activities in 1989.

112. *Final SEIS*, 2–4.

113. National Science Board's Committee on the NSF Role in Polar Regions, *The Role of the National Science Foundation in Polar Regions: A Report to the National Science Board*, NSB–87–128 (June 1987). In concurring with all of the board's fifteen recommendations, the Division of Polar Programs issued a plan describing current and future activities to implement each recommendation. See the Division of Polar Programs' "Implementation Plan for the National Science Board Report: The Role of the National Science Foundation in Polar Regions," June 1988.

114. In December 1986 the NSF commissioned a panel of experts to report on safety practices in Antarctica. The panel's report, recommending improvements for U.S. operations in the interrelated areas of safety, health, and the environment, was completed in June 1988. See *Safety in Antarctica: Report of the U.S. Antarctic Program Safety Review Panel*, NSF–88–78 (June 30, 1988).

115. *Final SEIS*, 1–5.

116. Ibid., 1–5. See also the testimony of Dr. Corell in House Committee on Science, Space, and Technology, *NSF Antarctic Environment Act of 1991: Hearing*, 22.

117. The NSF released the agenda for review and public comment in August 1988. See Robert M. Andersen and Lawrence Rudolph, "On Solid International Ground in Antarctica: A U.S. Strategy for Regulating Environmental Impact on the Continent," *Stanford Journal of International Law* 26, no. 1 (fall 1989): 96.

118. Ibid.

119. National Science Foundation, Division of Polar Programs, Office of the General Counsel, "Implementation of the National Science Foundation's Strategy for Compliance with Environmental Law in Antarctica," November 16, 1989.

120. "NSF Reports Focus on Improving USA Environmental Practices," *AJUS* 14, no. 2 (June 1990): 1.

121. See NSF's "Strategy for Compliance with Environmental Law in Antarctica," 1.

122. This discussion draws substantially from *Final SEIS*, 2–32, 2–33.

123. In 1989 the division also developed a videotape on environmental awareness for tour ship passengers, which is now used by all U.S. tour operators.

124. *Final SEIS*, 2–32, 2–33.

125. Ibid., 2–33. Such an award was presented to an Antarctic Support Associates' engineer for suggestions related to collection of earth fill materials during the 1990–91 season.

126. Ibid.

127. In January 1993, the director of NSF elevated the status of the Division of Polar Programs to the Office of Polar Programs and placed it in the Office of the Director. Under this change, which was effective immediately, the new office reports directly to the director. No staff or organizational changes were made. See *AJUS* 28, no. 1 (March 1993): 1.

128. *Final SEIS*, 2–33.
129. See statement of Corell, in House Committee on Science, Space, and Technology, *NSF Antarctic Environment Act of 1991*, 16, 17.
130. *Final SEIS*, 2–33.
131. Ibid.
132. See House Committee on Science and Technology, *Antarctic Conservation Act of 1978*, 11.
133. Mink, *American Foreign Policy Basic Documents, 1977–1980*, 386.
134. Letter from Douglas J. Bennett Jr., assistant secretary for congressional relations, to Thomas P. O'Neill Jr., Speaker of the House, in House Committee on Science and Technology, *Antarctic Conservation Act of 1978*, 17. A similar letter was sent to the president of the Senate.
135. Public Law 95–54116, *U.S. Code*, vol. 16, secs. 2401–12 (1982), signed into law by President Carter on October 28, 1978.
136. The full text of the act is reprinted in NSF, *Antarctic Conservation Act of 1978*, NSF 95–154 (October 1995).
137. NSF's issued regulations were published in the March 6, 1979, *Federal Register*. See "Regulations to Implement Antarctic Conservation Act of 1978 Now in Force," *AJUS* 14, no. 2 (June 1979): 1. The regulations only covered certain provisions of the act. The Environmental Defense Fund criticized the regulations and argued that they should extend protection "to any species that occurs in Antarctica and more vigorously enforce the substantive requirements of the Act." See Manheim, "On Thin Ice," 10. Also see "NSF Issues Final Regulations for Antarctic Conservation Act," *AJUS* 24, no. 1 (March 1989); and House Committee on Science and Technology, *Antarctic Conservation Act of 1978*, 4.
138. *U.S.Code*, vol. 16, secs. 2431–44 (Supp. V. 1987).
139. Andersen and Rudolph, "On Solid International Ground in Antarctica," 118.
140. Ibid., 118–19.
141. See Douglas R. Arnold, *Congress and the Bureaucracy: A Theory of Influence* (New Haven: Yale University Press, 1979). Also see Steven S. Smith and Christopher J. Deering, *Committees in Congress*, 2d ed. (Washington, D.C., 1990) for a fresh examination of the status of the U.S. Congress's committee system.
142. These weaknesses are evident in Peter Wilkniss's remark during hearings on the Antarctic Environment Act of 1991. Responding to a question from Congressman Boucher, Dr. Wilkniss stated that "a number of other bills have been introduced, and if yours would be the only legislation waiting while the other bills go forward which come from different perspectives and do not recognize the centrality of the research and its benefits in the Antarctic, that might not be a good way to go. . . ." See House Committee on Science, Space, and Technology, *NSF Antarctic Environment Act of 1991: Hearings*, 30.
143. Senate Committee on Foreign Relations, *Antarctica Legislation: Hearing on S. 2575*; S. J. Res. *206* and S. Res. *186*, 101st Cong., 2d sess., 27 July 1990, 7.
144. Ibid.
145. It also embodied similar goals and concerns to those reflected in S. R. 206.
146. See House Committee on Merchant Marine and Fisheries, *Antarctic Environmental Protection: Hearing before the Subcommittees on Fisheries and Wildlife Conservation and the Environment, Oceanography and Great Lakes, Coast Guard and Navigation*, 101st Cong., 2d sess., 26 June 1990.

147. Antarctic Protection Act, H. R. 3977, *Congressional Record*, 101st Cong., 2d sess., 1990, 136: H9605.
148. "U.S. Legislation and Antarctica," *AJUS* 26, no. 1 (March 1991): 11.
149. Interview by Ethel R. Theis with Mr. Lawrence Rudolph, assistant general counsel, NSF, on March 12, 1992. Also see Ibid., "U.S. Legislation and Antarctica."
150. House Committee on Interior and Insular Affairs, *Establish an Antarctica World Park: Hearing*, 5.
151. Ibid., 12.
152. Ibid., 18.
153. Ibid., 13.
154. Testimony by Bill Houston, deputy assistant secretary for territorial and international affairs, Department of the Interior, in ibid., 53.
155. Ibid., 63.
156. Testimony of R. Tucker Scully, in ibid., 62.
157. Statement of Dr. Peter Wilkniss, in ibid., 84.
158. Ibid., 84.
159. Protocol on Environmental Protection to the Antarctic Treaty, XIth Special Consultative Meeting in Madrid, Doc. XI ATSCM/2, June 2, 1991, adopted 4 October 1991.
160. *U.S. Code*, vol. 33, sec. 1901. See the discussion on Annex IV in chapter 8.
161. Protocol of 1978 relating to the International Convention for the Prevention of Pollution from Ships, 1973, with annexes. Done at London, February 17, 1978, entered into force for the United States October 2, 1983. For discussion, see Christopher C. Joyner, *Antarctica and the Law of the Sea* (Dordrecht, The Netherlands: Martinus Nijhoff, 1992), 151–56.
162. National Environmental Policy Act of 1969, Public Law No. 91-190, *U.S. Statutes at Large* 83: 852; codified at *U.S. Code*, vol. 42, secs. 4321–47.
163. See "Statement of Christopher Joyner," in House Committee on Merchant Marine and Fisheries, *Antarctic Treaty Protocol on Environmental Protection, Hearing*, 143, 150–51.
164. See Manheim, "On Thin Ice."
165. "Statement of the Antarctica Project, Greenpeace, Friends of the Earth-U.S., Environmental Defense Fund . . .", in House, Hearings before the Subcommittee on Science of the Committee on Science, Space, and Technology on H. R 3532, The Antarctic Environmental Protection Act, 103d Cong., 2d sess., February 8, 1994, p. 8 (mimeograph).
166. "Hearings on U.S. Implementing Legislation for Protocol," *The Antarctica Project* 3, NO. 1 (winter 1994): 1. See also ibid., "Statement of the Antarctica Project," 17–19.
167. "Hearings on U.S. Implementing Legislation." See also "Statement of the Antarctic Project," 10–11.
168. "Hearings on U.S. Implementing Legislation." See also "Statement of the Antarctica Project," 13–14.
169. See "Statement of the Antarctic Project," 12–13.
170. Ibid., 15.
171. Ibid., 20–21.
172. This study uses the terms environmentalists, NGOs, and public interest groups interchangeably.
173. As used here, the term "NGOs" refers specifically to nonprofit organiza-

tions that seek to influence the direction of U.S. Antarctic environmental policy and actions in accordance with their goals and objectives. NGOs obtain financial support through memberships fees, foundations, special gifts, government grants, and educational activities.

174. ASOC is an alliance of over two hundred organizations concerned with protecting Antarctica. The Antarctica Project, headquartered in Washington, D.C., serves as the northern hemisphere secretariat of ASOC.

175. For a general assessment of environmental interest groups, see Robert Cameron Mitchell, "From Conservation to Environmental Movement: The Development of the Modern Environmental Lobbies," in *Government and Environmental Politics*, ed. Michael J. Lacey (Washington, D.C.: The Woodrow Wilson Center Press, 1991), 91–92.

176. Walter A. Rosenbaum, *Environmental Politics and Policy* (Washington, D.C.: Congressional Quarterly Press, 1985), 37.

177. See Michael J. Lacey, "The Environmental Revolution and the Growth of the State: Overview and Introduction," in Lacey, ed., *Government and Environmental Politics*, 9. The Ford Foundation has played an important role in institutionalizing environmental law advocacy. EDF, the first environmental law group, received a total of $994,000 from the Ford Foundation from 1971 to 1977. See Mitchell, "From Conservation to Environmental Movement," in Lacey, ibid., 89, 90, and 109.

178. Court decisions in the 1960s broadened the class of interests entitled to seek judicial review and recognized injuries other than economic or direct physical damage as grounds for standing. "NEPA was used by environmental lawyers to challenge many federally funded projects on procedural grounds for not preparing acceptable environmental impact statements. In all, thousands of lawsuits have been filed by the environmental groups and the volume of litigation continues unabated." See Mitchell, "From Conservation to Environmental Movement," in Lacey, ibid., 101.

179. Lacey, "The Environmental Revolution," in Lacey, ibid., 9.

180. Markus G. Schmidt, *Common Heritage or Common Burden? The United States Position on the Development of a Regime for Deep Seabed Mining in the Law of the Sea Convention* (Oxford: Clarendon Press, 1989), 64. M. J. Peterson has pointed out that, although the governments of Australia, France, Japan, New Zealand, the United Kingdom, the United States, and Germany permit environmental groups a political role, other ATCPs like Chile and the former Soviet Union "are less approachable." See Peterson, *Managing the Frozen South*, 165. Philip Quigg has similar comments on the varying degrees of influence exerted on their governments by environmentalists. See Quigg, *A Pole Apart*, 178.

181. Schmidt, *Common Heritage or Common Burden?*, 64. This comment was made in the context of the Third United Nations Conference on the Law of the Sea negotiations.

182. "Oceans, Antarctic Resources and Environmental Concerns," statement by Patsy T. Mink, assistant secretary of state for oceans and international environmental and scientific affairs, before the Subcommittee on Arms Control, Oceans and International Environment of the Senate Foreign Relations Committee, February 6, 1978, in *American Foreign Policy Current Documents 1977–80*, Doc. 152, 383.

183. Two members of the Subcommittee on Fisheries and Wildlife Conserva-

tion of the House of Representatives's Merchant Marine and Fisheries Committee also attended the conference to advise the U.S. delegation. James E. Heg, "Conference on the Conservation of Antarctic Seals," *AJUS* 7, no. 3 (May June 1972): 45.
184. Ibid.
185. See Philip Quigg, *A Pole Apart*, 181.
186. Arnfinn Jørgersen-Dahl and Willy Østreng, "Introduction: The Antarctic Challenge," in Jørgersen-Dahl and Østreng, eds., *The Antarctic Treaty System*, 2.
187. The fact that executive branch agencies took no steps to encourage American companies to prospect in Antarctica seems to have been largely ignored. Quite the contrary, as F. M. Auburn has noted, "even a country as strongly committed to free access to Antarctic resources as the United States has taken active steps to discourage its oil companies from Antarctic prospecting." Auburn, *Antarctic Law and Politics*, 260.
188. Statement of Greenpeace, in House Committee on Science, Space, and Technology, *NSF Antarctic Environment Act of 1991: Hearing*, 109.
189. That conference called on the ATCPs to "negotiate to establish the Antarctic Continent and the surrounding seas as the first world park, under the auspice of the United Nations." See Greenpeace International, "The World Park Option for Antarctica: Background for a Fourth UN Debate," November 17, 1986, 1.
190. In addition to Greenpeace and ASOC, the International Union for the Conservation of Nature, the Environmental Defense Fund, and the Cousteau Society have been active in advocating a world park regime, and their proposals are all similar.
191. Statement of Greenpeace in House Committee on Science, Space, and Technology, *NSF Antarctic Environment Act of 1991: Hearing*, 110.
192. In an anonymous article, "Why Tempt the Devil? Protecting Antarctica," reprinted in Senate Committee on Foreign Relations, *Antarctica Legislation: Hearing*, 12, R. Tucker Scully is quoted as having said, "We are considered to be the potential pillager, which is totally false."
193. See House Committee on Interior and Insular Affairs, *Establish an Antarctica World Park: Hearing*, 73.
194. Ibid.
195. Statement of Susan Sabella of Greenpeace, Ibid., 38.
196. Statement of Lee Kimball, director, Antarctica Program, World Resources Institute, in House Committee on Merchant Marine and Fisheries, *Convention on the Regulation of Antarctic Mineral Resource Activities: Hearing*, 24.
197. Statement of Susan Sabella of Greenpeace in House Committee on Interior and Insular Affairs, *Establish a World Park: Hearing*, 38.
198. Ibid.
199. Gunther Handl, ed., *Yearbook of International Environmental Law*, vol 1, (Norwell: Kluwer Academic Publishers, 1990), 340.
200. Ethel R. Theis interview with Lawrence Rudolph, March 12, 1992. Also see *Environmental Defense Fund, Inc., v. Walter E. Massey, et al.*, Civil Action No. 91-1068, *Memorandum in Support of Motion to Dismiss and in Opposition to Plaintiff's Preliminary Injunction*, July 22, 1991.
201. *Environmental Defense Fund., Inc,. v. Massey*, 986 F2d 528 (U.S. Court of

Appeals, D.C. Circuit, January 29, 1993). For discussion, see David A. Wirth, "Environment-Extraterritorial Effects of Activities of Federal Agencies," *American Journal of International Law* 87 (October 1993), 626–35.

7. Geostrategic Interests

1. See, for example, Office of Technology Assessment, *Polar Prospects*; John Tinker, "Cold War Over Antarctica Wealth," *New Scientist* 83 (September 20, 1979); "Antarctic Oil is Estimated as Enormous," *The Washington Post*, March 3, 1975, reprinted in Senate Committee on Foreign Relations, *U.S. Policy with Respect to Mineral Exploration and Exploitation*, 33; J. Spivak, "Frozen Assets," *Wall Street Journal*, February 21, 1974, p. 1, col. 1; Deborah Shapley, "Antarctica: World Hunger for Oil Spurs Security Council Review," *Science* 184 (1974): 776–77; Stanley N. Wellborn, "Antarctic Riches: For Some or for All?" *U.S. News and World Report*, January 24, 1983; and Michael J. Berlin, "Global Pressure Builds for a Sharing of Antarctica's Wealth," *The Interdependent* (November/December 1983).

2. Evaluations of the petroleum potential of Antarctica are based on four main types of data: geological studies of Antarctic outcrops and plate tectonic reconstruction; geophysical studies including radio echo surveys and seismic surveys; the National Science Foundation–sponsored, worldwide Deep Sea Drilling Project; and research reports from oil companies. Egil Bergsager, "Basic Conditions for the Exploration and Exploitation of Mineral Resources in Antarctica: Options and Precedents," in *Antarctic Resource Policy: Scientific, Legal and Political Issues*, ed. Francisco Orrego Vicuña (New York: Cambridge University Press, 1983), 171.

3. See, for example, James E. Mielke, "Potential Mineral Resources in Antarctica and the Antarctic Minerals Convention," *Congressional Research Service*, Report 90-72 (February 5, 1990); Marteen J. de Wit, *Minerals and Mining in Antarctica: Science, Technology, Economics and Politics* (New York: Oxford University Press, 1985); James H. Zumberge, "Mineral Resources and Geopolitics in Antarctica," *American Scientist* 67, no. 1 (January/February 1979); Brewster, *Antarctica: Wilderness at Risk*; Zumberge, "Potential Mineral Resource Availability," in Charney, ed., *The New Nationalism*, 115, 124; and John F. Splettstoesser and Gisela A. M. Dreschhoff, eds., *Mineral Resources Potential of Antarctica* (Washington, D.C.: American Geophysical Union, 1990). As early as 1974, the U.S. Geological Survey reported that geological and geophysical data indicated that oil or gas or both were present in the Antarctic. U.S. Department of the Interior, Geological Survey Circular No. 703, *Mineral Resources of Antarctica* 15 (1974).

4. The terms petroleum, oil, and hydrocarbons are used interchangeably throughout this chapter.

5. The harvesting of whales has diminished steadily since the 1930s. In addition, the exploitation of the seal and penguin populations has ceased. For instructive discussions on the Antarctic marine living resources, see generally, N. D. Bankes, "Environmental Protection in Antarctica: A Comment on the Conservation of Antarctic Marine Living Resources," *The Canadian Yearbook of International Law* 19 (1981); Sayed El-Sayed, "Antarctic Marine Living Resources: The BIOMASS Program," *Oceanus* 31, no. 2 (summer 1988); and George A. Knox, "The Key Role of Krill in the Ecosystem of the Southern Ocean with Special Reference to the Convention on the Conservation of Antarctic Marine Living Resources," *Ocean Management* 9 (1984).

6. The total population of Antarctic birds is estimated at 200 million, with penguins being the largest single species. See Boleslaw A. Boczek, "The Protection of the Antarctic Ecosystem: A Study in International Environmental Law," *Ocean Development and International Law* 13 (1983): 366.
7. Ibid.
8. Bruce C. Parker and Ernest G. Angino, "Environmental Impact of Exploiting Mineral Resources and Effects of Tourism in Antarctica," in Splettstoesser and Dreschoff, eds. *Mineral Resources Potential of Antarctica*, 242; and Brewster, *Antarctica: Wilderness at Risk*, 71.
9. Parker and Angino, ibid., 242.
10. Ibid.
11. The notion of Gondwanaland as a supercontinent was introduced in 1883 by the Austrian geologist, Suess. De Wit, *Minerals and Mining in Antarctica*, 73. For a lucid discussion of the Gondwana hypothesis, see Campbell Craddock, "The Mineral Resources of Gondwanaland," in Splettstoesser and Dreschhoff, eds., *Mineral Resources Potential of Antarctica*, 1–6.
12. Zumberge, "Mineral Resources and Geopolitics in Antarctica," 50.
13. Geoff Mosley, *Antarctica: Our Last Great Wilderness* (Victoria: Australian Conservation Foundation, 1986), 39.
14. Pyne, *The Ice*, 353. The decision was made to cease drilling before the operation led to a blowout. The *Glomar Challenger* was part of the National Science Foundation's worldwide Deep Sea Drilling Project.
15. Brewster, *Antarctica: Wilderness at Risk*, 88.
16. Estimating resource potential is far more uncertain than determining proved oil resources. Proved resources are those quantities of naturally occurring hydrocarbons that geological and engineering studies demonstrate to be recoverable from known reservoirs under existing economic and operating conditions. Reserves are only proved by the drilling of wells. Undiscovered resources, on the other hand, are speculative estimates of the undrilled quantity of recoverable oil or gas based on geological analyses of particular areas. Nonetheless, estimates of undiscovered resources are important because they target which areas within a geologic province should receive priority in the search effort. John Norton Garrett, "Conventional Hydrocarbons in the United States Arctic: An Industry Appraisal," in *United States Arctic Interests: The 1980s and 1990s*, ed. William E. Westermeyer and Kurt M. Shusterich (New York: Springer-Verlag, 1984), 50.
17. Pyne, *The Ice*, 353.
18. G. D. Triggs, "Negotiation of a Minerals Regime," in Triggs, ed., *The Antarctic Treaty Regime*, 182.
19. Brewster, *Antarctica: Wilderness at Risk*, 89.
20. John C. Behrendt, ed., *Petroleum and Mineral Resources of Antarctica*, USGS Circular 909 (1983), 1.
21. Mielke, "Potential Mineral Resources in Antarctica," CRS-2. John F. Splettstoesser and Gisela A.M. Dreschhoff have noted that mineral deposits of economic value have not been discovered anywhere in Antarctica or its continental shelf. See their *Mineral Resources Potential of Antarctica*, xii.
22. Pyne, *The Ice*, 352.
23. John B. Anderson, "Geology and Hydrocarbon Potential of the Antarctic Continental Margin," in Splettstoesser and Dreschoff, eds., *Minerals Resource Potential of Antarctica*, 175.

24. Ibid.

25. Ibid., 197.

26. Ibid. The moratorium on mineral resource activities encouraged by the ATCPs in 1977 was largely due to the lack of a comprehensive database, necessary for adequately assessing the existence and range of mineral resources in the region.

27. E. W. Cole, "Claims of Sovereignty over the Antarctic," thesis, Judge Advocate General's Office, U.S. Army (1958), 85.

28. National Security Council report, "Statement of U.S. Policy on Antarctica," NSC 5905/1, April 7, 1959, reprinted in *FRUS 1958–1960*, 2:554. The report also listed among U.S. policy objectives that of maintaining "a leading position in Antarctica which would satisfy necessary U.S. political, scientific, and potential economic, military, and other interests." Ibid., 559.

29. Ibid., 560.

30. Dr. Ray's statement is in Senate Committee on Foreign Relations, *U.S. Policy with Respect to Mineral Exploration and Exploitation*, 16.

31. See Senate Committee on Foreign Relations, *The Antarctic Treaty: Hearings*, particularly pages 6, 112, and 63.

32. See Senate Committee on Foreign Relations, *U.S. Policy with Respect to Mineral Exploration and Exploitation*, 16.

33. During his testimony on April 13, 1965, John T. McNaughton, assistant secretary of defense for international security affairs, read the referenced sentence. He did not indicate the name of the agency that drafted the document. See House Committee on Interior and Insular Affairs, *Antarctica Report—1965: Hearings*, 65.

34. No elaboration was made on the nature and dimension of this "equitable" and "wide" use of resources. William D. McElroy, "Antarctic Research: A Pattern for Science Management," *Antarctic Journal of the United States (AJUS)*, 6, no. 2 (March/April 1971): 26.

35. Dr. Ray's statement, in Senate Committee on Foreign Relations, *U.S. Policy with Respect to Mineral Exploration and Exploitation*, 6.

36. Statement by Patsy Mink, assistant secretary of state and chairwoman of the Antarctic Policy Group, in Senate Committee on Foreign Relations, *U.S. Policy with Respect to Exploitation of Antarctic Resources: Hearing before the Subcommittee on Arms Control, Oceans, and International Environment*, 95th Cong., 2d sess., 6 February 1978, 18. These same principles were expressed by most of the U.S. government officials interviewed for this study in 1992.

37. The press release was emphatic in stating that full understanding of the environmental consequences of any mineral resource development was essential, and that Antarctic mineral resources, if ever developed, must be used wisely and developed only under effective environmental safeguards. Department of State Press Release 224, dated September 14, 1979, in *Department of State Bulletin* 79, no. 2032 (November 1979): 22.

38. Ibid.

39. Office of Management and Budget, *The U.S. Antarctic Program*, a report to the Committees on Appropriations of the Senate and House of Representatives, May 1983.

40. See R. Tucker Scully's statement in Senate Committee on Commerce,

Science, and Transportation, *Antarctica: Hearing*, 2. Mr. Scully restated these objectives during interviews with the authors.

41. Colson, "The United States Position on Antarctica," 292, 293. Ambassador Colson reaffirmed these objectives during interviews with the authors.

42. The statement of Ambassador Bohlen is in Senate Committee on Foreign Relations, *Antarctica Legislation: Hearing*, 32.

43. Mineral resource activities involve prospecting, exploration, and development. Prospecting aims at identifying areas of mineral resource potential for possible exploration and development. It includes geological, geochemical, and geophysical investigations and remote sensing and sampling. See Francis Auburn, "Convention on the Regulation of Antarctic Mineral Resource Activities," in Splettstoesser and Dreschoff, eds., *Mineral Resources Potential of Antarctica*, 263.

44. Ambassador Bohlen's statement is in Senate Committee on Foreign Relations, *Antarctica Legislation: Hearing*, 22.

45. Ibid. Also see R. Tucker Scully, "Antarctic Resources Focus of Two Important Meetings," *AJUS* 18, no. 1 (March 1983): 3.

46. In 1977 the ATCPs agreed to a policy of "voluntary restraint" on "all exploration and exploitation of Antarctic mineral resources," conditioned on "satisfactory progress being made towards a minerals regime." The policy was not legally binding. It was established in lieu of an indefinite moratorium on all mineral activities. See statement of James N. Barnes, in Senate Committee on Commerce, Science, and Transportation, *Antarctica: Hearing*, 66, 67.

47. Ambassador Bohlen's statement is in Senate Committee on Foreign Relations, *Antarctica Legislation: Hearing*, 22.

48. R. Tucker Scully made this observation during a hearing. See Senate Committee on Commerce, Science, and Transportation, *Antarctica: Hearing*, 6.

49. Ibid.

50. Scully's statement is in House Committee on Interior and Insular Affairs, *Establish an Antarctica World Park: Hearing*, 54.

51. See J. D. Heap, "Current and Future Problems Arising from Activities in the Antarctic," in Triggs, ed., *The Antarctic Treaty Regime*, 207.

52. See Susan R. Fletcher, *Antarctica: Environmental Protection Issues*, Congressional Research Service Report for Congress 89-272-ENR (April 10, 1989), CRS-34.

53. CRAMRA, Article 18.

54. Ibid., Articles 23–27.

55. Ibid., Article 33.

56. Fletcher, *Antarctica: Environmental Protection Issues*, CRS-34.

57. CRAMRA, Article 31.

58. Christopher Beeby, "The Convention on the Regulation of Antarctic Mineral Resource Activities and Its Future," in *Antarctica's Future: Continuity or Change*, ed. R. A. Herr, H. R. Hall, and M. G. Howard (Tasmania: Australian Institute of International Affairs, 1990), 53.

59. Krill has been touted as an answer to the world's food shortage and the protein deficiency. Yet only the major fishing nations such as Japan and the former Soviet Union have found krill economically and politically appealing. Such countries harvest krill not only because of its high protein content but also because overfishing and marine pollution have depleted commercial fishing stocks. See Brewster, *Antarctica: Wilderness at Risk*, 69.

60. Powell, "Antarctic Marine Living Resources and CCAMLR," in Herr, et al., eds., *Antarctica's Future*, 62.
61. Karl-Herman Kock, "Fishing and Conservation in Southern Waters," *Polar Record* 30 (1994): 5.
62. Ibid., 5.
63. Ibid. See also Raymond Arnaudo, "CAMLR Commission Meets in Hobart, Tasmania, for Eighth Annual Meeting," *AJUS* 25, no. 1 (March 1990): 19.
64. Friedheim and Akaha, "Antarctic Resources and International Law," 142.
65. Ibid.
66. Arnaudo, "CAMLR Commission Meets in Hobart, Tasmania," 19.
67. Friedheim and Akaha, "Antarctic Resources and International Law," 125.
68. Ibid.
69. Handl, *Yearbook of International Environmental Law*, 1:179.
70. Ibid.
71. Ibid. Also see Department of State, "Final Environmental Impact Statement on the Negotiation of an International Regime for Antarctic Mineral Resources" (1982), 5-10; Sayed El Sayed and Mary McWhinnie, "Antarctic Krill: Protein of the Last Frontier," *Oceanus* 22 (1979): 13.
72. Prior to 1904, the Antarctic whale population is estimated to have reached 975,000, but they are now believed to number no more than 338,000. See Zumberge, "Potential Mineral Resource," in Charney, ed., *The New Nationalism*, 122.
73. In May 1994 the International Whaling Commission adopted a special sanctuary for whales throughout the Southern and Indian Oceans. See Andrew Darby, "Sanctuary for Whales Stops Japanese Kill," *Sydney Morning Herald*, May 28, 1994, 2; and Andrew Darby, "Whale Sanctuary Nations Sink Japan," Ibid., 23.
74. John C. Behrendt, "Recent Geophysical and Geological Research in Antarctica Related to the Assessment of Petroleum Resources," in *Mineral Resources Potential of Antarctica*, ed. J. Splettstoesser and G. Dreschoff (Washington, D.C.: American Geophysical Union, 1990), 167.
75. Behrendt, "Recent Geophysical and Geological Research," 169.
76. Parker and Angino, "Environmental Impact of Exploiting Mineral Resources," 248.
77. Christopher C. Joyner, "The Southern Ocean and Marine Pollution: Problems and Prospects," *Case Western Resources Journal of International Law* 17 (1985): 174.
78. Parker and Angino, "Environmental Impact of Exploiting Mineral Resources," 249.
79. Joyner, "The Southern Ocean and Marine Pollution," 176.
80. Parker and Angino, "Environmental Impact of Exploiting Mineral Resources," 249.
81. Neff, "Composition and Fate of Petroleum and Spill-Treating Agents in the Marine Environment," in Geraci and St. Aubin, eds., *Sea Mammals and Oil*, 8.
82. Ibid.
83. M. Lynne Corn, Claudia Copeland, and Pamela Baldwin, "Arctic Resources Over a Barrel," Congressional Research Service Issue Brief, Order Code IB91011 (updated March 18, 1991), CRS-1. Although larger spills have occurred, the *Exxon Valdez* disaster was unique because it was in a body of water ringed by islands and relatively isolated from the open sea. The island enclosure

in Prince William Sound delayed dissipation of the spill, exposing animals to the oil for an extended time and allowing the oil to soak into the beaches and sediments. During congressional hearings on the *Exxon Valdez* spill, the cost of the cleanup and restoration activities was estimated at $1 billion. See Senate Committee on Commerce, Science and Transportation, *Exxon Oil Spill: Hearings before the National Ocean Policy Study and the Subcommittee on Merchant Marine,* 101st Cong., 1st sess., part 2, 10 May and 20 July 1989, 110.

84. *Exxon Oil Spill,* 110. The spill occurred at a particularly sensitive time of year—a breeding time for local species such as otters and shorebirds, the spawning season for herring, and a time of migration to the area by thousands of seabirds, waders, and waterfowl. Over 25,000 seabirds are known to have died, along with hundreds of marine mammals and terrestrial wildlife that fed on the carcasses of polluted animals. See Greenpeace testimony in House Committee on Merchant Marine and Fisheries, *Antarctic Environmental Protection: Hearing,* 140.

85. See Greenpeace testimony in House Committee on Merchant Marine and Fisheries, *Antarctic Environmental Protection: Hearing,* 134.

86. Polly A. Penhale, "Research Team Focuses on Environmental Impact of Oil Spill," *AJUS* 24, no. 2 (June 1989): 9. This issue of *AJUS* provides insightful discussions of the accident and actions taken in its aftermath.

87. See Greenpeace testimony in House Committee on Merchant Marine and Fisheries, *Antarctic Environmental Protection: Hearing,* 134.

88. *AJUS* 24, no. 2 (June 1989).

89. Neff, "Composition and Fate of Petroleum and Spill-Treating Agents," 18.

90. Robert C. Clark Jr. and John S. Finley, "Occurrence and Impact of Petroleum on Arctic Environments," in *The Arctic Ocean: The Hydrographic Environment and the Fate of Pollutants,* ed. Louis Rey (New York: John Wiley and Sons., 1982), 319.

91. Neff, "Composition and Fate of Petroleum and Spill-Treating Agents," 23.

92. Parker and Angino, "Environmental Impact of Exploiting Mineral Resources," 242.

93. Neff, "Composition and Fate of Petroleum and Spill-Treating Agents," 23.

94. Ibid., 24.

95. See Christopher C. Joyner, "Non-Militarization of the Antarctic: The Interplay of Law and Geopolitics," *Naval War College Review* 42, no. 4 (autumn 1989): 83–104.

96. The United States emphasized Arctic matters due to their geopolitical importance in the strategic competition between the two superpowers. See Gary Luton, "Strategic Issues in the Arctic Region," in *Ocean Yearbook 6,* ed. Elisabeth Mann Borgese and Norton Ginsburg (Chicago: University of Chicago Press, 1986), 399; and Olli-Pekka Jalonen, "The Strategic Significance of the Arctic," in *The Arctic Challenge: Nordic and Canadian Approaches to Security and Cooperation in an Emerging International Region,* ed. Kari Möttölä (Boulder: Westview Press, 1988), 157–81. With the collapse of the Soviet Union in 1990, it seems reasonable to assume that this threat is greatly diminished.

97. Obviously, the Antarctic does not impact U.S. national security considerations to the same degree that it does upon more proximate claimant nations. Argentina, Australia, Chile, and New Zealand do not view the Antarctic merely as a distant, frigid ice mass but rather as a nearby continent from which hostile military activity could threaten their national security. See Central Intelligence Agency (CIA), *Polar Regions Atlas,* 5; and Christopher C. Joyner, "Security Issues

and the Law of the Sea," *Ocean Development and International Law Journal* 15 (1985): 181–82. In Argentina, especially, serious concerns persist about the need to protect the southern flank of the mainland from possible attack. See Christopher C. Joyner, "Anglo-Argentine Rivalry: On the Road to Antarctica," in *The Falklands War: Lessons for Strategy, Diplomacy, and International Law*, ed. Alberto Coll and Anthony C. Arend (Winchester: Allen and Unwin, 1985), 190. For an insightful look at the geopolitical significance of Antarctica for Chile and Argentina, as well as for other South American countries with lesser Antarctic interests, see Jack Child's *Antarctica and South American Geopolitics*.

98. See Robert H. Rutford, in Polar Research Board, *Antarctic Treaty System: An Assessment*, proceedings of workshop held at Beardmore South Field Camp, Antarctica, January 7–13, 1985 (Washington, D.C.: National Academy Press, 1986), 56.

99. Carter, *Little America*, 217.

100. Carter, *Little America*, 204. Carter also notes that Germans had been in the Antarctic before; maps of the southernmost continent recalled those earlier visits in names like the Filchner Ice Shelf and Kaiser Wilhelm Land. But this was a new Germany. Ibid., 204–5.

101. Ibid., 205.

102. Ibid.; Walter Sullivan, *Quest for a Continent* (New York: McGraw Hill, 1957), 126; Laurence M. Gould, *The Polar Regions in Their Relation to Human Affairs* (New York: American Geographical Society, 1958), 16.

103. Quoted by Carter in *Little America*, 205.

104. Walter Sullivan, "Antarctica in a Two-World Power," *Foreign Affairs* 36 (October 1957): 159.

105. Ibid.; Phillip C. Jessup and Howard J. Taubenfeld, *Controls for Outer Space and the Antarctic Analogy* (New York: Columbia University Press, 1959), 162.

106. Jessup and Taubenfeld, ibid.; and Pyne, *The Ice*, 339.

107. With the German Antarctic expedition positioned along Princess Astrid Coast, King Haakon VII of Norway—thirty years after Amundsen had named the polar plateau in his honor—proclaimed title over Queen Maud Land. Pyne, ibid. Norway refused to apply the sector principle, which would have had undesirable complications for its Arctic claims. Instead, Norway's claim included the coast and an indefinite hinterland that, ironically, did not extend to the South Pole. Ibid.

108. Peterson, *Managing the Frozen South*, 72.

109. See CIA, *Polar Region Atlas*, 5; Sugden, *Arctic and Antarctic*, 394–95.

110. Latin American states, especially Argentina and Chile, regard the Drake Passage as their "gateway" to Antarctica, as well as a naval choke point of potential significance in the event the Panama Canal should be closed. See Jack Child, "Latin Lebensraum: The Geopolitics of Ibero-American Antarctica," *Applied Geography* 10, no. 4 (October 1990): 289. The United Kingdom has also traditionally assigned high priority to safeguarding the right of free passage through these waters.

111. The paper "Antarctica" is reprinted in *Foreign Relations of the United States (FRUS) 1948*, vol. 1, pt. 2 (1980), 977–83.

112. Ibid., 978.

113. In his testimony Captain Ronne emphasized the utmost importance of the Drake Passage to the United States should the Panama Canal be made inoperative during a war. Senate Committee on Armed Services, *Antarctic Expedition:*

Hearing on S. 3381, A Bill to Authorize the President to Provide Assistance to an Expedition to the Antarctic in Furtherance of the Interests of the United States, 83rd Cong., 2d sess., 1 July 1954, 7.

114. Sullivan, "Antarctica in a Two-Power World," 162. The following year, Senator Alexander Wiley of Wisconsin asserted in Congress that, "in the event of war, there is always a possibility of the destruction of the Panama Canal by enemy attack. This makes the maintenance of an open waterway around Cape Horn vital to our national security. One Russian submarine base properly located, without deterrent from the United States, might well close this vital route." *Congressional Record* 104 (January 27, 1958), 1032.

115. The former senator from California's remarks are in Senate Committee on Foreign Relations, *The Antarctic Treaty: Hearings*, 5.

116. Alan D. Hemmings, "Is Antarctica Demilitarized?" in Herr, Hall, and Haward, eds., *Antarctica's Future*, 224.

117. See Allison C. Hayes, "Antarctica: Challenge for the 1990s," Research Paper, U.S. Naval War College, Newport, Rhode Island (March 5, 1984), 4.

118. Jessup and Taubenfeld, *Controls for Outer Space*, 163. Regarding Antarctica's potential use as a launching site for intercontinental ballistic missiles, Walter Sullivan suggested in 1957 that, once ballistic missiles have sufficient range to reach any part of the planet, Antarctica would provide an advantageous base from which to launch thermonuclear weapons. Also, mobile launching sites would be hard to locate in that vast continent. See Sullivan, "Antarctica in a Two-Power World," 163.

119. Senate Committee on Foreign Relations, *The Antarctic Treaty: Hearings*, 5.

120. House Committee on Interior and Insular Affairs, *Antarctica Legislation—1960: Hearings*, 64.

121. Senate Committee on Foreign Relations, *The Antarctic Treaty: Hearings*, 53.

122. As evidence, Goldblat notes that deployment of naval vessels such as nuclear submarines would not make much sense, even if only because of the remoteness of the Antarctic from all possible strategic targets. He adds that the "situation could perhaps change if the range of submarine-launched ballistic missiles and their accuracy were to increase considerably, and if the survival of submarines in the areas of their present deployment were to be jeopardized as a result of some dramatic improvements in anti-submarine warfare techniques." Jozef Goldblat, "The Polar Regions: Their Strategic Significance and Arms Control Implications," in *The Polar Regions and Their Strategic Significance*, ed. L. Caflish and F. Tanner (Geneva: Programme for Strategic and International Security Studies, 1989, Special Study No. 2/1989), 57.

123. Peter J. Beck, "Antarctica as a Zone of Peace, A Strategic Irrelevance? A Historical and Contemporary Survey," in Herr, et al., eds., *Antarctica's Future*, 202.

124. "Antarctica," PPS-31 in *FRUS 1948*, vol. 1, pt. 2, 979.

125. Some cause for this posturing lies in the government of Juan Peron, elected in 1946, which spotlighted "Antartida Argentina" as a salient issue in Argentina's domestic politics.

126. To strengthen its title to the Antarctic Dependencies of the Falkland Islands, the British War Department decided in January 1943 to launch Operation Tabarin. Operation Tabarin (1943–45) was a secret naval expedition that was instructed to remove Argentine and Chilean marks of sovereignty (e.g., plaques) and initiate more or less continuous occupation of Antarctica. Antarc-

tica became a key element in the Anglo-Argentine-Chilean relationship, as evidenced by the acrimonious tone of diplomatic exchanges on the subject between Buenos Aires, London, and Santiago.

127. Jessup and Taubenfeld, *Controls for Outer Space*, 149.

128. One well-publicized incident occurred at Hope Bay, where on February 15, 1952, the British deployed a detachment of Royal marines to dismantle an Argentinean hut. In response, Argentine military personnel used machine guns to oppose the reconstruction of the British base at Hope Bay in "Argentine" territory. Another notable clash occurred on Deception Island in 1953 when the British dismantled two huts built by Argentine and Chilean forces and removed two Argentineans found in one of the huts. Robert D. Hayton, "The American Antarctic," *American Journal of International Law* 50, no. 3 (July 1956): 593. See also Peter Beck, "A Cold War: Britain, Argentina and Antarctica," *History Today* 37, no. 6 (1987): 18–21; and Sugden, *Arctic and Antarctic*, 406.

129. See generally *Antarctica Cases* (*U.K. v. Argentina*) (*U.K. v. Chile*) I.C.J. Pleadings 11 (1956).

130. Argentina Note to the United Kingdom (4 May 1955), in Ibid., 91–93.

131. Christopher C. Joyner, "U.S. Soviet Cooperative Diplomacy: The Case of Antarctica," in *U.S.-Soviet Cooperation: A New Future*, ed. Nish Jamgotch (New York: Praeger, 1989), 41.

132. As noted earlier, concern about the Soviet Union's intentions for Antarctica was expressed as early as 1948 by the U.S. Department of State paper "Antarctica," PPS-331, reprinted in *FRUS 1948*, vol. 1, pt. 2.

133. Ibid., 980.

134. Letter from Robert A. Lovett, acting secretary of state to the British ambassador (Inverchapel), reprinted in *FRUS 1948*, vol. 1, pt. 2, 974.

135. Such a concern by the U.S. government was plain: "[T]here is nothing to prevent the Russians from sending an expedition to the unclaimed sector of the Antarctic continent between 90° and 150° west longitude, establishing a permanent base there, conducting explorations and laying official claim to territory on the basis of these activities." "Antarctica," PPS-331, in *FRUS 1948*, 980.

136. As noted in Quigg, *A Pole Apart*, 137; and Beck, "Antarctica as a Zone of Peace," 200. Peter Beck has noted that redundant references to "friendly parties" made the anti-Soviet orientation of American policy glaringly transparent. He also mentioned that it remains debatable whether the Soviet Union actually possessed sufficient naval and air capacity during the 1950s and early 1960s to sustain any major facilities in Antarctica. Beck, ibid., 200–201.

137. Hayton, "The Antarctic Settlement of 1959," 353; and Steven J. Burton, "New Stresses on the Antarctic Treaty: Toward International Legal Institutions Governing Antarctic Resources," *Virginia Law Review* 65, no. 3 (April 1979): 476.

138. See Beck, "Antarctica as a Zone of Peace," 196.

139. Joyner, "Anglo-Argentine Rivalry: On the Road to Antarctica," in Coll and Arend, eds., *The Falklands War*, 208.

140. Donald W. McNemar points out that "policymakers must evince appreciation that law is . . . a policy instrument for structuring order, and building joint-benefit precedents." See his, "The Task of Education: Broadening Professional Training and Public Awareness," in *Proceedings of the 75th Anniversary Convocation, ASIL* (April 23–25, 1981): 176.

141. Ethel R. Theis's interviews with two Department of State officials: David A. Colson, deputy assistant secretary, Bureau of Oceans and International Envi-

ronmental Affairs (October 15, 1991), and R. Tucker Scully, director, Office of Oceans and Polar Affairs (November 5, 1991).

142. Joyner, "Non-Militarization of the Antarctic," 88.

143. U.S. policymakers often stress the treaty's significance in setting a precedent for agreements aimed at curbing the strategic arms race. Many people at the Department of State acknowledge that the Treaty has served as a model for other treaties such as the Nuclear Nonproliferation Treaty and the Outer Space Treaty. Ethel R. Theis's interview with R. Tucker Scully, director, Office of Oceans and Polar Affairs (November 5, 1991). The treaty also paved the way for two other agreements that curtailed the deployment of nuclear weapons in new areas and environments: the Treaty on the Prohibition of the Emplacement of Nuclear Weapons and Other Weapons of Self-Destruction on the Seabed and the Ocean Floor and the Subsoil Thereof. *Done* February 11, 1971, 480 UNTS 43 (1972), 23 UST 701, TIAS no. 7337; and the "Treaty for the Prohibition of Nuclear Arms in Latin America," February 14, 1967, 634 UNTS. 282 (1970), 21 UST 77, TIAS no. 1838. See also Protocol II to the Treaty, 634 UNTS. 364 (1971), 22 UST 754, TIAS no. 7137.

144. For example, the CCAMLR's preamble reaffirms the parties' commitment to support the treaty's goal of keeping peace and security in Antarctica. Convention on the Conservation of Antarctic Marine Living Resources (CCAMLR), *done* May 20, 1980, TIAS no. 10240 (entered into force April 7, 1982).

145. Of course, each party involved in such a dispute must give its consent before the dispute can be referred to the court.

146. See, for example, Giorgio Bosco, "Settlement of Disputes under the Antarctic Treaty," in *International Law for Antarctica*, ed. Francesco Francioni and Tullio Scovazzi (Milano: Giuffre Editore, 1987), 22–26.

147. Antarctic Treaty, Article VII.

148. Senate Committee on Foreign Relations, *The Antarctic Treaty: Hearings*, 41, 67; and Quigg, *A Pole Apart*, 147.

149. The United States has exercised the right of inspection most often: nine times between 1964 and 1993. Telephone interview on February 18, 1993, with Raymond Arnaudo, chief, Division of Polar Programs at the Department of State. See also S. V. Vinogradov, "Verification Machinery in the Antarctic Treaty System," in *Control over Compliance with International Law*, ed. W. E. Butler (Norwell: Kluwer Academic Publishers, 1991), 100. See also, Boleslaw A. Bozcek, "The Soviet Union and the Antarctic Regime," *American Journal of International Law* 78 (October 1984): 855; and Auburn, *Antarctic Law and Politics*, 110. No violations of the treaty provisions have ever been reported by the consultative parties. Greenpeace has also inspected Antarctic bases. Greenpeace's inspections have not revealed any violations of the treaty itself, but they have recorded numerous violations of Antarctic Treaty–related rules and recommendations. See Kelly Rigg, "Environmentalists' Perspectives on the Protection of Antarctica," in Cook, ed., *The Future of Antarctica*, 75.

150. The Soviet Union had never before agreed to any on-site inspection, and the treaty was negotiated at the height of the Cold War. The observation by a Soviet jurist that, in the Antarctic verification by inspection does not compromise national security, explains the shift in Soviet policy. See Bozcek, "The Soviet Union and the Antarctic Regime," 855. Also, it was the first inspection agreement ever negotiated between the United States and the Soviet Union. Miller, "Antarctica: White Continent of Promise," 55.

151. Ambassador Phleger made this comment during the hearings in 1960 on ratification of the treaty. See, Senate Committee on Foreign Relations, *The Antarctic Treaty Hearings*, 38.

152. Statement of R. Tucker Scully in House Committee on Science and Technology, *U.S. Antarctic Program: Hearings*, 3.

153. During those negotiations, the U.S. delegation proposed a system of international inspection in the form of permanent foreign observers on vessels engaged in sealing. This proposal was rejected by other states because, in the absence of commercial sealing, it was considered unnecessary to establish a detailed verification system. Vinogradov, "Verification Machinery," 100.

154. Joyner, *Antarctica and the Law of the Sea*, 246.

155. Ibid.

156. Article XII of the convention established the inspection system. The text of the convention is reprinted in OTA's, *Polar Prospects*, 185–211.

157. Quigg, *A Pole Apart*, 148.

158. "Oil and National Security," *The Washington Post*, March 30, 1995, A26.

159. Ibid.

160. "Oil Firms Look to Deep Water, *The Washington Post*, April 10, 1995, A4.

161. Ibid.

162. Ibid. Companies such as Texaco, Amoco, Mobil, and Shell Oil are banding together to develop a common strategy.

163. "Oil and National Security," *The Washington Post*, March 30, 1995, A26.

164. Ibid.

165. Ibid.

166. James W. Curlin, Peter Johnson, William Westermeyer, and Candice Stevens, "Arctic Offshore Petroleum Technologies," *Oceanus* 29, no. 1 (spring 1986): 74.

167. The crude oil price for February 1996 delivery closed at $20.26 on the New York Mercantile Exchange. See *The New York Times*, January 6, 1996, 32.

168. The average cost of drilling exploratory and appraisal wells near the shore in the Gulf of Mexico is about $5 to $6 million per well. The average cost for Arctic exploratory wells ranges from $50 to $60 million each, with significant variations among regions. Curlin et al., "Arctic Offshore Petroleum Technologies," 74.

169. Ibid.

170. Ibid.

171. John Norton Garrett, "The Economics of Antarctic Oil," in Alexander and Hanson, eds., *Antarctic Politics and Marine Resources*, 188.

172. With negligible supplies of mineral resources and a large population, Japan is searching for greater access to sources of energy. Friedheim and Akaha, "Antarctic Resources and International Law," 163.

173. Lee A. Kimball, "Special Report on the Antarctic Minerals Convention," in Splettstoesser and Dreschoff, eds., *Mineral Resources Potential of Antarctica*, 275. Environmental groups and some government officials have suggested that Japan's activities around Antarctica in the 1987–1988 season closely resembled minerals prospecting and exploration. Fueling the controversy was the statement of Japan's ambassador to the CCAMLR in 1987. He said, "While environmental concerns are of unquestionable importance, they have been stressed to the point of unduly deterring human activities which seek to investigate the resource poten-

tial of this vast region. What is needed is a balanced management strategy which emphasizes the importance of development of the resource potential of the Antarctic for the benefit of mankind." See Friedheim and Akaha, "Antarctic Resources and International Law."

174. Kimball, "Special Report on the Antarctic Minerals Convention," 275.

175. Ibid.

176. See, for example, Mort D. Turner, John F. Splettstoesser, and Joseph J. McClelland Jr., "Antarctic Logistic Support for the Earth Science," in Splettstoesser and Dreschoff, eds., *Mineral Resources Potential of Antarctica*, 232; and Behrendt, "Recent Geophysical and Geological Research," in ibid., 166.

177. Turner et al., "Antarctic Logistic Support for the Earth Science," 232.

178. Ibid. The authors note that the geodesic dome constructed at the South Pole by the U.S. is an example of what might be used to enclose any mining operation in Antarctica.

179. In comparison to other continental shelves, that of Antarctica is particularly deep, on average about 500 meters below sea level, 800 meters in the Ross Sea. See Friedheim and Akaha, "Antarctic Resources and International Law," 124; and Triggs, ed., *The Antarctic Treaty Regime*, 9. Also in the Antarctic the shelf ice is moving seaward at about two kilometers a year, a factor that would preclude on-site drilling, except possibly where ice rises exist. See, F. G. Larminie, "Mineral Resources: Commercial Prospects for Antarctic Minerals," in Triggs, ed., *The Antarctic Treaty Regime*, 180.

180. Behrendt, "Recent Geophysical and Geological Research," 170.

181. Splettstoesser and Dreschoff, eds., *Mineral Resources Potential of Antarctica*, xii.

182. Turner et al., "Antarctic Logistic Support for the Earth Science," 234.

183. Ibid.

184. M. W. Holdgate, "The Use and Abuse of Polar Environmental Resources," *Polar Record* 22, no. 136 (January 1984): 35.

185. See, for example, K. R. Croasdale, "Arctic Offshore Technology and Its Relevance to the Antarctic," in Polar Research Board, *Antarctic Treaty System: An Assessment*, 245–63.

186. Behrendt, "Recent Geophysical and Geological Research," 170.

187. Turner et. al., "Antarctic Logistic Support for the Earth Science," 234.

188. Robert Miller, *Liability or Asset: A Policy for the Falkland Islands* (London: Institute for European Defence and Strategic Studies, 1986), 47.

189. Ibid.

190. Ibid.

191. Curlin et al., "Arctic Offshore Petroleum Technologies," 73.

192. William E. Westermeyer, "The Transportation of Arctic Energy Resources," in Westermeyer and Shusterich, eds., *United States Arctic Interests*, 127.

193. Westermeyer has also noted that, although the *Polar Star* and the *Polar Sea* are the world's most powerful nonnuclear icebreakers, even more powerful icebreakers will be needed to support extensive, commercial tanker operations. Ibid., 128.

194. Behrendt, "Recent Geophysical and Geological Research," 170.

195. Larminie, "Mineral Resources: Commercial Prospects for Antarctic Minerals," 179.

196. Kimball, "Special Report on the Antarctic Minerals Convention," 275.

8. Ideological Interests and Ecopolitics

1. The 1979 Agreement Governing the Activities of States on the Moon and Other Celestial Bodies was opened for signature on December 18, 1979; it entered into force following the deposit of the fifth instrument of ratification on July 11, 1984. See Carl Q. Christol, "The Moon Treaty Enters into Force," *American Journal of International Law (AJIL)* 79 (1985): 163.

2. "United Nations Convention on the Law of the Sea," UN Doc. A/CONF.62/122 (1982), reprinted in *International Legal Materials* 21 (1982): 1261.

3. In particular, these newly independent states challenged the traditional doctrine that norms of customary law are automatically binding on new states. For an examination of the new states' attitudes toward international law, see, Louis Henkin, *How Nations Behave*, 2d ed. (New York: Columbia University Press, 1979), 119–34; Anthony J. Dolman, *Resources, Regimes and World Order* (New York: Pergamon Press, 1981), 79–105; and Boleslaw A. Boczek, "Ideology and the Law of the Sea: The Challenge of the New International Economic Order," *Boston College International and Comparative Law Review* 7 (spring 1984): 1–30.

4. Boczek, "Ideology and the Law of the Sea," 2.

5. The Group of 77 has worked to further policies designed to correct inequities in the world economy in favor of developing states. The group (formed at the first meeting of the UN Conference on Trade and Development in 1964) now has some 130 members. See Allan Young, "Antarctic Resource Jurisdiction and the Law of the Sea: A Question of Compromise," *Brooklyn Journal of International Law* 9 (winter 1985): 61–62; and Lung-Chu Chen, *An Introduction to Contemporary International Law: A Policy Oriented Perspective* (New Haven, Yale University Press, 1989), 29.

6. The common heritage of mankind principle is embodied in the 1974 UN Declaration on the Establishment of a New International Economic Order. G.A. Res. 3201, S-VI UN GAOR Supp. (no. 1), 3 UN Doc. A/9559 (1974), reprinted in *U.N. Year Book* (1974), 326.

7. Address by President Johnson at the launching of the *Oceanographer*, a new research vessel, on July 13, 1966 (cited in Quigg, *A Pole Apart*, 280, note 47; and in James K. Sebenius, *Negotiating the Law of the Sea*, vol. 154 (Cambridge: Harvard Economics Studies, 1984), 7.

8. The initial introduction of the CHM concept to the United Nations was actually in the context of space negotiations. Aldo Armando Cocca, ambassador-at-large for Argentina, speaking before the UN Committee on Outer Space, used the expression in June 1967, two months before the famed "*note verbale*" of Ambassador Pardo. See Bradley Larschan and Bonnie C. Brennan, "The Common Heritage of Mankind Principle in International Law," *Columbia Journal of Transnational Law* 21 (1983): 318, note 45.

9. United Nations General Assembly, 22 UN GAOR, C.1 (15th Mtg.), I, UN Doc. A/C.1/p.v. 1525 (1967). [Hereinafter cited as Pardo Statement on Common Heritage.] Also see Louis Henkin, "Politics and the Changing Law of the Sea," *Political Science Quarterly* 89 (March 1974): 53.

10. Larschan and Brennan, "The Common Heritage of Mankind Principle," 318.

11. Summary of draft proposed by the United States for a "United National Convention on the International Sea Bed Area," August 3, 1970, in *AJIL* 65

(1971): 179. Also see, Aaron L. Danzig, "A Funny Thing Happened to the Common Heritage on the Way to the Sea," *San Diego Law Review* 12 (April 1975): 658, note 10.

12. Declaration of Principles Governing the Seabed and the Ocean Floor, and the Subsoil Thereof Beyond the Limits of National Jurisdiction," G.A. Res. 2749, 25 UN GAOR Supp. (no. 280), 24 UN Doc. A/8028 (1970). The Resolution was adopted by 104 voters to 0, with 14 abstentions. The United States voted in favor. See Gillian D. Triggs, "The Antarctic Treaty System: Some Jurisdictional Problems," in Triggs, ed., *The Antarctic Treaty Regime*, 108, note 58.

13. The Informal Single Negotiating Text is reprinted in *International Legal Materials* 14 (1975): 682.

14. See UN Convention on the Law of the Sea, Article 136.

15. Keith Suter, *Antarctica: Private Property or Public Heritage* (Leichhardt, Australia: Pluto Press of Australia, 1991), 70.

16. Remarks by Ian Brownlie in *ASIL Proceedings of the 75th Anniversary Convocation* (Washington, D.C., April 23–25, 1981), 36.

17. Christopher C. Joyner and Ethel R. Theis, "The United States and Antarctica," 94.

18. Articles 136 and 137 of the 1982 UN Convention on the Law of the Sea provides that the "area and its resources are the common heritage of mankind" and that the resources are "not subject to alienation" except in accordance with the rules, regulations, and procedures to be established by the International Seabed Authority.

19. The Moon Treaty of 1979 in Article 11, paragraph 1, provides that the "moon and its natural resources are the common heritage of mankind" and would establish, when feasible, an international mechanism to govern exploitation of these.

20. See Dolman, *Resources, Regimes and World Order*, 223–60.

21. See John Warren Kindt, "A Regime for Ice-Covered Areas: The Antarctic and Issues Involving Resource Exploitation and the Environment," in Joyner and Chopra, eds., *The Antarctic Legal Regime*, 199.

22. For the CHM doctrine to be regarded as an accepted principle of contemporary international law, at least three qualifying pre-conditions must be met. (1) The legal content of the CHM notion must be so distinct and well defined that the concept can be fully integrated into the corpus of international law. (2) Resultant state practice must comply with the development of the CHM notion, and, additionally, evidence of *opinion juris* (i.e, consensus) must be demonstrated and evident. (3) Customary acceptance of the CHM concept as determined by state conduct and behavior must be demonstrated, or at least sufficiently broad based, to attest to its widespread acceptance. See Christopher C. Joyner, "Legal Implications of the Concept of the Common Heritage of Mankind," *International and Comparative Law Quarterly* 35 (January 1986): 198. Also see Louis Sohn, "The Law of the Sea: Customary International Law Developments," *American University Law Review* 34 (1985): 271, 279–80.

23. Triggs, "The Antarctic Treaty System: Some Jurisdictional Problems," 99. Larschan and Brennan have suggested, however, that *res communis* and the CHM concept are fundamentally different. They credit Bing Cheng for this insight. According to Cheng, in *res communis*, "as long as a State respects the exclusive quasi-territorial jurisdiction of other States over their own ships, aircraft, and spacecraft, general international law allows it to use the area or even to abuse it

more or less as it wishes, including the appropriation of its natural resources. . . ." The CHM notion, on the other hand, "wishes basically to convey the idea that the management, exploitation, and distribution of the natural resources of the areas in question are matters to be decided by the international community, and are not to be left to the initiative and discretion of individual states or their nationals." See Larschan and Brennan, "The Common Heritage of Mankind Principle in International Law," 319.

24. Stephen Gorove, "The Concept of the Common Heritage of Mankind: A Political, Moral, or Legal Innovation," *San Diego Law Review* 9 (1970): 402.

25. John Warren Kindt, *Marine Pollution and the Law of the Sea*, vol. 4. (Buffalo, N.Y.: William S. Hein, 1986), 2110.

26. Doleman, *Resources, Regimes and World Order*, 226.

27. See "Comment," "UNCLOS III: The Remaining Obstacles to Consensus on the Deep Seabed Mining Regime," *Texas International Law Journal* 16 (winter 1981): 85.

28. Ibid.

29. Ibid.

30. Steven J. Burton,"New Stresses on the Antarctic Treaty, 500. Also see, Christopher C. Joyner, "Antarctica and the Law of the Sea: Rethinking the Current Legal Dilemmas," *San Diego Law Review* 18 (1981): 425; and "Note," "Thaw in International Law? Rights in Antarctica Under the Law of Common Spaces," *Yale Law Journal* 87 (March 1978): 826–27.

31. House Committee on Merchant Marine and Fisheries, *Law of the Sea: Hearings Before the Subcommittee on Oceanography*, 96th Cong., 1st sess., on "Law of the Sea Conference Oversight," February 27, 1979, "Deep Seabed Mining—H.R. 2759, H.R. 3268," May 22 and 23, June 7, 1979, 19.

32. Ibid., 18.

33. "Deep Seabed Hard Mineral Resources Act," Pub. Law 96–283, reproduced in *International Legal Materials* 19 (1980), 1003; 20 (1981), 1228. The act, section 2(b)(2), declares itself to be transitional, pending the entry into force of a comprehensive law of the sea treaty. It authorizes the administrator of NOAA to issue to U.S. citizens, or corporations and other entities controlled by U.S. citizens, licenses for exploration and permits for commercial recovery of the deep seabed resources. Ibid., sec. 102(a). For a brief but informative discussion of major provisions of the act, see Jon Van Dyke and Christopher Yuen, "Common Heritage v. Freedom of the High Seas: Which Governs the Seabed?," *San Diego Law Review* 19 (April 1982): 499–500, note 21.

34. Pub. Law 96–283, sec. 2(a)(3). The act includes a revenue-sharing fund, created by imposing a special tax on seabed miners of 0.75 percent of the processed value of the metals. Cited in Triggs, "The Antarctic Treaty System," in Triggs, *The Antarctic Treaty Regime*, 108 note 69.

35. Kindt, "A Regime for Ice-Covered Areas," 101.

36. Ibid.

37. Elizabeth Mann Borgese, "The Law of the Sea Treaty," *Scientific American* 248 (March 1983): 47.

38. United Nations General Assembly Official Records, 43rd Session, UN Doc. A/43/C 1/PV 44 (1988), 23.

39. See United Nations General Assembly Official Records, 40th Session, UN Doc. A/C.1/40/PV.48 (1985), 43–50 (Statement of Mr. Wijewardane); Ibid., 41st Session, UN Doc. A/C.1/41/PV.49 (1986), 36–39 (Statement of Mr. Wijew-

ardane); Ibid., 42d Session, UN Doc. A/C.1/42/PV.46 (1987), 26–31 (Statement of Mr. Wijewardane); Ibid., 43d Session, UN Doc. A/C.1/43/PV.44 (1988), 28–35 (Statement of Mr. Jayasinghe); Ibid., 44th Session, UN Doc A/C.1/44/PV.46 (1989), 6–12 (Statement of Mr. Jayasinghe).

40. See "Question of Antarctica," study requested under General Assembly Resolution 38/77, Report of the Secretary-General (Part Two: Views of States), Vol. III (1984), UN Doc. A/39/583 (Part II) (1984), 32–36 (Pakistan); United Nations General Assembly Official Records, 40th Session, UN Doc. A/C.1/40/PV.55 (1985), 42–45 (Statement of Mr. Saeed); Ibid., 41st Session, UN Doc. A/C.1/41/PV.50 (1986), 20–25 (Statement of Mr. Mohiuddin); Ibid., 42d Session, UN Doc. A/C.1/42/PV.46 (1987), 31–36 (Statement of Mr. Chohan); Ibid., 43d Session, UN Doc. A/C.1/43/PV.45 (1988), 18–22 (Statement of Mr. Chohan); Ibid., 44th Session, UN Doc. A/C.1/44/PV.42 (1989), 22–27 (Statement of Ahmad Kamal).

41. See "Question of Antarctica," Report of the Secretary-General, UN Doc. A/39/583 (Part II, Vol. I) (1984), 92 (Bangladesh); United Nations General Assembly Records, 40th Session, UN Doc. A/C.1/40/PV.50 (1985), 12–19 (Statement of Mr. Ali); Ibid., 42d Session, UN Doc. A/C.1/42/PV.48 (1987), 2–12 (Statement of Mr. Siddiky); Ibid., 44th Session, UN Doc. A/C.1/44/PV.45 (1989), 8–12 (Statement of Mr. Mohiuddin).

42. See Report of the Secretary-General, UN Doc. A/39/583 (Part II, Vol. II), 93 (Indonesia); United Nations General Assembly Official Records, 40th Session, UN Doc. A/C.1/40/PV.52 (1985), 42–48 (Statement of Mr. Wisnoemoerti); Ibid., 41st Session, UN Doc. A/C.1/41/PV.49 (1986), 56–63 (Statement of Mr. Alatas); Ibid., 42d Session, UN Doc. A/C.1/42/PV.48 (1987), 12–22 (Statement of Mr. Alatas); Ibid., 43d Session, UN Doc. A/C.1/43/PV.45 (1988), 36–42 (Statement of Mr. Sutresna); Ibid., 44th Session, UN Doc. A/C.1/44/PV.44 (1989), 2–7 (Statement of Mr. Poernomo).

43. See United Nations General Assembly Official Records, 42d Session, UN Doc. A/C.1/42/PV.48 (1987), 41–45 (Statement of Mrs. Namgyel); Ibid., 43d Session, UN Doc. A/C.1/43/PV.45 (1988), 14–17 (Statement of Mr. Penjor); Ibid., 44th Session, UN Doc. A/C.1/44/PV.45 (1989), 12–15 (Statement of Mr. Tshering).

44. See United Nations General Assembly Official Records, 40th Session, UN Doc. A/C.1/40/PV.53 (1985), 26–28 (Statement of Mr. Josse); Ibid., 41st Session, UN Doc. A/C.1/41/PV.49 (1986), 63–67 (Statement of Mr. Josse); Ibid., 42d Session, UN Doc. A/C.1/42/PV.47 (1987), 22–28 (Statement of Mr. Josse); Ibid., 43d Session, UN Doc. A/C.1/43/PV.44 (1988), 24–27 (Statement of Mr. Rana); Ibid., 44th Session, UN Doc. A/C.1/44/PV.42 (1989), 6–11 (Statement of Mr. Josse).

45. See United Nations General Assembly Official Records, 43d Session, UN Doc. A/C.1/43/PV.44 (1988), 43–48 (Statement of Mr. Tiogson); Ibid., 44th Session, UN Doc. A/C.1/44/PV.44 (1989), 26–27 (Statement of Mrs. Reyes).

46. These conditions are extrapolated from the description of the "common heritage of mankind" prepared by Ambassador Pardo in his historic statement, "Declaration and Treaty Concerning the Reservation Exclusively for Peaceful Purposes of the Seabed and of the Ocean Floor, Underlying the Seas Beyond the Limits of Present National Jurisdiction, and the Use of Their Resources in the Interests of Mankind," UN Doc. A/AC.105/C.2/SR.75 (17 August 1967).

47. See Rudiger Wolfrum, "The Principle of the Common Heritage of Man-

kind," *Zeitschrift fur auslandisches offentliches Recht und Volkerrecht* (1983): 312, 316-319.

48. Gorove, "The Concept of the Common Heritage of Mankind," 390.

49. See Elizabeth Mann Borgese, "The New International Economic Order and the Law of the Sea," *San Diego Law Review* 14 (1977): 584, 590.

50. Pardo Statement on Common Heritage.

51. Ibid.

52. See Christopher Pinto, "Toward a Regime Governing International Public Property," in *Global Planning and Resource Management: Toward International Decision-Making in a Divided World*, ed. Anthony Dolman (New York: Pergamon, 1980), 202-24.

53. See Boczek, "Ideology and the Law of the Sea," 1-31.

54. See Anthony Dolman, "The Common Heritage of Mankind and Global Reform," in his *Resources, Regimes and World Order*, 223, 241-53.

55. The formal decision to begin detailed negotiations on minerals was taken in 1981 and was recorded in Recommendation XI-1, reprinted in Heap, ed., *Handbook of the Antarctic Treaty System*, 197. The negotiations began in 1982 and were concluded in 1988.

56. The political instability of several Middle Eastern and African oil-producing countries during the 1970s also contributed to wide fluctuations in world energy supplies, leading many governments to perceive Antarctica as a potentially stable source of hydrocarbon fuels.

57. See Richard B. Bilder, "The Present Legal and Political Situation in Antarctica," in Charney, ed. *The New Nationalism*, 184; and Note, "Thaw in International Law?" 848-53.

58. Todd J. Parriott has noted that, since developing states "could not otherwise exploit Antarctic resources without some international regime governing Antarctica, these nations obtain a 'free ride' from developed nations who could exploit Antarctic resources without sharing the benefits of such exploitation." Under these conditions, there is a disincentive for developed nations to exploit resources. See his "Territorial Claims in Antarctica", 98.

59. Fernando Zegers Santa Cruz, "El Sistema Antartico y la Utilizacion de los Recursos," *University of Miami Law Review* 33 (1978): 431.

60. Address by David A. Colson, then assistant legal adviser, Department of State, "United Nations and the Antarctic Treaty System: The Consultative Parties," in *ASIL Proceedings of the 80th Annual Meeting* (Washington, D.C., April 9-12, 1986), 276-79.

61. Most countries have expressed opposition to abandoning their claims. See, for example, Peter J. Beck, "Britain's Antarctic Dimension," *International Affairs* 59 (summer 1983); Edward S. Milensky and Steven I. Schwak, "Latin America and Antarctica," *Current History* (February 1983); and Luis H. Mericq, *Antarctica: Chile's Claim* (Washington, D.C.: National Defense University, 1987).

62. Brigadier Luis H. Mericq of the Chilean army includes "the defense and maintenance of the rights of sovereignty" as one of the goals of Chile's Antarctic policy. Mericq, *Antarctica: Chile's Claim*, 104.

63. John Warren Kindt has raised the credible possibility that, if development of a "common rights" approach seemed imminent, consultative parties holding historical rights, together with those governments who have held their potential claims in abeyance, would be thrust into reviving old claims and asserting new ones. In other words, the "common rights" approach could precipitate a "race"

for establishing claims to Antarctic areas. See his "A Regime for Ice-Covered Areas," 99.
64. Henkin, "Politics and the Changing Law of the Sea," 54.
65. Per Magnus Wikman, "Managing the Global Commons," *International Organization* 36 (summer 1982): 512.
66. Beck, "Britain's Antarctic Dimension," 436.
67. Auburn, *Antarctic Law and Politics*, 295.
68. Anthony Parsons, *Antarctica: The Next Decade*, 21
69. This view is not universal. During an interview with Ethel R. Theis on January 23, 1992, Eric Melby, National Security Council staff member, indicated that he did not perceive increases in the number of treaty members necessarily as a threat. He felt that it would require, however, more skill, more diplomacy, and more time to work out issues and reach consensus. Raymond Arnaudo, chief, Division of Polar Affairs, Department of State, expressed similar views during an interview with Ethel R. Theis on November 20, 1991. Mr. Arnaudo indicated that he found the greater the representation, the better.
70. World Resources Institute, "Blueprint for Antarctica," the Tinker Foundation Workshop held in Luzarches, France, October 22–25, 1989, 22.
71. Interview by Ethel R. Theis with John B. Talmadge, section head, Polar Coordination and Information Section, NSF, February 2, 1992.
72. Interview by Ethel R. Theis with Wesley Scholz, director, Office of International Commodities, Bureau of Economic and Business Affairs, Department of State, January 25, 1992.
73. Ibid.
74. Ibid.
75. Protocol on Environmental Protection to the Antarctic Treaty, XIth Special Consultative Meeting in Madrid, Doc. XI ATSCM/2, June 21, 1991, adopted October 4, 1991. [Hereinafter Madrid Environmental Protocol]. The protocol was signed by twenty-three of the twenty-six Antarctic Treaty Consultative Parties and by eight non-consultative parties. Statement of James Neil Barnes and Beth Claudia Marks to the Committee on Foreign Relations of the U.S. Senate on the Antarctic Environmental Protection Protocol, Hearings before the Senate Foreign Relations Committee, 102d Cong., 2d sess. (May 4, 1992) (mimeograph).
76. For discussion of the negotiations that produced this agreement, see Christopher C. Joyner, "The Antarctic Minerals Negotiating Process," 888–905; and Vicuña, *Antarctic Mineral Exploitation*.
77. For an analysis of the minerals treaty and its operational provisions, see generally Christopher C. Joyner, "1988 Antarctic Minerals Convention," 69–85; and Rudiger Wolfrum, *The Convention on the Regulation of Antarctic Mineral Resource Activities* (Berlin: Springer-Verlag, 1992).
78. See Antarctic Docs. XV ATCM/WP/2 and XV ATCM/WP/3, in *Final Report of Fifteenth Antarctic Treaty Consultative Meeting* (1989), 202–13.
79. See Antarctic Docs. XV ATCM/WP/7 (Chile), in Ibid., 227; XV ATCM/WP/4 (New Zealand), in Ibid., 214; XV ATCM/WP/8 (United States), in Ibid., 237; XV ATCM/WP/14 (Sweden), in Ibid., 243.
80. Madrid Environmental Protocol, Article 2.
81. Ibid., Article 3, paragraph 1.
82. Ibid.
83. Ibid., Article 7.
84. Ibid., Article 25. However, pursuant to a U.S. proposal, any state will have

the right to withdraw from the protocol (presumably giving it the right to mine without regulation) if an amendment lifting the ban is enacted but not ratified within five years of its proposal. Ibid., Article 25, paragraph 6.

85. Ibid., Article 25, paragraph 4.
86. Ibid., Article 14.
87. Ibid., Articles 18, 19, and 20.
88. Ibid., Article 8.
89. Ibid., Article 9.
90. Annex I to the Protocol on Environmental Protection to the Antarctic Treaty: Environmental Impact Assessment, XI ATSCM/2, 21 June 1991.
91. Ibid., Articles 1, 2, and 3.
92. Annex II to the Protocol on Environmental Protection to the Antarctic Treaty: Conservation of Antarctic Fauna and Flora, XI ATSCM/2, 21 June 1991.
93. Ibid., Article 6.
94. Ibid., Article 4, paragraph 2.
95. Ibid., Article 1, paragraph (h)(v).
96. Annex III to the Protocol on Environmental Protection to the Antarctic Treaty: Waste Disposal and Waste Management, XI ATSCM/2, 21 June 1991.
97. Recommendation XII-4, "Man's Impact on the Antarctic Environment, Code of Conduct for Antarctic Expeditions and Station Activities," reprinted in Heap, ed., *Handbook of the Antarctic Treaty System*, 2062.
98. "Recommendation XV-3, "Human Impact on the Antarctic Environment: Waste Disposal," reprinted in Heap, ed., *Handbook of the Antarctic Treaty System*, 2063.
99. Annex III, Article 8.
100. See Ibid., Article 5, paragraph 1(b) (Disposal of Waste in the Sea).
101. Ibid., Article 5, paragraph 1.
102. Antarctic and Southern Ocean Coalition, "A Critique of the Protocol to the Antarctic Treaty on Environmental Protection, ASOC Information Paper No. 1, XVI ATCM (October 8, 1991), 8.
103. Annex IV to the Protocol on Environmental Protection to the Antarctic Treaty: Prevention of Marine Pollution, XI ATSCM/2, June 21, 1991.
104. International Convention for Preventing Pollution from Ships, done at London, November 2, 1973, IMCO Doc. MP/CONF/WP.35, amended by Protocol of 1978 Relating to the International Convention for the Prevention of Pollution from Ships, done at London, February 17, 1978, entered into force October 2, 1983, IMCO Doc. TSPP/CONF/11.
105. Annex IV, Article 2.
106. Ibid., Article 9, paragraph 1.
107. Ibid., Article 9, paragraph 2.
108. The problem here seems obvious. Most U.S. vessels operating in circumpolar Antarctic waters are government owned or operated and will qualify for this exception of sovereign immunity. To the extent that determination of "appropriate measures" and "reasonable and practicable" conditions for compliance is principally left to the discretion of American vessel operators, violations become more possible and prospects for enforcement by the U.S. government are narrowed.
109. Annex to Recommendation XVI: Annex V to the Protocol on Environmental Protection to the Antarctic Treaty: Area and Management, in Antarctic

Treaty, *Final Report of the Sixteenth Antarctic Consultative Meeting* (7–18 October 1991), 116–125.
110. Ibid., Article 3.
111. Ibid., Article 4.
112. Ibid., Article 5.
113. Ibid., Article 5, paragraph 2.
114. Ibid., Article 5.
115. States that have ratified the protocol include the following: Argentina, Australia, Brazil, Chile, China, Ecuador, France, Germany, Italy, Netherlands, New Zealand, Norway, Peru, Poland, South Africa, South Korea, Spain, Sweden, United Kingdom, and Uruguay. "Protocol Ratification Update," *The Antarctica Project* 4, no. 3 (August 1995): 1; "Protocol Ratification Update," *The Antarctica Project* 4, no. 43 (November 1995): 1; "Protocol Ratification Update," *The Antarctica Project* 5, no. 1 (March 1996): 1.
116. National Environmental Policy Act of 1969, *U.S. Code*, vol. 42, Secs. 4321–70 (1988).
117. The Environmental Defense Fund initially brought suit against the NSF because that agency failed to conduct an environmental impact statement prior to operating an incinerator at McMurdo Station, the largest U.S. research facility in the Antarctic. The NSF argued that NEPA did not apply in Antarctica, on the grounds that the continent was not a sovereign foreign country and NEPA's jurisdictional reach was therefore excluded. See *Environmental Defense Fund, Inc. v. Walter E. Massey, Director, National Science Foundation*, 986 F2d 528, U.S. Court of Appeals, D.C. Circuit, January 29, 1993, reprinted in ILM 32 (1993): 505.
118. Ibid., 986 F2d at 535.
119. "Protocol Ratification Update," *The Antarctica Project* 4, no. 43 (November 1995): 2.
120. Ibid.

Conclusion

1. Testimony of R. Tucker Scully, director, Office of Oceans and Polar Affairs, Department of State. See Senate Committee on Commerce, Science and Transportation, *Antarctica: Hearing*, 3. Mr. Scully reiterated the importance of the Antarctic Treaty for the pursuit of U.S. interests in Antarctica in interviews with both authors.
2. "Question of Antarctica," Study Requested Under General Assembly Resolution 38/77, Report of the Secretary General, UN GAOR Annex (Agenda Item 66), UN Doc. A/39/583, vol. III (1984), 126.
3. National Science Foundation, *Facts About the U.S. Antarctic Program* (Washington, D.C.: NSF, 1988), 9.
4. On November 5, 1991, the director of NSF dedicated a new $23 million McMurdo Station Science and Engineering Laboratory. The center will primarily support NSF-funded investigators and provide state-of-the-art laboratory space, analytical instrumentation, and staging areas for a variety of scientific disciplines. *Antarctic Journal of the United States* 27, no. 1 (March 1992): 1.

Selected Bibliography

Primary Sources

United States Congress

House Committee on Appropriations. *Expedition to the Antarctic Regions: Hearing before the Subcommittee on Deficiencies.* 76th Cong., 1st sess., 2 June 1939.

———. *Report on the International Geophysical Year: Hearing before the Subcommittee of the Committee on Appropriations.* 85th Cong., 1st sess., 1 May 1957.

———, *Review of First Eleven Months of the IGY: Hearing before the Subcommittee on Independent Offices.* 85th Cong., 2d sess., 2 June 1958.

———. *Report on the International Geophysical Year: Hearing before the Subcommittee on Independent Offices.* 86th Cong., 1st sess., February 1959.

House Committee on Energy and Commerce. *Ozone Layer Depletion: Hearing before the Subcommittee on Health and Environment.* 100th Cong., 1st sess., 9 March 1987.

———. *Ozone Layer Depletion: Hearing before the Subcommittee on Oversight and Investigations.* 101st Cong., 1st sess., 15 May 1989.

House Committee on Foreign Affairs. *President's Special Report on United States Policy and International Cooperation in Antarctica.* Referred to the Committee on Foreign Affairs, HR Document No. 358. Report prepared by the Historical Studies Division, Bureau of Public Affairs, Department of State. 88th Cong., 2d sess., May 1964.

———. *The Political Legacy of the International Geophysical Year.* Study prepared for the Subcommittee on National Security Policy and Scientific Developments. Report prepared by Harold Bullis, Congressional Research Service, Washington, D.C: Government Printing Office, November 1973.

———. *Preserving Antarctica's Ecosystem: Hearing before the Subcommittee on Human Rights and International Organization.* 101st Cong., 2d sess., 2 May and 19 July 1990.

———. *Implementing Legislation for the Protocol on Environmental Protection to the Antarctic Treaty: Joint Hearings before the Subcommittee on Economic Policy, Trade and Environment and the Committee on Merchant Marine and Fisheries.* 103d Cong., 1st sess., 16 November 1993.

House Committee on Interior and Insular Affairs. *Antarctica Legislation—1960: Hearings before the Subcommittee on Territorial and Insular Affairs.* 86th Cong., 2d sess., 13, 14 June 1960.

House Committee on Interior and Insular Affairs. *Antarctica Legislation—1961: Hearing before the Subcommittee on Territorial and Insular Affairs.* 87th Cong., 1st sess., 24, 25 August 1961.

———. *DeepFreeze 1962 Operations: Hearings before the Subcommittee on Territorial and Insular Affairs.* 87th Cong., 2d sess., 25 May 1962.

———. *DeepFreeze 1963–64 Operations: Hearings before the Subcommittee on Territorial and Insular Affairs.* 88th Cong., 2d sess., 28 May and 10 August 1964.
———. *Antarctica Report–1965: Hearings before the Subcommittee on Territorial and Insular Affairs.* 89th Cong., 1st sess., 12, 13 April, 6, 7 May, and 15 June 1965.
———. *Establish an Antarctica World Park: Hearing before the Subcommittee on Insular and International Affairs.* 101st Cong., 2d sess., 18 September 1990.
———. *Commemorating the 30th Anniversary of the Ratification of the Antarctic Treaty, and Encouraging U.S. Support of an International Agreement to Close the Antarctic to Commercial Mineral Development for 99 Years.* Report 102–45, Part 1. 102d Cong., 1st sess., 30 April 1991.
House Committee on Interstate and Foreign Commerce. *International Geophysical Year. The Arctic and Antarctic: A Report Submitted by the Committee on Interstate and Foreign Commerce to the Committee of the Whole House.* 85th Cong., 2d sess., 17 February 1958.
House Committee on Merchant Marine and Fisheries. *Antarctic Fauna and Flora Conservation: Hearing before the Subcommittee on Fisheries and Wildlife Conservation and the Environment on H.R. 7749 (Bill to implement Agreed Measures).* 95th Cong., 1st sess., 1977.
———. *Antarctic Marine Living Resources: Hearings before the Subcommittee on Fisheries and Wildlife Conservation and the Environment on H.R. 3416, a bill to implement the CCAMLR.* 98th Cong., 1st sess., 30 June 1983.
———. *Antarctic Marine Living Resources Convention Act of 1983: Report to Accompany H.R. 3416 Submitted by the Committee on Merchant Marine and Fisheries.* 98th Cong., 1st sess., 13 October 1983.
———. *Convention on the Regulation of Antarctic Mineral Resource Activities: Hearing before the Subcommittee on Oversight and Investigations.* 101st Cong., 2d sess., 14 March 1990.
———. *Antarctic Environmental Protection: Hearing before the Subcommittees on Fisheries and Wildlife Conservation and the Environment, Oceanography and Great Lakes, Coast Guard and Navigation.* 101st Cong., 2d sess., 26 June 1990.
———. *Antarctic Protection Act of 1990.* Report to accompany H.R. 3977. 101st Cong., 2d sess., 10 September 1990.
———. *Antarctic Treaty on Environmental Protection: Joint Hearing before the Subcommittee on Fisheries and Wildlife Conservation and the Environment and Subcommittee on Oceanography, Great Lakes and the Outer Continental Shelf and the Subcommittee on Coast Guard and Navigation.* 102d Cong., 1st sess., 5 March 1991.
———. *Oil Spill Response Technology: Hearing before the Subcommittee on Oceanography, Great Lakes and the Outer Continental Shelf and the Subcommittee on Coast Guard and Navigation.* 102d Cong., 1st sess., 18 June 1991.
———. *Antarctic Treaty Protocol on Environmental Protection: Hearing before the Subcommittees on Oceanography, Great Lakes, and the Outer Continental Shelf, Coast Guard and Navigation, and Fisheries and Wildlife Conservation and the Environment.* 102d Cong., 2d sess., 30 June 1992.
House Committee on Science, Space, and Technology. *Review of the Results of the Antarctic Ozone Expedition: Hearing before the Committee on Science, Space, and Technology.* 100th Cong., 1st sess., 29 October 1987.
———. *Antarctic Minerals Policy: Hearing before the Subcommittee on Science, Research and Technology and the Subcommittee on Transportation, Aviation and Materials.* 101st Cong., 2d sess., 12 July 1990.

———. *NSF Antarctic Environment Act of 1991: Hearing before the Subcommittee on Science*. 102d Cong., 1st sess., 14 May 1991.

———. *The Antarctic Environmental Protection Act of 1993: Hearing before the Subcommittee on Science*. 103d Cong., 1st sess., 23 February 1993.

———. *H.R. 3532—The Antarctic Environmental Protection Act: Hearing before the Subcommittee on Science, Space and Technology*. 103d Cong., 2d sess., 8 February 1994.

House Committee on Science and Technology. Subcommittee on Energy Research, Development and Demonstration. *Polar Energy Resources Potential.* Report prepared by the Congressional Research Service. 94th Cong., 2d sess., September 1976.

———. *Antarctic Conservation Act of 1978: Report submitted by the Committee on Science and Technology Committed to the Whole House.* Report 95–1031, pt. 2. 95th Cong., 2d sess., 18 May 1978.

———. *U.S. Antarctic Program: Hearings before the Subcommittee on Science, Research, and Technology.* 96th Cong., 1st sess., 1, 3 May 1979.

Senate. *Protocol on Environmental Protection to the Antarctic Treaty.* Message from the President transmitting the Protocol on Environmental Protection to the Antarctic Treaty, with annexes, done at Madrid October 4, 1991, and an additional annex done at Bonn October 17, 1991. 102d Cong., 2d sess. Treaty Doc. 102–22. 18 February 1992.

Senate Committee on Armed Services. *Antarctic Expedition: Hearing before the Committee on Armed Services, on S. 3381, A Bill to Authorize the President to Provide Assistance to an Expedition to the Antarctic in Furtherance of the Interests of the United States.* 83d Cong., 2d sess., 1 July 1954.

Senate Committee on Commerce, Science, and Transportation. *Antarctic Living Marine Resources Negotiations: Hearing before the National Ocean Policy Study.* 95th Cong., 2d sess., 14 June 1978.

———. *Coast Guard Polar Icebreaking Operations: Hearing before the Subcommittee on Merchant Marine.* 98th Cong., 2d sess., 11 April 1984.

———. *Antarctica: Hearing before the Subcommittee on Science, Technology and Space.* 98th Cong., 2d sess., 24 September 1984.

———. *Protecting Antarctica's Environment: Hearing before the Subcommittee on Science, Technology and Space.* 101st Cong., 1st sess., 8 September 1989.

———. *Monitoring the Arctic and Antarctic Environments: Hearings before the Subcommittee on Science, Space and Technology.* 102 Cong., 1st sess., 13 May 1991.

———. *S. 1427: The Antarctic Scientific Research, Tourism and Marine Resources Act of 1993 to Implement the Protocol on Environmental Protection to the Antarctic Treaty: Hearing before the Committee on Commerce, Science and Transportation.* 103d Cong., 1st sess., 20 October 1993.

———. *Antarctic Scientific Research, Tourism and Marine Resources Act of 1993: A Report on S. 1427.* 103d Cong., 2d sess., 25 January 1994.

Senate Committee on Energy and Natural Resources. *Oversight on U.S. Activities in Antarctica: Hearing before the Committee on Energy and Natural Resources.* 96th Cong., 1st sess., 23 April 1979.

Senate Committee on Environment and Public Works. *Implications of the Findings of the Expedition to Investigate the Ozone Hole over the Antarctic: Hearing before the Subcommittee on Environmental Protection and Hazardous Wastes and Toxic Substances.* 100th Cong., 1st sess., 27 October 1987.

Senate Committee on Foreign Relations, *The Antarctic Treaty: Hearings before the Committee on Foreign Relations.* 86th Cong., 2d sess., 14 June 1960.

———. *U.S. Policy with Respect to Mineral Exploration and Exploitation in the Antarctic: Hearing before the Subcommittee on Oceans and International Environment.* 94th Cong., 1st sess., 15 May 1975.

———. *U.S. Policy with Respect to Exploitation of Antarctic Resources: Hearing before the Subcommittee on Arms Control, Oceans, and International Environment.* 95th Cong., 2d sss., 6 February 1978.

———. *Antarctica Legislation: Hearing before the Committee on Foreign Relations on S. 2575; S.J. Res. 206, and S. Res. 186.* 101st Cong., 2d sss., 27 July 1990.

———. *Protocol on Environmental Protection to the Antarctic Treaty (Treaty Doc. 102–22): Hearing before the Committee on Foreign Relations.* 102d Cong. 2d sss., 4 May 1992.

U.S. Congressional Agencies

Congressional Research Service. *The Role of Advisory Committees in U.S. Foreign Policy.* Report prepared for the Senate Committee on Foreign Relations and the House Committee on International Relations by the Foreign Affairs Division. 94th Cong., 1st sess., April 1975.

———. *Effects of Recent Activities in Antarctica.* Report for Congress prepared by M. Lynne Corn, Marjorie Ann Browne, Eugene H. Buck, and James E. Mielke. Report 88–439 SPR, 15 June 1988.

———. *Antarctica: Environmental Protection Issues.* Report for Congress prepared by Susan R. Fletcher. Report 89–272 ENR, 10 April 1989.

———. *Potential Mineral Resources in Antarctica and the Antarctic Minerals Convention.* Report for Congress prepared by James E. Mielke. Report 90–72 SPR, 5 February 1990.

———. *Antarctic Mineral Resources Regime: Diplomacy and Development.* Issue Brief prepared by James E. Mielke and Marjorie Ann Browne, 28 January 1991.

———. *Antarctic Mineral Resources: Diplomacy and Dilemma.* Issue Brief prepared by James E. Mielke and Marjorie Ann Browne, 7 March 1991.

General Accounting Office. Comptroller General of the United States. *Financing Research in Antarctica: Tighter Control of Logistic Support Costs Needed.* Report to the Senate Committee on Foreign Relations. ID–77–59, 30 December 1977.

Office of Technology Assessment, *Polar Prospects: A Minerals Treaty for Antarctica.* OTA-O-428. Washington, D.C.: GPO, September 1989.

U.S. Executive Branch Agencies

Bureau of the Budget. "Planning and Conduct of the U.S. Program for Antarctica." Circular N. A-51, from Maurice H. Stans, Director, to the heads of executive branch agencies. 3 August 1960.

Central Intelligence Agency. *Polar Regions Atlas.* Washington, D.C.: GPO, May 1978, reprinted November 1981.

———. "National Intelligence Survey—Antarctica" (NIS) 69. January 1956.

Department of State. Memorandum from President Roosevelt to the Secretary of State dated July 13, establishing the United States Antarctic Services. In *Foreign Relations of the United States 1939.* Vol. 2.

———. Letter from the Secretary of State to President dated August 1, 1940, regarding territorial claims in the Antarctic advanced by certain governments. In *Foreign Relations of the United States 1940*. Vol. 2.

———. Department of State Policy and Information Statement dated January 27, 1947, regarding United States policy with regard to the polar regions. In *Foreign Relations of the United States 1947*. Vol. 1.

———. "Antarctica." Paper prepared by the Policy Planning Staff (PPS-31) and Draft Agreement prepared by the Department of State, June 9, 1948. *Foreign Relations of the United States, 1948*. Vol. 1, pt. 2.

———. "U.S. Policy with Regard to the Antarctic." *Foreign Relations of the United States, 1948*. Vol 1.

———. Report to the National Security Council (NSC) by the National Security Council Planning Board. June 28, 1954. NSC 5424; and "Antarctica: Draft Statement of Policy Proposed by the NSC." *Foreign Relations of the United States, 1952–1954*. Vol. 1. pt. 2.

———. "Question of Antarctica." *Foreign Relations of the United States, 1955–1957*. Vol. 11.

———. "Progress Report on NSC 5424/1, Antarctica," Memorandum from the Assistant Secretary for International Organization Affairs dated December 5, 1956. *Foreign Relations of the United States, 1955–1957*. Vol. 11.

———. "Establishment of a Condominium over Antarctica," and "Application of UN Trusteeship over Antarctica." *Foreign Relations of the United States, 1955–1957*. Vol. 11.

———. "Status of the Antarctic Region: Memorandum of the Department of State." March 30, 1956. *Basic Documents American Foreign Policy, 1950–1956*. Vol. 1. Department of State Publication 6446, 1957.

———. "Statement of U.S. Policy on Antarctica, NSC 5804/1." *Foreign Relations of the United States, 1958–1960*. Vol. 2.

———. "United States Proposes Conference on Antarctica." *Department of State Bulletin* 38, no. 988 (June 2, 1958).

———. "Statement Issued by the Department of State on the Occasion of the Signing of the Antarctic Treaty." Press Release 827 (December 1, 1958).

———. "Statement by President Eisenhower on the Occasion of the Signing of the Antarctic Treaty." Press Release 828 (December 1, 1958).

———. "Conference on Antarctica to Meet at Washington in October." *Department of State Bulletin* 40, no. 1042 (June 15, 1959).

———. "Statements by Americans Favoring an International Solution to the Antarctic Question." Historical Division, Bureau of Public Affairs (1959).

———. "Twelve Nations Sign Treaty Guaranteeing Nonmilitarization of Antarctica and Freedom of Scientific Investigation." Press Release 827 (December 1, 1959).

———. "Antarctic Treaty Enters into Force." Statement by President Kennedy. *Department of State Bulletin* 65, no. 1150 (July 10, 1961).

———. "Measures in Furtherance of Principles and Objectives of the Antarctic Treaty: Recommendations adopted at the First Consultative Meeting under Art. IX., Canberra, July 24, 1961." *American Foreign Policy Current Documents, 1961*, Doc. 163.

———. "Antarctica: Measures in Furtherance of Principles and Objectives of the Antarctic Treaty: Recommendations adopted at the Second Consultative Meeting under Art. IX., Buenos Aires, July 28, 1962." *American Foreign Policy Current Documents, 1962*, Doc. III.149.

———. "U.S. to Conduct Inspection in Antarctica." *Department of State Bulletin* 69, no. 1266 (September 30, 1963).
———. "A Framework for Assessing Environmental Impacts of Possible Antarctic Mineral Development." Prepared by the Ohio State University for the Department of State (1977).
———. Memorandum. "Implementation of Agreed Measures for the Conservation of Antarctic Fauna and Flora." April 5, 1978.
———. "Oceans: Antarctic Resources and Environmental Concerns." Statement by Patsy T. Mink, Assistant Secretary for Oceans and International Environmental and Scientific Affairs. *Department of State Bulletin* 78, no. 2013 (April 1978).
———. "Final Environmental Impact Statement on the Negotiation of a Regime for Conservation of Antarctic Marine Living Resources" (June 1978).
———. "Final Environmental Impact Statement on the Negotiation of an International Regime for Antarctic Mineral Resources" (August 1982).
———. "Antarctica: 10th Meeting of Treaty Consultative Parties." Text of Press Release 224, September 14, 1979. Opening remarks by Lucy Wilson Benson, Undersecretary for Security Assistance, Science and Technology, and press communique issued by the chairman of the meeting on October 10, 1979. *Department of State Bulletin* 79, no. 2032 (November 1979).
———. "Parties to Antarctic Treaty Meet in London." Statement by Deputy Assistant Secretary for Oceans and International Environmental and Scientific Affairs at the opening session of the Ninth Antarctic Treaty Consultative Meeting in London on September 19. *Department of State Bulletin* 79, no. 2032 (November 21, 1979).
———. "Antarctica: Tenth Meeting of Antarctic Treaty Consultative Parties." Press Release 224 (September 14, 1979).
———. "The Success of the Antarctic Treaty." Address by John D. Negroponte, Assistant Secretary for Oceans and International Environmental and Scientific Affairs. *Department of State Bulletin* 87, no. 2123 (June 1987).
———. "State Department Perspectives on Environmental Issues." Address by John D. Negroponte, Assistant Secretary for Oceans and International Environmental and Scientific Affairs. *Department of State Bulletin* 87, no. 2129 (December 1987).
Johnson, Lyndon. "Statement by the President in Response to a Progress Report by the Antarctic Policy Group." *Public Papers of the Presidents of the U.S.: Lyndon B. Johnson*. Vol 1. (Washington, D.C.: GPO, [May 1] 1965).
———. "Remarks Following a Meeting with Members of the Antarctic Policy Group," *Public Papers of the Presidents of the U.S.: Lyndon B. Johnson*. Vol. 1. (Washington, D.C.: GPO, [May 20] 1965).
Kennedy, John F. "Statement by the President upon Entry into Force of the Antarctic Treaty." *Public Papers of the Presidents of the U.S.: John F. Kennedy. Containing the Public Messages, Speeches, and Statements of the President January 20 to December 31, 1961* (June 23, 1961). U.S. G.P.O., Washington, D.C., 1962.
National Academy of Sciences. National Research Council. *Report by the Polar Research Board. Science in Antarctica: The Life Sciences in Antarctica, pt. 1; The Physical Sciences in Antarctica, pt. 2*. Publications 839 and 878, respectively. Washington, D.C., 1961.
———. *An Evaluation of Antarctic Marine Ecosystem Research*. Washington, D.C., 1981.

———. *U.S. Research in Antarctica in 2000 A.D. and Beyond: A Preliminary Assessment.* Washington, D.C., 1986.
National Science Foundation. *Final Regulations Implementing the Antarctic Conservation Act of 1978,* Pub. L. 95–45. *Federal Register* 44, no. 111 (7 June 1979), reprinted October 1984.
———. *U.S. Antarctic Program Final Environmental Impact Statement.* June 1980, reprinted October 1984.
———. Committee on the NSF Role in Polar Regions. *The Role of the National Science Foundation in Polar Regions: A Report to the National Science Board.* NSB-87-128, June 1987.
———. Division of Polar Programs. "Implementation Plan for the National Science Board Report: The Role of the National Science Foundation in Polar Regions." June 1988.
———. *Report of the U.S. Antarctic Program Safety Review Panel.* June 1988.
———. Division of Polar Programs. *Facts about the United States Antarctic Research Program.* July 1988.
———. Division of Polar Programs. *Facts about the United States Antarctic Research Programs.* October 1994.
———. Division of Polar Programs. *U.S. Antarctic Program Environmental Protection Agenda.* 31 August 1988.
———. "Regulations Pursuant to the Antarctic Conservation Act of 1978." Amending Title 45 of the *Code of Federal Regulations,* by adding part 670. June 1989.
———. Division of Polar Programs. Office of the General Counsel. *Implementation of the National Science Foundation's Strategy for Compliance with Environmental Law in Antarctica.* 16 November 1989.
———. Office of the General Counsel. Report to the Director. "A National Science Foundation Strategy for Compliance with Environmental Law in Antarctica." 29 December 1989.
———. Division of Polar Programs. *Survival in Antarctica.* NSF 90–91, 1990 edition.
———. "A Long-Range Science Plan for the Division of Polar Programs." Recommendations by the Division's Advisory Committee for Polar Programs. April 1990.
———. "Environmental Defense Fund v. Walter E. Massey, Director, National Science Foundations: Memorandum in Support of Motion to Dismiss and in Opposition to Plaintiff's Motion for Preliminary Injunction." 22 July 1991.
———. Memorandum from General Counsel to Director regarding Environmental Defense Fund v. Massey. 29 August 1991.
———. Division of Polar Programs. Office of Safety, Environment and Health. *Final Supplemental Environmental Impact Statement for the United States Antarctic Program.* October 1991.
———. "Permit Issued under the Antarctic Conservation Act of 1978." *Federal Register* 57, no. 195 (7 October 1992).
National Security Council. National Security Decision Memorandum 71 (NSDM-71). "U.S. Antarctic Policy and Program." 10 July 1970.
———. National Security Decision Memorandum 318. "U.S. Policy for Antarctica." 25 February 1976.
Office of Management and Budget. Circular No. A-51 Revised, "Planning and Conduct of the U.S. Program in Antarctica." 4 August 1971.

———. "The U.S. Antarctic Program." Report submitted by OMB to the Committees on Appropriations of the Senate and the House of Representatives. May 1983.
President, Executive Order 12114, "Environmental Effects Abroad of Major Federal Actions." Title 3, 4 January 1979. *Federal Register* 44, no. 6 (9 January 1979).
Reagan, Ronald. "Statement by Press Secretary Speakes on U.S. Antarctic Policy." *Public Papers of the Presidents of the U.S.: Ronald Reagan.* Vol. 1 (Washington, D.C.: GPO, [March 29] 1982).
———. "Message from the President on the Commemoration of the 25th Anniversary of the Antarctic Treaty. In *Antarctic Journal of the United States* 19, no. 4 (December 1984).

Dissertations, Research Papers, and Other Unpublished Documents

Bing, R. N. "The Role of the Developed States in the Formulation of International Controls for Unoccupied Regions, Outer Space, the Ocean Floor and Antarctica." Ph.D. diss., Tufts University, 1972.
Council of Managers of National Antarctic Programs. "Report to the XVI Antarctic Treaty Consultative Meeting." Bonn, October 1991.
Hayes, Allison C. "Antarctica: Challenge for the 1990s." Research Paper. U.S. Naval War College, March 5, 1984.
Keho, Jeffrey D. "Factors Affecting U.S. Naval Operations in Antarctica." Research Paper. U.S. Naval War College, 1987.
———. "Frozen Assets: Antarctica and the United States in 1990." Research Paper. U.S. Naval War College, 1989.
Laughlin, Thomas L. "Minerals Regime in Antarctica." Paper presented at the Center for Oceans, Law and Policy Polar Regions Seminar. University of Virginia, March 26, 1987.
———. "The Environmental Aspects of the New Antarctic Mineral Convention." Paper presented at the International Studies Association Conference, London, April 1, 1989.
Mehaffey, Donald C. "The Strategic Importance of Antarctica." Research Paper. U.S. Naval War College, June 1972.
Oerding, James B. "Frozen Friction Point: A Geopolitical Analysis of Sovereignty in the Antarctic Peninsula." M.A. thesis, University of Florida at Gainesville, 1977.
Plott, B. M. "The Development of US Antarctic Policy." Ph.D. diss., Tufts University, 1969.
SCAR, Steering Committee for the International Geosphere-Biosphere Programme. "The Role of Antarctica in Global Change." April 1989.
Sherman, Kenneth. "The Antarctic Marine Ecosystem: A Model for Science Management During Global Change." NOAA, National Marine Fisheries Service, Northeast Fisheries Center, Narragansett, Rhode Island, November 10, 1989.

Secondary Sources

Books

Adams, Harry. *Beyond the Barrier with Byrd: An Authentic Story of the Byrd Antarctic Exploring Expedition.* New York: M. A. Donohue, 1932.

Akehurst, Michael. *A Modern Introduction to International Law.* 5th ed. Winchester: George Allen and Unwin, 1985.
Alexander, Lewis M., and Lynne Carter Hanson, eds. *Antarctic Politics and Marine Resources: Critical Choices for the 1980s.* Kingston: University of Rhode Island Press, 1985.
Andresen, Steinar, and Willy Østreng. *International Resource Management: The Role of Science and Politics.* New York: Belhaven Press, 1989.
Arbuet, Jeber V., Roberto Puceiro R., and Belter Garre C. *Antartida: Continente de los Mas, para los Menos.* Buenos Aires: Fundacion de Cultura Universitaria, 1979.
Auburn, Francis M. *Antarctic Law and Politics.* Bloomington: University of Indiana Press, 1982.
Baird, Patrick D. *The Polar World.* New York: John Wiley and Son, 1964.
Barnes, James N. *Let's Save Antarctica.* Australia: Greenhouse, 1982.
Beard, Charles A. *The Idea of National Interest: An Analytical Study in American Foreign Policy.* New York: MacMillan, 1966.
Beck, Peter. *The International Politics of Antarctica.* London: Croom Helm, 1986.
———. *The Falklands as an International Problem.* London: Routledge, 1988.
Beeby, Christopher. *The Antarctic Treaty.* Wellington: Institute of International Affairs, 1972.
Behrendt, John C. ed. *Petroleum and Mineral Resources of Antarctica,* United States Geological Survey Circular 909 (1983).
Beitz, Charles R. *Political Theory and International Relations.* Princeton: Princeton University Press, 1979.
Bertrand, Kenneth J. *Americans in Antarctica: 1775-1948.* New York: Special Publication No. 39, American Geographic Society, 1971.
Bond, Creina, and Roy Siegfried. *Antarctica, No Single Country, No Single Sea.* New York: Mayflower Books, 1979.
Bonner, W. N., and D. W. Walton, eds. *Key Environments: Antarctica.* Oxford: Pergamon Press, 1985.
Boyle, Francis A. *World Politics and International Law.* Durnham: Duke University, 1985.
Brewster, Barney. *Antarctica: Wilderness at Risk.* San Francisco: Friends of the Earth, 1982.
Brierly, J. L. *The Law of Nations.* 6th ed. by Sir Humphrey Waldock. London: Oxford University Press, 1963.
Brown, Seyom. *The Forces of Power: Constancy and Change in U.S. Foreign Policy from Truman to Johnson.* New York: Columbia University Press, 1968.
Brown, Seyom, Nina W. Cornell, Larry L. Fabian, and Edith B. Weiss. *Regimes for the Ocean, Outer Space and Weather.* Washington, D.C.: The Brookings Institution, 1977.
Brownlie, Ian. *Principles of Public International Law.* 3rd ed. New York: Oxford University Press, 1979.
Bull, Hedley. *The Anarchical Society: A Study of Order in World Politics.* New York: Columbia University Press, 1977.
Bulmer-Thomas, Victor, ed. *Britain and Latin America: A Changing Relationship.* England: Cambridge University Press, 1989.
Burns, Robert A. *Diplomacy, War and Parliamentary Democracy: Further Lessons from the Falklands or Advice from Academe.* Lanham: University Press of America, 1980.

Bush, W. M., ed. *Antarctica and International Law: A Collection of Interstate and National Documents.* Vols. 1, 2, and 3. New York: Oceana Publications, 1982.
Butterfield, H., and Martin Wight, eds. *Diplomatic Investigations: Essays in the Theory of International Politics.* London: Allen and Unwin, 1966.
Byrd, Richard E. *Little America.* New York: G.P. Putnam's Sons, 1930.
———. *Discovery: The Story of the Second Byrd Expedition.* New York: G.P. Putnam's Sons, 1935.
———. *Alone.* New York: G.P. Putnam's Sons, 1938.
Caflish L., and F. Tanner, eds. *The Polar Regions and Their Strategic Significance.* Geneva: Programme for Strategic and International Security Studies, Special Study No. 2/1989, 1989.
Calvert, P. *The Falklands Crisis: The Rights and Wrongs.* London: Frances Pinter, 1982.
Cameron, Ian [Donald Gordon Payne]. *Antarctica: The Last Continent.* Boston: Little, Brown, 1974.
Caras, Roger A. *Antarctica: Land of Frozen Time.* Philadelphia: Chilton Books, 1962.
Carter, Paul A. *Little America: Town at the End of the World.* New York: Columbia University Press, 1979.
Chapman, S. *IGY.: Year of Discovery: The Story of the International Geophysical Year.* Ann Arbor: Michigan, 1959.
Chapman, Walter. *The Loneliest Continent.* New York: Geographical Society, 1964.
Charney, Jonathan I., ed. *The New Nationalism and the Use of Common Spaces: Issues in Marine Pollution and the Exploitation of Antarctica.* Totowa: Allenheld, Osmund, 1982.
Chaturvedi, Sanjay. *Dawning of Antarctica: A Geopolitical Analysis.* New Delhi: Segment Books, 1990.
Child, Jack. *Antarctica and South American Geopolitics: Frozen Lebensraum.* New York: Praeger Publishers, 1988.
Christie, E. W. Hunter. *The Antarctic Problem: An Historical and Political Study.* London: Allen and Unwin, 1951.
Clinton, David W. *Deliberating in the Bazaar: The Two Faces of the National Interest.* Baton Rouge: Louisiana State University Press, 1993.
Coll, Alberto, and Anthony C. Arend, eds. *The Falklands War: Lessons for Strategy, Diplomacy, and International Law.* Winchester: Allen and Unwin, 1985.
Cook, Grahame. *The Future of Antarctica: Exploitation versus Preservation.* Manchester: Manchester University Press, 1990.
Coplin, William D. *Introduction to International Politics: A Theoretical Overview.* Chicago: Markham Publishing, 1971.
Couloumbis, Theodore A., and James H. Wolfe. *Introduction to International Relations: Power and Justice.* Englewood Cliffs: Prentice-Hall, 1978.
Crabb, Cecil B. Jr. *Policy-Makers and Critics: Conflicting Theories of American Foreign Policy.* New York: Praeger, 1976.
Deacon, George. *The Antarctic Circumpolar Ocean.* Cambridge: Cambridge University Press, 1984.
Debenham, Frank. *Antarctica: The Story of a Continent.* New York: MacMillan, 1961.
Deese, David A., and Joseph A. Nye, eds. *Energy and Security.* Cambridge: Ballinger Publishing, 1980.
Deutsch, Karl W. *The Analysis of International Relations.* 2d ed. Englewood Cliffs: Prentice-Hall, 1978.

Dewart, Gilbert. *Antarctic Comrades: An American with the Russians in Antarctica.* Columbus: Ohio State University Press, 1989.
De Wit, Marteen J. *Minerals and Mining in Antarctica: Science, Technology, Economics and Politics.* New York: Oxford University Press, 1985.
Dillon, G. M. *The Falklands, Politics and War.* New York: St. Martin's Press, 1989.
Dolman, Antony J. *Resources, Regimes, World Order.* New York: Pergamon Press, 1981.
Dougherty, James E., and Robert L. Pfaltzgraff Jr. *Contending Theories of International Relations.* 2d ed. New York: Harper and Row, 1981.
Dufek, George J. *Operation Deepfreeze.* New York: Harcourt Brace, 1957.
Elliott, Lorraine M. *International Environmental Politics: Protecting the Antarctic.* New York, St. Martin's Press, 1994.
Ellsworth, Lincoln. *Beyond Horizons.* New York: Book League of America, 1938.
Fifield, Richard. *International Research in the Antarctic.* New York: Oxford University Press, 1987.
Fogg, G. E., and David Smith. *The Explorations of Antarctica: The Last Unspoilt Continent.* New York: Sterling Publishing, 1990.
Fox, Robert. *Antarctica and the South Atlantic: Discovery, Development, and Dispute.* London: British Broadcast Corp., 1985.
Francioni, Francesco, and Tullio Scovazzi, eds. *International Law for Antarctica.* Milan: Giuffre Editore, 1987.
Frankel, Joseph. *National Interest.* London: Pall Mall, 1970.
Friis, Herman R., and Shelby G. Bale Jr., eds. *United States Polar Exploration.* Athens: Ohio University Press, 1970.
George, Alexander L. *Presidential Decisionmaking in Foreign Policy: The Effective Use of Information and Advice.* Boulder: Westview Press, 1980.
Gifford, Prosser. *The National Interests of the United States in Foreign Policy.* Lanham: University Press of America, 1981.
Gould, Laurence M. *The Polar Regions in Their Relation to Human Affairs.* New York: American Geographical Society, 1958.
Gribbin, John. *The Hole in the Sky: Man's Threat to the Ozone Layer.* New York: Bantam Books, 1988.
Halle, Louis J. *The Sea and the Ice: A Naturalist in Antarctica.* Boston: Houghton Mifflin, 1973.
Hamzah, B. A. *Antarctica in International Affairs.* Malaysia: Institute of Strategic and International Studies, 1987.
Handl, Gunther, ed. *Yearbook of International Environmental Law.* Vol. 1. Norwell: Kluwer Academic Publishers, 1990.
Harris, Colin M., and Bernard Stonehouse. *Antarctica and Global Climatic Change.* Boca Raton: Lewis Publishers, 1991.
Harrison, Albert A., Yvonne A. Clearwater, and Christopher P. McKay, eds. *From Antarctica to Outer Space: Life in Isolation and Confinement.* New York: Springer-Verlag, 1991.
Hayes, Margaret Daly. *Latin America and the U.S. National Interest: A Basis for U.S. Foreign Policy.* Boulder: Westview Press, 1984.
Henkin, Louis. *How Nations Behave: Law and Foreign Policy.* 2d ed. New York: Columbia University Press, 1979.
Herr, R. A., H. R. Hall, and M. G. Haward, eds. *Antarctica's Future: Continuity or Change.* Tasmania: Australian Institute of International Affairs, 1990.
Hoffmann, Fritz L., and Olga Mingo Hoffmann. *Sovereignty in Dispute: The Falklands/Malvinas, 1943–1982.* Boulder: Westview Press, 1984.

Hoffman, Stanley. *Primacy or World Order.* New York: McGraw Hill, 1978.
Jennings, R. Y. *The Acquisition of Territory in International Law.* Manchester: Manchester University Press, 1963.
Jessup, Phillip C., and Howard J. Taubenfeld. *Controls for Outer Space and the Antarctic Analogy.* New York: Columbia University Press, 1959.
Johansen, Robert C. *The National Interest and the Human Interest: An Analysis of U.S. Foreign Policy.* Princeton: Princeton University Press, 1980.
Johnson, Stanley. *Antarctica: The Last Great Wilderness.* London: Weidenfeld and Nicolson, 1985.
Jones, A. G. E. *Antarctica Observed: Who Discovered the Antarctic Continent.* North Yorkshire: Caedmon Whitby, 1982.
Jørgersen-Dahl, A., and W. Østreng. *The Antarctic in World Politics.* New York: St. Martin's Press, 1991.
Joyner, Christopher C. *Antarctica and the Law of the Sea.* Dordrecht: Martinus Nijhoff, 1992.
Joyner, Christopher C., and Sudhir K. Chopra, eds. *The Antarctic Legal Regime.* Dordrecht: Martinus Nijhoff Publishers, 1988.
Kelly, Philip, and Jack Child. *Geopolitics of the Southern Cone and Antarctica.* Boulder: Lynne Rienner, 1988.
Keohane, Robert O., ed. *Neorealism and Its Critics.* New York: Columbia University Press, 1986.
King, H. G. R. *The Antarctic.* London: Blandford Press, 1969.
Kinney, Douglas. *National Interest, National Honor: The Diplomacy of the Falkland Crises.* New York: Praeger Publishers, 1989.
Kish, John. *The Law of International Spaces.* Leiden: A. W. Sijthoff, 1973.
Klotz, Frank G. *America on the Ice: Antarctica Policy Issues.* Washington, D.C.: National Defense University Press, 1990.
Krasner, Stephen D. *Defending the U.S. National Interest: Raw Materials, Investments and U.S. Foreign Policy.* Princeton: Princeton University Press, 1978.
Laws, R. M., ed. *Antarctica: The Last Frontier.* London: Boxtree, 1989.
Levi, Werner. *Law and Politics in the International Society.* Beverly Hills: Sage Publications, 1976.
———. *Contemporary International Law: A Concise Introduction.* Boulder: Westview Press, 1979.
Lewis, R. *A Continent for Science: The Antarctic Adventure.* New York: The Viking Press, 1965.
Lewis, Richard S., and Phillip M. Smith, eds. *Frozen Future: A Prophetic Report from Antarctica.* New York: Quadrangle Books, 1973.
Lindsey, G. R. *Strategic Aspects of the Polar Regions.* Toronto: Canadian Institute of International Affairs, 1977.
Longone, John. *Life at the Bottom: The People of Antarctica.* Boston: Little, Brown and Co., 1977.
Mawson, Douglas. *The Home of the Blizzard.* London: Heinemann, 1915.
May, John. *The Greenpeace Book of Antarctica: A New View of the Seventh Continent.* New York: Doubleday Publishing, October 1989.
McWhinnie, M. A., ed. *Polar Research: To the Present and the Future.* Boulder: Westview Press, 1978.
Mear, Roger, and Robert Swan. *In the Footsteps of Scott.* London: Jonathan Cape, 1987.

Mericq, Luis H. *Antarctica: Chile's Claim.* Washington, D.C.: National Defense University, 1987.
Mickleburgh, Edwin. *Beyond the Frozen Sea: Visions of Antarctica.* London: Butler and Tanner Ltd., 1987.
Milia, Fernando A., ed. *La Atlantartida: Un Espacio Geopolitico.* Buenos Aires: Editorial Pleamar, 1978.
Miller, Lynn H. *Global Order: Values and Power in International Politics.* Boulder: Westview Press, 1985.
Miller, Robert. *Liability or Asset: A Policy for the Falkland Islands.* London: Institute for European Defence and Strategic Studies, 1986.
Mitchell, Barbara. *Frozen Stakes: The Future of Antarctic Minerals.* London: International Institute for Environment and Development, 1983.
Mitchell, Barbara, and John Tinker. *Antarctica and Its Resources.* London: Earthscan Press, 1980.
Moneta, C., ed. *La Antartida en el Sistema Internacional del Futuro.* Buenos Aires: Grupo Editor Latinoamericano, 1988.
Morgan, Patrick M. *Theories and Approaches to International Politics.* 3rd. ed. New Brunswick: Transaction Books, 1981.
Morgenthau, Hans J. *Politics Among Nations: The Struggle for Power and Peace.* 5th ed. New York: Alfred Knopf, 1978.
Myhre, Jeffrey D. *The Antarctic Treaty System: Politics, Law, and Diplomacy.* Boulder: Westview Press, 1986.
Nagel, Ernest. *The Structure of Science: Problems in the Logic of Scientific Explanations.* 2d ed. Indianapolis: Hackett Publishing, 1979.
Neider, Charles. *Edge of the World: Ross Island, Antarctica.* Garden City: Doubleday, 1974.
———, ed. *Antarctica: Authentic Accounts of Life and Exploration in the World's Highest, Driest, Windiest, Coldest and Most Remote Continent.* New York: Random House, 1972.
Nuechterlein, Donald E. *United States National Interests in a Changing World.* Lexington: University of Kentucky Press, 1973.
———. *National Interests and Presidential Leadership: The Setting of Priorities.* Boulder: Westview Press, 1978.
———. *America Overcommitted: United States National Interests in the 1980s.* Lexington: University of Kentucky Press, 1985.
Orrego, Vicuña Francisco, and Augusto Salinas Aramayo, eds. *El Desarrollo de la Antarctica.* Santiago: Editorial Universitaria, 1977.
Orrego, Vicuña Francisco. *Antarctic Bibliography.* Santiago: Instituto de Estados Internacionales, Universidad the Chile, 1987.
———. *Antarctic Mineral Exploitation: The Emerging Legal Framework.* Cambridge: Cambridge University Press, 1988.
———, ed. *Antarctic Resource Policy: Scientific, Legal and Political Issues.* Cambridge: Cambridge University Press, 1983.
Orrego, Vicuña Francisco, Maria Teresa Infante, and Pilar Armanet, eds. *Politica Antarctica de Chile.* Santiago: Editorial Universitaria, 1984.
Osgood, Robert Endicott. *Ideals and Self-Interest in America's Foreign Relations: The Great Transformation of the 20th Century.* Chicago: University of Chicago Press, 1953.
Parker, Bruce C., and Mary C. Holliman, eds. *Environmental Impact in Antarctica.* Blacksburg: Virginia Polytechnic Institute and State University, 1978.

Parsons, Anthony. *Antarctica: The Next Decade*. New Rochelle: Cambridge University Press, 1987.
Peterson, M. J. *Managing the Frozen South: The Creation and Evolution of the Antarctic Treaty System*. Berkeley: University of California Press, 1988.
Pinochet de la Barra, Oscar. *La Antartida Chilena*. 1st ed. Santiago: Editorial del Pacifico, 1948.
Polar Research Board. *Antarctic Treaty System: An Assessment*. Proceedings of Workshop held at Beardmore South Field Camp, Antarctica, January 7–13, 1985. Washington, D.C.: National Academy Press, 1986.
Porter, Eliot. *Antarctica*. New York: E. P. Dutton, 1978.
Potter, Neil. *Natural Resource Potentials of the Antarctic*. American Geographical Society, Occasional Publication No. 4. Burlington: The Lane Press, 1969.
Pyne, Stephen J. *The Ice: A Journey to Antarctica*. Iowa City: University of Iowa Press, 1986.
Quadri, Ricardo P. *La Antartida en la Politica Internacional*. Buenos Aires: Editorial Pleamar, 1983.
Quam, Louis O., ed. *Research in the Antarctic*. Washington, D.C.: American Association for the Advancement of Science, 1971.
Quigg, P. *A Pole Apart: The Emerging Issue of Antarctica*. New York: McGraw Hill, 1983.
Rodgers, Eugene. *Beyond the Barrier: The Story of Byrd's First Expedition to Antarctica*. Annapolis: Naval Institute Press, 1990.
Ronne, Finn. *Antarctic Conquest: The Story of the Ronne Antarctic Expedition, 1946–1948*. New York: G.P. Putnam's Sons, 1949.
———. *Antarctic Command*. Indianapolis: Bobbs-Merrill, 1961.
———. *Antarctica, My Destiny: A Personal History by the Last of the Great Polar Explorers*. New York: Hastings House, 1979.
Rose, Lisle A. *Assault on Eternity: Richard E. Byrd and the Exploration of Antarctica, 1946–47*. Annapolis: Naval Institute Press, 1980.
Rosenau, James N. *The Scientific Study of Foreign Policy*. Rev. ed. New York: Nichols Publishing Co., 1980.
———, ed. *International Politics and Foreign Policy: A Reader in Research and Theory*. Rev. ed. New York: The Free Press, 1969.
Santis, H., and R. Riesco. *Las Fronteras Antarticas de Chile*. Santiago: Imprenta de Gendarmeria, September 1986.
Schatz, Gerald S., ed. *Science, Technology and Sovereignty in the Polar Regions*. Lexington: D.C. Heath, 1974.
Schwerdtfeger, W. *Weather and Climate of the Antarctic*. New York: Elsevier Science Publishing, 1984.
Shapley, Deborah. *The Seventh Continent: Antarctica in a Resource Age*. Washington, D.C.: Resources for the Future, 1985.
Sharma, R. C., ed. *Growing Focus on Antarctica*. Delhi: Rajesh, 1986.
Siple, Paul A. *90° South: The Story of the American South Polar Conquest*. New York: G.P. Putnam's Sons, 1959.
Splettstoesser, John F., and Gisela A. M. Dreschhoff, eds. *Mineral Resources Potential of Antarctica*. Washington, D.C.: American Geophysical Union, 1990.
Stonehouse, B. *Polar Ecology*. New York: Chapman and Hall, 1989.
Sugden, David. *Arctic and Antarctic: A Modern Geographical Synthesis*. Totowa: Barnes and Noble, 1982.
Sullivan, Walter. *Quest for a Continent*. New York: McGraw Hill, 1957.

288 / Selected Bibliography

———. *Assault on the Unknown: The International Geophysical Year.* New York: McGraw Hill, 1961.
Suter, Keith D. *Antarctica: World Law and the Last Wilderness.* Sydney: Friends of the Earth, 1978.
———. *Antarctica: Private Property or Public Heritage.* Atlantic Highlands: Zed Books, 1991.
Triggs, G. *The Antarctic Treaty Regime: Environment and Resources.* Cambridge: Cambridge University Press, 1987.
Urrutia, Cecilia. *La Antartida Chilena.* Santiago: Empresa Editorial Nacional Quimantu, Ltda., 1972.
Vasquez, John A. *The Power of Power Politics: A Critique.* New Brunswick: Rutgers University Press, 1983.
Vaughan, Norman D. *With Byrd at the Bottom of the World: The South Pole Expedition of 1928–1930.* Harrisburg: Stackpole Books, 1990.
Von Glahn, Gerhard. *Law Among Nations: An Introduction to Public International Law.* 5th ed. New York: MacMillan Publishing, 1986.
van Mleghem, J., and P. Van Oye, eds. *Biodegraphy and Ecology in Antarctica.* The Hague: Dr W. Junk Publishers, 1965.
Wall, Patrick. *The Southern Ocean and the Security of the Free World.* London: Stacey International, 1977.
Walton, D. H., ed. *Antarctic Science.* Cambridge: Cambridge University Press, 1987.
Waltz, Kenneth N. *Theory of International Politics.* Reading: Addison-Wesley, 1979.
Westermeyer, William E. *The Politics of Mineral Resource Development in Antarctica: Alternative Regimes for the Future.* Boulder: Westview Press, 1984.
Wilson, Tuzo. *IGY: The Year of the New Moons.* New York: Alfred Knopf, 1961.
Wolfrum, Rudiger, ed. *Antarctic Challenge III: Conflicting Interests, Cooperation, Environmental Protection, Economic Development.* Proceedings of an Interdisciplinary Symposium July 7–12, 1987. Berlin: Duncker and Humblot, 1987.

Articles

Ahluwalia, K. "The Antarctic Treaty: Should India Become a Party to It?" *Indian Journal of International Law* 1 (1960–1961).
Alexander, Frank C. Jr. "Legal Aspects: Exploitation of Antarctic Resources: A Recommended Approach to the Antarctic Resource Problem." *University of Miami Law Review* 33 (December 1978).
Almond, Gabriel. "The Return to the State." *American Political Science Review* 82, no. 3 (September 1988).
Almond, Harry H. Jr. "Demilitarization and Arms Control: Antarctica." *Case Western Reserve Journal of International Law* 17, no. 2 (spring 1985).
Alverson, Dayton L. "Tug-of-War for the Antarctic Krill." *Ocean Development and International Law Journal* 8, no. 2 (1980).
Amstrup, Niels. "The Early Morgenthau. A Comment on the Intellectual Origins of Realism." *Cooperation and Conflict* 13 (1978).
Andersen, Robert M., and Lawrence Rudolph. "On Solid International Ground in Antarctica: A U.S. Strategy for Regulating Environmental Impact on the Continent." *Stanford Journal of International Law* 26, no. 1 (fall 1989).
Anderson, David. "The Conservation of Wildlife under the Antarctic Treaty." *Polar Record* 14, no. 88 (January 1968).
Anderson, Peter. "How the South Was Won." *The Wilson Quarterly* (autumn 1981).

Archdale, H. E. "Claims to the Antarctic." *Yearbook of World Affairs* 12 (1958).
Auburn, Francis M. "The White Desert." *International and Comparative Law Quarterly* 19 (1970).
———. "Legal Implications of Petroleum Resources Off the Antarctic Continental Shelf." *Ocean Yearbook* 1 (1978).
———. "United States Antarctic Policy." *Marine Technology Society Journal* 12, no. 1 (1978).
———. "Consultative Status Under the Antarctic Treaty." *International and Comparative Law Quarterly* 28 (1979).
Argon, Daniel. "Energy Security in the 1990s." *Foreign Affairs* 67, no. 1 (fall 1988).
Argon, R. D. "United States Inspects Four Peninsula Stations." *Antarctic Journal of the United States* 10, no. 3 (1975).
Auburn, Francis M. "The Antarctic Environment." In *Yearbook of World Affairs*. London: Stevens, 1981.
Baldwin, Gordon W. "The Dependence of Science on Law and Government: The International Geophysical Year, A Case Study." *Wisconsin Law Review* (January 1964).
Bankes, N. D. "Environmental Protection in Antarctica: A Comment on the Conservation of Antarctic Marine Living Resources." *The Canadian Yearbook of International Law* 19 (1981).
Barcelo, John J. III. "The International Legal Regime for Antarctica." *Cornell International Law Journal* 19, no. 2 (summer 1986).
Barnes, James N. "The Emerging Convention on the Conservation of Antarctic Marine Living Resources." In *The New Nationalism and the Use of Common Spaces: Issues in Marine Pollution and the Exploitation of Antarctica*, edited by Jonathan I. Charney. Totowa: Allenheld, Osmund, 1982.
Barrie, G. N. "The Antarctic Treaty: Example of Law and Its Sociological Infrastructure." *Comparative and International Law Journal of South Africa* 8 (July 1975).
Beck, Peter J. "Cooperative Confrontation in the Falkland Island Dispute: The Anglo-Argentine Search for a Way Forward, 1968–1981." *Journal of Interamerican World Affairs* 24, no. 1 (February 1982).
———. "Britain's Antarctic Dimension." *International Affairs* 59, no. 3 (summer 1983).
———. "British Antarctic Policy in the Early 20th Century." *Polar Record* 21, no. 134 (1983).
———. "Securing the Dominant Place in the Wan Antarctic Sun for the British Empire: The Policy of Extending British Control over Antarctica." *Australian Journal of Politics and History* 29, no. 3 (1983–1984).
———. "Britain's Role in the Antarctic: Some Recent Changes in Organization." *Polar Record* 22, no. 136 (January 1984).
———. "The United Nations and Antarctica." *Polar Record* 22, no. 137 (May 1984).
———. "Preparatory Meetings for the Antarctic Treaty, 1958–1959." *Polar Record* 22, no. 141 (1985).
———. "The Future of the Falkland Islands: A Solution Made in Hong Kong?" *International Affairs* 61, no. 4 (autumn 1985).
———. "The United Nations Study on Antarctica, 1984." *Polar Record* 22, no. 140 (May 1985).

———. "Antarctica Enters the 1990s: An Overview." *Applied Geography* 10, no. 4 (October 1990).

———. "Regulating One of the Last Tourism Frontiers: Antarctica." *Applied Geography* 10, no. 4 (October 1990).

Beeby, Christopher. "Towards an Antarctic Mineral Resources Regime." *New Zealand International Review* 7, no. 3 (May/June 1982).

———. "The Antarctic Treaty System and a Minerals Regime." *New Zealand Foreign Affairs Review* 35 (January/March 1985).

Bergin, Patrick T. "The Antarctic Treaty Regime and Legal and Political Implications." *Florida International Law Journal* 4, no. 1 (fall 1988).

Bernhardt, J. P. A. "Sovereignty in Antarctica." *California Western International Law Journal* 5 (1975).

Boczek, Boleslaw A. "The Protection of the Antarctic Ecosystem: A Study in International Environmental Law." *Ocean Development and International Law Journal* 13, no. 3 (1983–1984).

———. "The Soviet Union and the Antarctic Regime." *American Journal of International Law* 78 (October 1984).

Bogart, Paul S. "Environmental Threats in Antarctica." *Oceanus* 31, no. 2 (summer 1988).

Brigham, Lawson W. "The Soviet Antarctic Program." *Oceanus* 31, no. 2 (summer 1988).

Burmester, Henry C. "Liability for Damage from Antarctic Mineral Resource Activities." *Virginia Journal of International Law* 29, no. 3 (spring 1989).

Burton, Steven J. "New Stresses on the Antarctic Treaty: Toward International Legal Institutions Governing Antarctic Resources." *Virginia Law Review* 65, no. 3 (April 1979).

Butler, Shirley Oakes. "Owning Antarctica: Cooperation and Jurisdiction at the South Pole." *Journal of International Affairs* 31, no. 1 (spring/summer 1977).

Carl, Beverly May. "Claims to Sovereignty: Antarctica." *Southern California Law Review* 28 (1954–1955).

Carrol, James F. "Of Icebergs, Oil Wells, and Treaties: Hydrocarbon Exploitation Offshore Antarctica." *Stanford Journal of International Law* 19, no. 1 (spring 1983).

Cerone, Rudy J. "Survival of the Antarctic Treaty: Economic Self-Interest versus Enlightened International Cooperation." *Boston College International and Comparative Law Journal* 2, no. 1 (1978–1979).

Child, Jack. "South American Geopolitical Thinking and Antarctica." *International Studies Notes* 11, no. 3 (1985).

Colson, David A. "The Antarctic Treaty System: The Mineral Issue." *Law and Policy in International Business* 12 (1980).

———. "The United States Position on Antarctica." *Cornell International Law Journal* 19, no. 2 (summer 1986).

Conforti, Benedetto. "Territorial Claims in Antarctica: A Modern Way to Deal with an Old Problem." *Cornell International Law Journal* 19, no. 2 (summer 1986).

Daniels, Paul C. "The Antarctic Treaty." *Science and Public Affairs* 26, no. 10 (1970).

De Blij, H. J. "Physical Aspects: Resources, Environment, and Ecology." *University of Miami Law Review* 33 (December 1978).

Deporov, H. "Antarctica: A Zone of Peace and Cooperation." *International Affairs* (USSR), no. 11 (1983).

Dryden, Hugh L. "The IGY: Man's Most Ambitious Study of His Environment." *National Geographic Magazine* 109, no. 2 (February 1956).
Dufek, G. J. "What We Have Accomplished in Antarctica." *National Geographic Magazine* (October 1959).
Dugger, John A. "Exploiting Antarctic Mineral Resources—Technology, Economics and the Environment." *University of Miami Law Review* 33 (December 1978).
Elliot, David H. "Antarctica: Is There Any Oil or Natural Gas?" *Oceanus* 31, no. 2 (summer 1988).
Ellsworth, Lincoln. "My Flight Across Antarctica." *National Geographic Magazine* 70, no. 1 (July 1936).
El-Sayed, Sayed Z. "Biology of the Southern Ocean." *Oceanus* 18, no. 4 (summer 1975).
———, and M. A. McWhinnie. "Antarctic Krill: Problem of the Last Frontier." *Oceanus* 22 (1979).
Fowler, Alfred N. "Antarctic Logistics." *Oceanus* 31, no. 2 (summer 1988).
Francioni, Francesco. "Legal Aspects of Mineral Exploitation in Antarctica." *Cornell International Law Journal* 19 (summer 1986).
Frank, Ronald F. "The Convention on the Conservation of Antarctic Marine Living Resources." *Ocean Development and International Law Journal* 13 (1983–1984).
Friedheim, Robert, and Tuseneo Akaha. "Antarctic Resources and International Law: Japan, the United States and the Future of Antarctica." *Ecology Law Quarterly* 16, no. 1 (1989).
Gardam, Judith. "Management Regimes for Antarctic Marine Living Resources: An Australian Perspective." *Melbourne University Law Review* 15, no. 2 (December 1985).
Gordon, Arnold L. "The Southern Ocean and Global Climate." *Oceanus* 31, no. 2 (summer 1988).
Gould, Laurence M. "Some Geographical Results of the Byrd Antarctic Expedition." *Geographical Review* 21, no. 2 (April 1931).
———. "Strategy and Politics in the Polar Regions." *Annals American Academy Political Science* 255 (January 1948).
———. "Emergency of Antarctica: The Mythical Land." *Science and Public Affairs* 26, no. 10 (December 1970).
Graber, D. A. "Struggle for a Continent: Who Will Rule Antarctica." *World Affairs* (spring 1950).
Greenpeace International. "The Future of the Antarctic: Background for a Second UN Debate." October 22, 1984.
———. "Antarctica's Unspoiled Wilderness." *Greenpeace Examiner* 10, no. 4 (December 1985).
———. "The World Park Option for Antarctica: Background for a Fourth UN Debate." November 17, 1986.
Hall, H. R. "The Open Door into Antarctica: An Explanation of the Hughes Doctrine." *Polar Record* (April 1989).
Hambro, Edvard. "Some Notes on the Future of the Antarctic Treaty and Collaboration." *American Journal of International Law* 68 (April 1974).
Hanevold, Truls. "The Antarctic Treaty Consultative Meetings: Form and Procedure." *Cooperation and Conflict* 3/4 (1971).
———. "Inspections in Antarctica." *Cooperation and Conflict* 2 (1971).

Harding, Peter X. "Parallels in Isolation: Managing Space Problems in Antarctica." *Fletcher Forum* 5, no. 2 (summer 1981).
Hayton, Robert D. "The American Antarctic." *American Journal of International Law* 50, no. 3 (July 1956).
———. "Polar Problems and International Law." *American Journal of International Law* 52 (1958).
———. "The Antarctic Settlement of 1959." *American Journal of International Law* 54 (1960).
Heim, Barbara Ellen. "Exploring the Last Frontiers for Mineral Resources: A Comparison of International Law Regarding the Deep Seabed, Outer Space and Antarctica." *Vanderbilt Journal of Transnational Law* 23, no. 4 (1990).
Heimsoeth, Harold. "Antarctic Mineral Resources." *Environmental Policy and Law* 11, no. 3 (November 1983).
Honnold, Edward. "Draft Provisions of a New International Convention on Antarctica." *Yale Studies in World Public Order* 4, no. 1 (fall 1977).
———. "Thaw in International Law? Rights in Antarctica Under the Law of Common Spaces." *Yale Law Journal* 87, no. 4 (March 1978).
Howard, M. "The Convention on the Conservation of Antarctic Living Resources: A Five Year Review." *International and Comparative Law Quarterly* 38 (1989).
Ivanhoe, L. F. "Antarctica: Operating Conditions and Petroleum Prospects." *Oil Gas Journal* 78 (December 29, 1980).
Jacobsen, Mark P. "Recent Developments on the Convention on the Regulation of Antarctic Mineral Resource Activities." *Harvard International Law Review* 30, no. 1 (winter 1989).
Joyner, Christopher C. "Oceanic Pollution and the Southern Ocean: Rethinking the International Legal Implications for Antarctica." *Natural Resources Journal* 24, no. 1 (January 1984).
———. "Anglo-Argentine Rivalry After the Falklands/Malvinas War: Laws, Geopolitics and the Antarctic Connection." *Lawyer of the Americas* 15, no. 3 (winter 1984).
———. "Security Issues and the LOS: The Southern Ocean." *Ocean Development and International Law Journal* 15 (1985).
———. "Polar Politics in the 1980s: Some Preliminary Thoughts on Polar Contrasts and Geopolitical Considerations." *International Studies Notes* 11, no. 3 (1985).
———. "The Southern Ocean and Marine Pollution: Problems and Prospects." *Case Western Reserve Journal of International Law* 17 (1985).
———. "Protection of the Antarctic Environment: Rethinking the Problems and Prospects." *Cornell International Law Journal* 19, no. 2 (summer 1986).
———. "The Antarctic Minerals Negotiating Process." *American Journal of International Law* 81, no. 4 (October 1987).
———. "The Exclusive Economic Zone and Antarctica: The Dilemmas of Non-Sovereign Jurisdiction." *Ocean Development and International Law Journal* 19 (1988).
———. "The Evolving Antarctic Minerals Regime." *Ocean Development and International Law Journal* 19 (1988).
———. "The Antarctic Legal Regime and the Law of the Sea." *Oceanus* 31, no. 2 (summer 1988).
———. "The Evolving Antarctic Legal Regime." *American Journal of International Law* 83, no. 3 (July 1989).
———. "1988 Antarctic Minerals Convention." *Marine Policy Reports* 1 (1989).

———. "Non-Militarization of the Antarctic: the Interplay of Law and Politics." *Naval War College Review* 42 (autumn 1989).
———. "Japan and the Antarctic Treaty System," *Ecology Law Quarterly* 16, no. 1 (1989).
———. "Antarctica and the Indian Ocean States: The Interplay of Law, Interests, and Geopolitics." *Ocean Development and International Law Journal* 21, no. 1 (1990).
———. "Ice-Covered Regions in International Law." *Natural Resources Journal* 31 (winter 1991).
———. "The 1991 Antarctic Environmental Protocol: What Prospects for Antarctica as a World Park?" *Review of European Community and International Environmental Law* 1, no. 3 (1992).
———. "Fragile Ecosystems: Preclusive Restoration in the Antarctic." *Natural Resources Journal* 34 (fall 1994).
———. "The Antarctic Treaty Systems and the Law of the Sea: Competing Regimes in the Southern Ocean?" *The International Journal of Marine and Coastal Law* 10, no. 2 (May 1995).
———, and Blair Ewing. "Antarctica and the Latin American States: The Interplay of Law, Geopolitics, and Environmental Priorities." *Georgetown International Environmental Law Review* 4, no. 1 (spring/summer 1991).
———, and Ethel R. Theis. "The United States and Antarctica: Rethinking the Interplay of Law and Interests." *Cornell International Law Journal* 20 (winter 1987).
Kimball, Lee A. "Critical Antarctic Issues Emerging." *Oceanus* 26, no. 3 (fall 1983).
———. "Whither Antarctica?" *International Studies Notes* 11, no. 3 (spring 1985).
———. "Report on Antarctica: The Antarctic Minerals Convention." Special Report, *World Resources Institute* (February 1988).
———. "The Antarctic Treaty System." *Oceanus* 31, no. 2 (summer 1988).
———. "Report on Antarctica: The Antarctic Minerals Convention." Special Report, *World Resources Institute* (July 1988).
———. "Report on Antarctica, November 1989." *World Resources Institute* (1989).
———. "The Policy Framework for Environmental Monitoring in Antarctica." *Antarctic Journal of the United States* 24, no. 4 (December 1989).
———. "Southern Exposure: Deciding Antarctica's Future." *World Resources Institute* (November 1990).
Koch, Michael. "The Antarctic Challenge: Conflicting Interests, Cooperation, Environmental Protection, and Economic Development." *Journal of Maritime Law and Commerce* 15, no. 1 (January 1984).
Knudsen, Olav. "National Interests and Foreign Policy: On the National Pursuit of Material Interests." *Cooperation and Conflict* 14, no. 1 (1979).
Kratochwil, Friedrich. "On the Notion of 'Interest' in International Relations." *International Organization* 36 (winter 1982).
Macknis, Joseph. "United States Policy in Antarctica." *Marine Policy Reports* 2, no. 2 (May 1979).
Manheim, Bruce S. "On Thin Ice: The Failure of the National Science Foundation to Protect Antarctica." Report for Environmental Defense Fund Wildlife Program. August 17, 1988.
Marcoux, Michel J. "Natural Resource Jurisdiction on the Antarctic Continental Margin." *Virginia Journal of International Law* 11, no. 3 (May 1971).

McKenzie, Garry D. "National Security and the Austral Crescent." *Mershon Center Quarterly Report* 8, no. 4 (spring 1984).
Milenky, Edward S., and Steven I. Schwab. "Latin America and Antarctica." *Current History* 82 (February 1983).
Miller, A. C. "Antarctica: White Continent of Promise." *U.S. Naval Institute Proceedings* 88 (August 1962).
Molina, Mario J. "The Antarctic Ozone Hole." *Oceanus* 31, no. 2 (summer 1988).
Morgenthau, Hans J. "Another 'Great Debate': The National Interest of the United States." In *American National Security: A Reader in Theory and Policy*, edited by M. Berkowitz and P. G. Bock. New York: The Free Press, 1965.
Nuechterlein, Donald E. "The Concept of the 'National Interest': A Time for New Approaches." *Orbis* (spring 1979).
———. "Foreign Policy in the 1980s: In Search of the National Interest." *Foreign Service Journal* (May 1981).
———. "National Interests and National Strategy: The Need for Priority." In *Understanding U.S. Strategy: A Reader*, edited by Terry L. Heyns. Washington, D.C.: National Defense University Press, 1985.
Nye, Joseph. "Energy and Security in the 1980s." *World Politics* (October 1982).
Overholt, Deborah H. "Environmental Protection in the Antarctic: Past, Present, and Future." *The Canadian Yearbook of International Law* 28 (1990).
Oxman, Bernard H. "The Antarctic Regime: An Introduction." *University of Miami Law Review* 33 (December 1978).
———. "Evaluating the Antarctic Minerals Convention: The Decision Making System." *Interamerican Law Review* (fall 1989).
Pallone, F. "Resource Exploitation: The Threat to the Legal Regime of Antarctica." *Connecticut Law Review* 10 (1978).
Parriott, Todd J. "Territorial Claims in Antarctica: Will the United States Be Left in the Cold?" *Stanford Journal of International Law* 22, no. 1 (spring 1986).
SCAR. Steering Committee for the International Geosphere-Biosphere Programme. "The Role of Antarctica in Global Change." April 1989.
Scott, Ronald W. "Protecting United States Interests in Antarctica." *San Diego Law Review* 26, no. 3 (1989).
Scully, R. Tucker. "The Marine Living Resources of the Southern Ocean." *University of Miami Law Review* 33, no. 2 (December 1978).
———. "Inspection of Non-US Stations in Antarctica." *Antarctic Journal of the United States* 15, no. 5 (1980).
———. "Antarctic Resources Focus of Two Important Meetings." *Antarctic Journal of the United States* 18, no. 1 (March 1983).
———. "The Antarctic Mineral Resources Negotiations." *Oceanus* 31, no. 2 (summer 1988).
———. "The Antarctic Minerals Agreement." In *Proceedings of the Annual Meeting of the American Society of International Law* (April 5–8, 1989).
———. "Protecting Antarctica: Progress in Chile." *Antarctic Journal of the United States* 26, no. 1 (March 1991).
———, and L. A. Kimball. "Antarctica: Is There Life after Minerals." *Marine Policy Reports* (April 1989).
Sherman, Kenneth, and Alan F. Ryan. "Antarctic Marine Living Resources." *Oceanus* 31, no. 2 (summer 1988).
Shusterich, Kurt M. "The Antarctic Treaty System: History, Substance and Speculation." *International Journal* 39 (1984).

Simma, Bruno. "The Antarctic Treaty as a Treaty Providing for an Objective Regime." *Cornell International Law Journal* (summer 1986).
Simoes, Jefferson. "Brazilian Antarctic Research Program." *Polar Record* 22, no. 138 (September 1984).
Simpson, John A. "The International Geophysical Year—A Study of Our Planet." *Science and Public Affairs* 13, no. 10 (December 1957).
Simsarian, J. "The Acquisition of Legal Title to Terra Nullius." *Political Science Quarterly* 53 (1938).
———. "Inspection Experience under the Antarctic Treaty and the International Atomic Agency." *Antarctic Journal of the United States* 10, no. 3 (July 1966).
Siniff, Donald B. "Antarctic Marine Living Resources: Seals." *Oceanus* 31, no. 2 (summer 1988).
Siple, Paul A. "We Are Living at the South Pole." *National Geographic Magazine* 112, no. 1 (July 1957).
Sisco, Joseph J. "The United States Program in Antarctica." *Antarctic Journal of the United States* 1, no. 1 (1966).
Skagestad, Gunnar. "The Frozen Frontier Models for International Cooperation." *Cooperation and Conflict* 10 (1975).
Smith, Philip M. "International Cooperation in Antarctica: The Next Decade." *Science and Public Affairs* 26, no. 10 (December 1970).
———, and Rodney W. Johnson. "From the South Pole to the Moon: Parallels in Exploration." *Science and Public Affairs* 24, no. 10 (December 1968).
Sollie, Finn. "The Political Experiment in Antarctica." *Science and Public Affairs* 26, no. 10 (December 1970).
———. "The New Developments in the Polar Regions." *Cooperation and Conflict* 2/3 (1974).
Sondermann, Fred. "The Concept of the National Interest." *Orbis* 21, no. 1 (1977).
Stone, Jeffrey E. "International Agreements: Antarctic Resources." *Harvard International Law Journal* 22 (winter 1981).
Stuhlinger, Ernst. "Antarctic Research: A Prelude to Space Research." *Antarctic Journal of the United States* 4, no. 1 (January/February 1969).
Sullivan, Walter. "Antarctica in a Two-World Power." *Foreign Affairs* (October 1957).
———. "The International Geophysical Year." *International Conciliation* 521 (January 1959).
Waller, Deborah Cook. "Death of a Treaty: The Decline and Fall of the Antarctic Minerals." *Vanderbilt Journal of Transnational Law* 22, no. 3 (1989).
Walton, David W. H., and Elizabeth M. Morris. "Science, Environment and Resources in Antarctica." *Applied Geography* 10, no. 4 (October 1990).
Wasilewski, Peter. "NASA Participation in the 25th Japanese Antarctic Research Expedition." *Antarctic Journal of the United States* 19, no. 5 (1984).
———. "NASA Scientist Joins the 25th Japanese Antarctic Research Expedition, 1983–1984." *Antarctic Journal of the United States* 20, no. 1 (1985).
Wasserman, Ursula. "The Antarctic Treaty and Natural Resources." *Journal of World Trade Law* 12 (1978).
Westermeyer, William E. "Energy from the Polar Regions." *International Journal* 39 (autumn 1984).
Wijkman, Per Magnus. "Managing the Global Commons." *International Organization* 36, no. 3 (summer 1982).

Wilkniss, Peter. "The Polar Regions: Research in a Changing World." *Bulletin American Meteorological Society* 70, no. 2 (February 1989).
———. "Fuel Spill Clean in the Antarctic." *Antarctic Journal of the United States* 25, no. 4 (December 1990).
Wilson, Gregory P. "Antarctica, the Southern Ocean, and the Law of the Sea." *The Judge Adjutant General* 30 (summer 1978).
Wolfers, Arnold. "National Security as an Ambiguous Symbol." In *American National Security: A Reader in Theory and Policy*, edited by M. Berkowitz and P. G. Bock. New York: The Free Press, 1965.
Wright, Charles. "The Antarctic and the Upper Atmosphere." *Scientific American* (September 1962).
Wright, N. A., and P. L. Williams. "Mineral Resources of Antarctica." U.S. Geological Survey, Circular 705, 1974.
Young, Allan. "Antarctic Resource Jurisdiction and the Law of the Sea: A Question of Compromise." *Brooklyn Journal of International Law* 11, no. 1 (winter 1985).
Zegers, Santa Cruz F. "The Antarctic System and the Utilization of Resources." *University of Miami Law Review* 33, no. 2 (December 1978).
Zeller, E. J., D. F. Saunders, and E. E. Angino. "Putting Radioactive Wastes on Ice: A Proposal for an International Radionuclide Depository in Antarctica." *Science and Public Affairs* 29, no. 1 (January 1973).
Zorn, Stephen A. "Antarctic Minerals: A Common Heritage Approach." *Resources Policy* 10, no. 1 (March 1984).
Zumberge, James H. "Mineral Resources and Geopolitics in Antarctica." *American Scientist* 67, no. 1 (January/February 1979).
———. "18th Meeting of SCAR Focuses on Conservation and International Cooperation." *Antarctic Journal of the United States* 20, no. 2 (June 1985).

Index

Acheson, Dean, 39, 226
Advisory Committee on Oceans and International Environmental and Scientific Affairs. *See* Department of State
Agreed Measures for the Conservation of Antarctic Fauna and Flora, 44, 103, 104, 118, 181, 228, 243; Sites of Special Scientific Interest, 43, 103, 104; Specially Protected Areas, 43, 103, 104, 184
Amoco Cadiz, 144
Amundsen-Scott Station. *See* South Pole Station
Andersen, Rolfe Trolle, 56
Anderson, John B., 136, 255
Antarctic climate, 12, 13, 15, 17, 216; blizzards, 12, 13, 16; snowdrifts, 13, 16, 17; temperatures, 10, 12, 13, 15; "whiteout," 16
Antarctic Commission, 27
Antarctic Conservation Act of 1978, 111, 124, 243, 250
Antarctic continental shelf, 265
Antarctic Convergence, 14, 107, 112
Antarctic Environmental Protection, Clean-Up and Liability Act of 1990, 53, 64–66, 120
Antarctic Journal of the United States, 85
Antarctic Marine Living Resources Convention Act of 1984 (AMLRCA), 84, 112, 118
Antarctic Policy Group (U.S.), 48, 49, 50, 53–55, 58–61, 72, 75, 104, 106, 137, 139; Antarctic Policy Working Group (U.S.), 53, 54, 55
Antarctic and Southern Ocean Coalition (ASOC), 70–72, 126, 128–30, 140, 252
Antarctic Treaty, 20–44, 52, 78, 97, 188–93; amendments, 35; Antarctica as a nuclear-free zone, 224; Article IV, 34, 36, 37, 107; claims, 225; claims to territory, 30, 32, 36–40, 149, 150, 171, 173; consultative meetings, 34, 41–43, 54, 56–60, 102–10, 138–44, 173, 178, 179, 183; decisionmaking, 41–43; dispute settlement, 34, 52, 53; entry into force, 33; geographic scope, 33; inspection provision, 34, 153, 154, 224, 263; join, 35; major features, 33–36, 78, 97, 98; negotiation, 32, 33; origins and U.S. role, 29–33; ratification, 33, 223; recommendations, 42, 43, 54, 102, 103; resource development, 34, 35, 140; review conference, 35; secrecy problem, 41, 42; secretariat, 227; signatories, 175; withdraw, 35
Antarctic Treaty Consultative Parties (ATCPs), 42–44, 55–60, 67, 88, 100–10, 131, 132, 138–44, 174–76, 180, 181, 242
Antarctic Treaty System, 1, 35, 45, 77, 100–10, 131, 132, 177–87, 227, 228; defined, 43, 44; importance to the United States, 44, 100–10
Antarctica: as condominium, 30; ecosystem, 10–19; flora and fauna, 13; as international trusteeship, 30; territorial claims to the continent, 36–40
Antarctica Project, 126, 127, 128, 140
Arctic, 11, 157, 158, 258, 259; *Exxon Valdez* oil spill, 57, 144; strategic significance, 3, 33, 38, 97, 145, 156, 157
Argentina, 5, 30, 33, 38, 50, 97, 141, 151, 171, 173, 245, 259, 260; *Bahia Paraiso* oil spill, 57, 117, 144; claims to territory, 30, 32, 149–50, 171, 173, 221, 261, 262
Arnaudo, Raymond, 231–32, 235, 258, 263
ASOC. *See* Antarctic and Southern Ocean Coalition (ASOC)
Atkinson, Richard C., 47, 50, 89, 90, 217, 239, 240
Auburn, Francis M., 39, 108, 216, 228, 229, 245
Australia, 33, 41, 57, 58, 96, 108, 109, 148, 178, 179, 245, 249; CCAMLR Commission, Scientific Committee and Secretariat, 143; claims to territory, 36; reversal on CRAMRA, 57, 58, 108, 120, 140, 178, 179, 232, 233

Baker, James, 60
Ball, George, 49, 54

298 / Index

Barnes, James N., 257
Bay of Whales, 17
Beck, Peter J., 49, 214, 217, 222, 228, 261, 262, 270, 271
Beeby, Christopher, 257
Belgium, 105, 245
Benson, Lucy Wilson, 106
Bentley, Charles, 239
Berkner, L. V., 149, 221
Bernhardt, J. Peter, 214
Bertrand, Kenneth J., 40, 217, 218, 226
Biological Investigations of Marine Antarctic Systems and Stocks (BIOMASS), 71, 93, 94, 96, 241
Bohlen, E. U. Curtis, 59, 63, 65, 139, 232, 257
Boucher, Rick, 122
Brazil, 105, 175, 245
Brewster, Barney, 215
Bulgaria, 245
Bullis, Harold, 212
Bureau of the Budget (U.S.) (now Office of Management and Budget), 27, 28, 47
Burke, Arleigh, Adm., 29
Burton, Steven, 165, 268
Bush, George, 52, 53, 59, 66, 120, 140, 179
Bush, W. M., 217, 220, 231
Byrd, Adm. Richard E., 16, 17, 22–25, 37, 38, 79, 215–18, 226
Byrd Station (U.S.), 16

Cameron, Ian, 13, 214–16
Canada, 105, 158, 245
Carter, Jimmy, 110, 114
CCAMLR. *See* Convention on the Conservation of Antarctic Marine Living Resources
CCAS. *See* Convention on the Conservation of Antarctic Seals
Central Intelligence Agency, 75, 214, 216
Charney, Jonathan, 254
Child, Jack, 215
Chile, 5, 30, 33, 38, 97, 142, 149, 151, 171, 179, 245, 259, 260, 270; claims to territory, 30, 32, 36, 149, 150, 171, 221, 261, 262
China, 175
CHM. *See* Common heritage of mankind
Christie, E. W. Hunter, 223
Circular A-51 (OMB), 47, 50
Claims to Antarctic territory, 30, 32, 36, 140–50, 171; and common heritage, 170–74; "freezing" the legal status, 31, 36, 37; nonrecognition, 170; overlapping, 149, 150, 222; U.S. attitude toward claims, 30, 37, 170–74; *uti possidetis*, 36, 225
Cleveland, Harlan, 46, 49, 229, 237
Clinton, David, 212
Clinton, William J., 89, 122, 247
Code of Conduct for Antarctic Expeditions and Stations' Activities (1975), 181, 182
Cold War, 4, 150–53, 263
Colson, David A., 138, 225, 227, 228, 231, 242, 257, 270
Committee on Polar Research. *See* Polar Research Board
Committee on Science (U.S.), 62
Common heritage of mankind (CHM), 161–74, 266–68; common heritage and U.S. interests, 170–74; defined, 162–66; proposed regime for Antarctica, 166–70; status in international law, 167–70
Conference on Antarctica. *See* Washington Conference
Congressional Research Service, 136
Consultative Meetings. *See* Antarctic Treaty
Consultative Status. *See* Antarctic Treaty
Conte, Silvio, 64, 66, 120
Continental Shelf (Antarctica), 143
Convention on the Conservation of Antarctic Marine Living Resources (CCAMLR), 14, 44, 71, 105–108, 112, 127, 128, 141, 142, 153, 154, 215, 228, 245, 263, 264; Commission, 55, 107, 108, 143; Final Act, 106; origin, 106; provisions, 107; Scientific Committee, 107, 108, 143
Convention on the Conservation of Antarctic Seals (CCAS), 44, 104, 105, 143, 154, 228, 244
Convention on the Regulation of Antarctic Mineral Resource Activities (CRAMRA), 44, 55–57, 59, 63–67, 73, 108, 109, 119–22, 129, 130, 139–42, 154, 159, 160, 177, 178, 233, 245, 257; environmental considerations, 65, 109, 140–42, 159, 160; inspection regime, 141; policymaking commission, 141, 142; provisions, 109; ratification, 140, 223; regulatory committee, 142; resource development, 109, 140–42; Scientific, Technical and Environmental Advisory Committee, 142; U.S. Congress opposition, 140; Wellington Minerals Convention, 56, 57, 58, 65, 178
Convention on the Law of the Sea (1982), 266, 267, 268
Cook, James, 80

Corell, Robert, 250
Council on Environmental Quality, U.S., 111, 113
Council of Managers of Antarctic Programs, 88, 239
Cousteau, Jacques, 65, 234
Cousteau Society, 126, 129, 130, 140
Crevasses, 16
CRAMRA. *See* Convention on the Regulation of Antarctic Mineral Resource Activities

Daniels, Paul C., 32, 223
Dater, Henry M., 220
Deacon, George, 216
de Blij, Harm J., 215
Deep Seabed Hard Mineral Resources Act of 1980, 166
Department of Commerce (U.S.): National Oceanic and Atmospheric Administration (NOAA), 51, 54, 64, 84, 86, 92, 94, 95, 111–13, 125, 247, 248
Department of Defense (U.S.), 48, 49, 53, 82, 113
Department of the Interior (U.S.), 121; Geological Survey, 51, 84, 135, 136
Department of Justice (U.S.), 113–15
Department of the Navy (U.S.), 25, 27, 51, 82
Department of State (U.S.), 22, 23, 27, 45, 48, 53, 74, 75, 106, 111, 112, 118, 121, 136; Advisory Committee on Oceans and International Environmental and Scientific Affairs, 70, 127, 128; Bureau for Oceans and International Environment and Scientific Affairs (OES), 49; Bureau of International Organization Affairs, 49; Bureau of International Scientific and Technological Affairs, 49; Policy Planning Staff, 26, 147; position on claims, 37, 40
Department of Transportation: Coast Guard, 51, 52, 84, 111, 124, 125
Division of Polar Programs. *See* National Science Foundation
Drake Passage, 11, 147, 148, 213, 214
Dry Valley Drilling Project, 96
Dudeney, J. R., 12, 214
Dufek, Admiral George, 18, 27, 216

East Antarctica, 11
Eisenhower, Dwight, 26–29, 32, 33, 47, 48, 220
Ellsworth, Lincoln, 22, 23, 37, 38, 47, 218, 226
El-Sayed, Sayed, 14, 15, 254, 258

Eltanin. See Icebreakers
Endangered Species Act of 1976, 100
Environmental Action Memorandum (EAM), 114
Environmental Defense Fund (EDF), 112, 126, 128, 130, 131, 140, 252
Environmental Defense Fund, Inc., v. Walter E. Massey, 185, 253, 254, 273
Environmental Impact Assessment (EIA), 114, 123–25, 181
Environmental Impact Statement (EIS), 113–15
Environmental Protection Agency, 54, 111–13, 116, 125, 248
Esperanza Bay, 57
Executive Order 12114 (U.S.), 113, 114, 248
Expeditions (Antarctic): Byrd, 22, 23; Ellsworth, 22, 23; Germany, 24, 146; Great Britain, 96, 173; Norway, 147; Operation Deepfreeze I (U.S.), 28, 38; Operation Deepfreeze II (U.S.), 28, 39; Operation Highjump (U.S.), 25, 38, 219; Operation Windmill (U.S.), 38; Ronne, 22; United States Exploring Expedition, 21, 24, 25
Exxon Valdez oil spill. *See* Arctic

Falkland/Malvinas Islands War, 151, 173, 190, 261; Argentina claim, 151; British claim, 151
FIBEX. *See* First International Biological Experiment
Final Supplemental Environmental Impact Statement for the U.S. Antarctic Program (Final SEIS), 114, 115, 248, 250
Finland, 245
First International Biological Experiment (FIBEX), 93
Fogg, G. E., 214
Ford, Gerald, 52, 105
Fowler, Alfred E., 228, 236, 240
France, 21, 57, 58, 108, 142, 156, 178, 179, 245; claims, 36; reversal on CRAMRA, 57, 58, 108, 120, 140, 178, 179, 233, 245

Germany, 24, 146, 147, 245
Glaciology, 91, 92, 240
Glomar Challenger, 135
Gondwanaland, 135
Gore, Albert, 63–66, 120
Gould, Laurence M., 226, 235
Grady, Robert E., 59
Great Britain. *See* United Kingdom (U.K.)
Greece, 245

300 / Index

Greenpeace, 69, 71, 126, 128–30, 140
Guthridge, Guy, 241

Halle, Louis J., 215
Halley Bay Station (U.K.), 16
Handl, Gunther, 253, 258
Hanessian, John, 221
Haworth, Leland J., 213, 236–38
Hawke, Robert, 57, 58
Hayton, Robert D., 217, 226, 262
Heap, John, 228, 243, 245, 257, 271, 272
Helms, Jesse, 120
Herter, Christian, 223
Hollings, Ernest F., 62
House Committee on Foreign Affairs, 65
House Committee on Merchant Marine and Fisheries, 65
House Committee on Science, Technology and Space (now Committee on Science), 62, 89
Hughes, Charles Evans, 37, 38
Hull, Cordell, 146
Hydrocarbons, 45, 133–36, 143–45, 154–57, 254, 255, 264

Ice, 11, 12, 144; Antarctic ice sheet, 11; ice sheet, 11, 214; ice shelves, 11; pack ice, 11, 17
Icebergs, 11, 17, 143, 144, 157, 214
Icebreakers (U.S.), 158; *USNS Eltanin*, 85, 92, 93; *U.S. Polar Duke*, 82, 83, 95
ICSU. *See* International Council of Scientific Unions
IGY. *See* International Geophysical Year
Inderbitzen, Anton, 231
India, 175, 245; United Nations proposal, 30
Interagency Committee on Antarctica, 48
Interest groups. *See* Nongovernmental organizations (NGOs)
International Convention for the Prevention of Pollution from Ships, 1973/78 (MARPOL), 123, 182, 183, 231
International Council of Scientific Unions (ICSU), 87
International Court of Justice, 150, 152, 153
International Geophysical Year (IGY) 1957/1958, 1, 18, 21, 26–32, 78, 80, 137, 221; scientific achievements, 31
International law, 152, 153
International Whaling Commission, 143
Italy, 245

Jackson, Andrew, 21
Japan, 4, 41, 109, 142, 143, 245

Jennings, R. Y., 224
Johnson, Lyndon B., 48, 49, 110, 162, 163, 229, 246, 266
Jones, T. O., 28, 243
Jones, Walter, 66
Joyner, Christopher C., 221, 225, 232, 233, 234, 235, 243, 244, 251, 258, 259, 260, 262, 263, 267, 271

Kennedy, John F., 33, 48, 49
Kerry, John, 64, 66, 120, 122
Kimball, Lee, 103, 130, 234, 253, 264, 265
King George Island, 57
King, H. G. R., 216
Krasner, Stephen D., 213
Krill, 15, 134, 140–42, 254, 257; catches, 142; population, 142, 143

Lacey, Michael J., 252
Lane, Neal, 89
Laughlin, Thomas, 235, 237, 241
Law of the Sea Convention (1982), 162–64, 166, 169
Laws, R. M., 243
Lewis, Richard S., 12, 214
Little America Base (U.S.), 22

Malaysia, 166–70, 186
Madrid Conference, 58, 59, 109, 110
Malta, 162–64
Malvinas Islands. *See* Falkland/Malvinas Islands War
Manheim, Bruce S., Jr., 247, 250
Marine Mammal Commission, 54, 111, 113
Marine Mammal Protection Act of 1976, 100
Marie Byrd Land, 17, 22, 37–39
Mawson, Douglas, 12
Mayhew, Richard, 62
McMurdo Station (U.S.), 94, 96, 115–17
Meteorology, 92
Miller, A.C., 226
Minerals Convention. *See* Convention on the Regulation of Antarctic Mineral Resource Activities (CRAMRA)
Minerals Resources (Antarctic), 55–60, 133–46, 156, 157; mining operations, 140
Mink, Patsy T., 110, 127, 244, 246, 250, 252, 256
Monroe Doctrine, 146
Moon Treaty (1979), 162, 266, 267
Morgenthau, Hans, 212
Multiple condominium, 30
Myhre, Jeffrey D., 223

Index / 301

National Academy of Sciences (U.S.), 24, 80, 85, 86, 238; Committee on Polar Research, 28, 85; National Committee for the IGY, 28, 80
National Aeronautics and Space Administration (NASA), 53, 54
National Environmental Policy Act (NEPA), 66, 100, 113–15, 120, 121, 123–26, 130, 185, 186
National Geographic Magazine, 71
National Interests, United States, 1–6, 130, 131; defined, 3, 4; described in the Antarctic, 47, 52, 53, 65, 79, 98; limitations and strengths of the concept, 5, 6, 9
National Oceanic and Atmospheric Administration (NOAA). *See* Department of Commerce
National Research Council, 24, 86
National Science Board, 115
National Science Foundation, U.S. (NSF), 27–29, 45, 47–53, 62, 72, 80, 84, 85, 91, 92, 110–18, 121, 122–25, 135, 184, 185, 236, 237, 240, 249, 273; Advisory Committee for Polar Programs, 72, 86, 87; creation, 27–29, 81; Office of the Director, 50, 76; Office of the General Counsel, 116, 249; Office of Polar Programs (former Division of Polar Programs), 51, 72, 81–83, 86, 116, 117, 184, 249 (functions, 81–83, 91; funding [Antarctic program], 89; logistics support, 81; responsibilities, 51, 82, 114, 115)
National Security Council, 26, 28, 45, 49, 55, 73, 75, 136, 256; Operations Coordinating Board, 27, 48
National Security Decision, 1, 53
National Security Decision Memorandum No. 71, 50, 52, 230
National Security Decision Memorandum No. 318, 52, 230, 231
Naval Support Force, Antarctica (U.S.), 83, 84
Negroponte, John, 110
NEPA. *See* National Environmental Policy Act
Netherlands, 245
New International Economic Order (NIEO), 162–74
New Zealand, 11, 23, 56, 57, 173, 179, 244, 245; Claims, 36; CRAMRA headquarters, 177; CRAMRA negotiations, 56, 57
NIEO. *See* New International Economic Order

Nixon, Richard M., 50, 110, 137
NOAA. *See* Department of Commerce
Nongovernmental organizations (NGOs), 68–72, 126–31
Norway, 143, 147, 245; claims, 36
NSF. *See* National Science Foundation

O'Brien, Leo W., 52
Oceanography, 92
Office of Antarctic Programs, 229
Office of Management and Budget (OMB) (U.S.), 59, 73, 138, 219, 256
Office of Oceans and Polar Affairs, 53–54. *See also* Department of State
Office of Technology Assessment, 156, 158, 213, 223
Oil spills, 143–45, 156, 157, 258, 259
OMB. *See* Office of Management and Budget
O'Neill, Thomas P., 250
Operation Deepfreeze I. *See* Expeditions
Operation Deepfreeze II. *See* Expeditions
Operation Highjump. *See* Expeditions
Operation Tabarin, 261
Operation Windmill. *See* Expeditions
Oxman, Bernard H., 225
Owens, Wayne, 64, 65, 120, 206
Ozone, 63; depletion, 58, 63, 92, 98

Pakistan, 167
Palmer, Nathaniel, 21, 217
Palmer Peninsula, 37
Palmer Research Station (U.S.), 57, 94, 95, 116, 117
Panama Canal, 147, 148, 261
Pardo, Arvid, 162–64, 269, 270
People's Republic of China, 175
Peru, 245
Peterson, M. J., 105, 212, 222, 224, 226, 234, 242, 243, 248, 252, 260
Phleger, Herman, 149, 153, 222, 223, 235, 264
Phytoplankton, 15
Plott, Barry, 224
Poland, 142, 146
Polar Duke. *See* Icebreakers
Polar Experiment (POLEX)–South, 96
Polar Marine Biology, 93, 94
Polar Research Board (former Committee on Polar Research), 28, 29, 71, 83, 85, 86, 220
Polar Star, 217
Policy Coordinating Committee, 53, 55, 59
Porter, Roger, 59
PPS-31, 26, 147

Presidential Memorandum No. 6646, 80, 81
Principle of discovery and occupation, 36
Programmatic Environmental Impact Statement (PEIS), 114, 115
Protocol on Environmental Protection to the Antarctic Treaty (1991), 10, 19, 44, 55–61, 99, 104, 109, 110, 122–26, 155, 156, 176–87, 227, 246, 271–73; Antarctic Specially Managed Areas, 184; Antarctic Specially Protected Areas, 184, 122–26, 155, 176–87
Pyne, Stephen, 255

Queen Maude Land, 147
Quigg, Philip, 108, 109, 215, 220, 226, 229, 230, 232, 243, 245, 246, 253, 262, 263

Rational Decision Making Model, 7
Ray, Dixie Lee, 46, 137, 224, 256
Reagan, Ronald, 52, 81, 190, 245, 247
Recommendations of Consultative Parties. *See* Antarctic Treaty
Republic of Korea, 142
Res communis, 164
Research. *See* Scientific research
Resources (Antarctic): marine, 134, 135 (krill, 254, 257; whales, 254, 258); mineral (particularly hydrocarbons), 13, 133–46, 254–56
Richardson, Elliot, 165, 166
Ronne, Finn, 22, 147, 148, 260–61
Roosevelt, Franklin D., 24, 25, 38, 79, 218, 219
Ross Ice Shelf, 11, 37
Ross Sea, 135
Rudolph, Lawrence, 229, 249, 251, 253
Russia, 40
Rutford, Robert H., 220, 229, 238, 260
Rutherford, Robert H., 247

Sabella, Susan, 234, 243, 245, 253
Safety, Environment and Health Initiative (SEHI), 115
Safety in Antarctica, 115
Sastrugi, 16
SCAR. *See* Scientific Committee on Antarctic Research
Schmidt, Markus G., 127, 252
Schwarzenberger, Georg, 224
Scientific Committee on Antarctic Research (SCAR), 43, 86–88, 102, 105, 117, 228
Scientific research (U.S.), 78–99, 90–95; funding, 62, 63, 89; International Geophysical Year, 26–32, 78, 80; political significance, 31, 90; research vessels, 82, 92, 93, 95; scientific cooperation, 95–99
Scully, R. Tucker, 51, 52, 97, 138, 140, 141, 228, 232, 233, 235, 236, 241, 242, 246, 247, 256, 257, 263, 273
Second International BIOMASS Experiment (SIBEX), 93, 94
Sector principle, 36
Senate Committee on Commerce, Science and Transportation, 89
Senate Committee on Foreign Relations, 61, 64, 65
Senate Committee on Labor and Human Resources, 63, 189
Shapley, Deborah, 242, 254
Siple, Paul A., 13, 26, 214, 217, 219, 225, 226
Sisco, Joseph J., 98, 218, 242
Sites of Special Scientific Interest. *See* Agreed Measures for the Conservation of Antarctic Fauna and Flora
Smedal, G. T., 224
South Africa, 109, 245
South Pole, 40, 149, 150
South Pole Station (Amundsen-Scott Station)(U.S.), 94, 95, 216, 217
Southern Ocean, 14, 15, 92, 93, 104–108, 134; productivity, 14; sailing conditions, 17
Sovereignty issue, 30, 31, 34; compromise, 31, 32; in CCAMLR, 107; CRAMRA, 65, 139
Soviet Ruskaya station, 214
Soviet Union (now Russian Federation), 4, 12, 30, 40, 96, 105, 142, 150–52, 173, 245, 263; claims, 36, 151; fishing, 105, 142; IGY participant, 151
Spain, 176, 177
Specially Protected Areas. *See* Agreed Measures for the Conservation of Antarctic Fauna and Flora
Sri Lanka, 167
Standing Committee on Antarctic Logistics and Operations (SCALOP), 88
State Department. *See* Department of State
Studds, Garry, 122
Sullivan, Walter, 148
Sununu, John, 59
Supplemental Environmental Impact Statement (SEIS), 115
Sweden, 179, 245

Talmadge, John B., 235
Taubenfeld, Howard J., 148, 222, 227

Theis, Ethel R., 228, 229, 231, 243, 251, 253, 262, 263, 271
Transantarctic Mountains, 11
Triggs, Gillian D., 164, 243, 257, 265, 267
Trusteeship (International), 30, 221
Tydings, Millard, 23
Tyree, David M., 90

Ukraine, 245
United Kingdom (U.K.), 21, 30, 57, 150, 171, 175, 178, 183; overlapping claims, 30, 36, 141, 149, 150, 171, 173, 262
United Nations, 30, 41, 161, 163, 164, 173; and Antarctic Treaty, 161–75, 186; Conference on the Human Environment (1972), 102; General Assembly, 73, 161, 164, 266–69
United States Antarctic Policy, 26, 45–78, 136, 137, 155; active presence considerations, 29, 155; Antarctic stations, 81, 94, 95, 221, 222; bases for claims, 22–24, 32, 36–40; budget process, 50, 51, 170–74; common heritage concept, 170–74; decisionmaking process, 6–9, 18, 19, 45–78, 53, 54, 60, 61, 75, 76, 77; economic interests, 154–56; environmental interests, 18, 19, 47, 100, 138, 142–45, 159, 189, 190; historic rights, 40, 46–78; IGY participation, 1, 26–33, 79–81, 235–36; logistics support, 51, 84, 89, 95; objectives, 29, 50, 52, 53, 65, 97, 98, 188–93; organizational structure, 53, 54; position on claims to territory, 39, 40; resource interests, 47, 133–45, 154–56, 189, 190; scientific interests, 2, 10, 18, 29, 47, 78–99, 189, 190; strategic security interests, 4, 29, 47, 52, 77, 145–54, 189, 190, 260, 261
United States Antarctic Program (USAP), 81–85. *See also* National Science Foundation
United States Antarctic Service (USAS), 23, 24, 25, 38

United States Coast Guard. *See* Department of Transportation
United States Congress, 25, 61–67, 117–22, 191; House authorizing committee for Antarctic program, 62, 89; limitations on Antarctic policymaking, 67, 191; position on CRAMRA, 63, 66, 119–22; Senate authorizing committee for Antarctic program, 62, 63
United States Exploring Expedition, 24
United States Geological Survey. *See* Department of the Interior
Uruguay, 245
USAS. *See* United States Antarctic Service

Viña del Mar, 58, 59, 109, 110, 179
Vostok (Soviet Station), 12, 23

Walton, D. H., 16,
Washington Conference (1959), 32, 33, 224
Waterman, Allan T., 29
Weddell Sea, 17
Wellington minerals convention. *See* Convention on the Regulation of Antarctic Mineral Resource Activities (CRAMRA)
West Antarctica, 11
Westermeyer, William, 158, 265
Whaling interests, 134
White House, 45
Wilkes, Charles, 21, 22, 217, 218
Wilkes expedition. *See* United States Exploring Expedition
Wilkes Land, 37
Wilkniss, Peter E., 81, 232, 248, 250
World Meteorological Organization, 95
World Park Proposal, 128–30
World Resources Institute, 271

Zablocki, Clement J., 27
Zoellick, Robert, 59
Zumberge, James H., 78, 215, 239, 254, 255

UNIVERSITY PRESS OF NEW ENGLAND
publishes books under its own imprint and is the publisher for Brandeis University Press, Dartmouth College, Middlebury College Press, University of New Hampshire, Tufts University, Wesleyan University Press, and Salzburg Seminar.

LIBRARY OF CONGRESS CATALOGING-IN-PUBLICATION DATA
Joyner, Christopher C.
    Eagle over the ice : the U.S. in the Antarctic / by Christopher C. Joyner and Ethel R. Theis.
       p. cm.
    Includes bibliographical references and index.
    ISBN 0-87451-778-8 (cloth : alk. paper)
    1. Antarctica—International Status. 2. Antarctica—Discovery and exploration—American. 3. United States—Foreign relations.
    I. Theis, Ethel R. II. Title.
KWX465.J69 1997
341.2′9—dc20   96-22352